MW00574356

The New Theory of Time

Contributors

MICHELLE BEER is associate professor of philosophy at Florida International University.

BRIAN J. GARRETT is Queen Elizabeth II Research Fellow at the Australian National University.

H. SCOTT HESTEVOLD is associate professor of philosophy at the University of Alabama.

DAVID KAPLAN is professor of philosophy at the University of California, Los Angeles.

DELMAS KIERNAN-LEWIS is assistant professor of philosophy at Morehouse College, Georgia.

MURRAY MACBEATH is lecturer in philosophy at the University of Stirling, Scotland.

D. H. MELLOR is professor of philosophy at Cambridge University and a fellow of the British Academy.

L. NATHAN OAKLANDER is professor of philosophy at the University of Michigan–Flint.

GEORGE SCHLESINGER is professor of philosophy at the University of North Carolina at Chapel Hill.

QUENTIN SMITH is associate professor of philosophy at Western Michigan University.

CLIFFORD WILLIAMS is associate professor of philosophy at Trinity College, Illinois.

DAVID ZEILICOVICI is professor of philosphy at Bar Ilan University, Israel.

The New Theory of Time

EDITED AND WITH INTRODUCTIONS
BY L. NATHAN OAKLANDER
AND QUENTIN SMITH

Yale University Press
New Haven and London

Designed by Sally Harris / Summer Hill Books.
Set in Galliard type by DEKR Corp.,
Woburn, Massachusetts.

Printed in the United States of America by BookCrafters,
Chelsea, Michigan.

Library of Congress Cataloging-in-Publication Data

The new theory of time / edited and with introductions by L. Nathan Oaklander and Quentin Smith.
p. cm.
Includes bibliographical references and index.
ISBN 0-300-05796-2
1. Time. 2. Tense (Logic) I. Oaklander, L. Nathan, 1945– . II. Smith, Quentin, 1952–
BD638.N45 1994
115—dc20 93-47500
CIP

A catalogue record for this book is available from the British Library.

The paper in this book meets the guidelines for permanence and durability of the Committee on Production Guidelines for Book Longevity of the Council on Library Resources.

10 9 8 7 6 5 4 3 2 1

To Lillian Pierson Lovelace, for her support of philosophical research, and to the memory of Rose and Stanley Zoobris

Contents

PART III: *Time and Experience*

Preface

Philosophical problems arise when we try to understand the fundamental features of experience. One of the most basic experiences, and certainly one that has led to considerable philosophical perplexity, is that of time. This experience is reflected in such statements as "I can't wait until the Matisse exhibit opens," "Hurrah, I am finally graduating," "Thank goodness the exam is over!" "Here today, gone tomorrow," "Time heals all wounds," and the like. These expressions convey the flow, or passage, of time; the experience, or "perception," of events moving from the future into the present and then receding from the present into the more and more distant past. It is hard to think of a more familiar experience than temporal passage or temporal becoming. If time heals all wounds, it is because, with its passing, various present sorrows, frustrations, feelings of loss, and other unpleasant experiences recede into the past and, in so doing, fade in strength, if not in effect. To anticipate with eagerness the beginning of the Matisse exhibit is to look forward to opening day moving from the distant to the near future and from the near future to the present, the lived moment of the NOW.

Philosophers who have reflected on the experience of temporal passage and the concepts of past, present, and future have become puzzled, and it is easy to see why. Consider, for example, the birth of my first grandchild. Since that event is in the relatively distant future, it does not yet exist. Nevertheless, it will (I hope) eventually take place—that is, the event in question will come into existence. Subsequently, the event will cease to exist as it recedes into the past. It would thus appear that the passage of time involves events moving from the nonexistent future to the existing present to the nonexistent past. But if the future does not exist, how can it become present? How can what is not become what is? How can something (the present) come from nothing (the future), and what is (the present) become what is not (the past)? Questions like these have led some people to wonder how our experience of time could ever reflect the real nature of time.

To think of time as involving the concepts of past, present, and future gives rise to further questions. Augustine believed that neither future nor past but only the present exists or is real. But what is the present? If it is temporally extended, if it has any duration, then it can be divided into a nonexistent future and a nonexistent past. Thus the present must be a moment without extension. But how can the essence of time consist in a temporally unextended point? How can the durational aspect of real time be built up from past, present, and future?

Perhaps the most difficult problem facing those who seek to understand the experience of time is that of temporal passage itself. The notion of "passage" is essentially a spatial one. We speak of a quarterback *passing* a football through the air or of a train *passing* through a town. In these instances, as in many others, passage is understood in terms of movement through space. Thus something passes through space by being at different places at different times. But what sense can be made of the idea of the movement of time itself? Can time itself, or the events that occur in time, pass from one time to another? What might this mean? How could it be true?

Some of those who have thought about these questions have maintained that past, present, and future are real, do all exist, but that they differ in their temporal characteristics, or properties. Thus it has been argued that *pastness, presentness,* and *futurity* are nonrelational properties which events possess and that the passage of time consists in events changing with respect to these properties. In this way, temporal becoming, or passage, is taken to be a species of qualitative change. Just as an apple can change from green to red, so an apple's being green can change from being in the future to being in the present.

The analogy of temporal passage to qualitative change can help us, but only up to a point. Things change through time, but to suppose that time itself changes through time is circular. What is this time through which time moves? It is clear that we experience the passage of time, and some languages certainly reflect this by their use of tenses. But could our experience be deceiving us? Is it possible that the most pervasive aspect of our experience of ourselves and the world is an illusion? Could it be that time does not *really* pass, that time is unreal, and that all tensed judgments are false? The goal of a sound philosophy of time is to avoid such a conclusion by rendering the phenomena in question comprehensible. Moreover, in seeking to elaborate an adequate philosophy of time, we will need to consider another familiar feature of our experience of time, namely, that of succession. We experience time as involving not only passage but also succession. To say that time involves succession means that we experience events as occurring one after another, as temporally related. My typing on the computer keyboard is (roughly) simultaneous with the letters appearing on the screen, and the words themselves occur on the screen in succession, one after the other. But what metaphysical reality underlies our experience of succession? Does that experience depend on our experience of the transitory aspect of time (temporal passage), or does our experience of temporal passage depend on our experience of succession (temporal relations)? Or are both experiences equally fundamental and distinct? After all, it could be that each reflects some basic truth about the nature of temporal reality.

One way in which philosophers of time have approached these questions is through a consideration of the language of time. Many of them have argued that if we can determine what we mean or intend to express by the use of temporal language, then we will have an accurate picture of what reality must be like if our words and thoughts about time are to be true. Does our use of tensed language (which reflects temporal passage), as in the sentence "It is now 1994," depend, metaphysically speaking, on our use of tenseless language (which reflects temporal relations), as in the sentence "1994 is later than 1993; or does our use of tensed language capture an aspect of time that cannot be captured by tenseless language? It used to be thought that the question as to whether tensed discourse could be translated into tenseless discourse determined which theory of time is true. But there has been a reaction to this way of thinking, the outcome of which is a *new* tenseless theory of time, and it is this that constitutes the subject of this book.

The aim of these prefatory remarks is not to answer any of the questions raised, but rather to set the stage for the essays that follow, all of which, in one way or another, seek to make the experience of time intelligible and the language of time true. The essays in this volume contain most of the principal contributions to the new tenseless theory of time since its development in the early 1980s. In addition, there are several new essays by Smith and one by Oaklander. Together, they present the latest development in a philosophical debate between the tensed and the tenseless theories of time that is still ongoing.

The book is divided into three parts. Part I is devoted to the role (or lack of it) of language in the debate between these two fundamentally different ways of viewing time: the theory that takes *earlier than, later than,* and *simultaneity* as fundamental and the theory that takes the transitory temporal properties of *pastness, presentness,* and *futurity* as fundamental. Part II focuses on the logical difficulties inherent in what the defender of the tensed theory of time, perhaps prejudiciously, calls the "commonsense conception of time." Here what many have considered to be the strongest argument against the tensed theory—McTaggart's paradox—is debated. Part III returns to the nature of our experience of time. It is this that provides the defender of the tensed theory with perhaps the strongest argument against the tenseless theory. For, allegedly, the defender of the tenseless theory cannot explain the presence of experience and the experience of temporal passage. Here, as in the preceding parts, the reader is left to decide whether the theory being criticized is adequately defended against the objections raised.

The importance of the debate can hardly be overestimated, given the role played by time in numerous other philosophical issues. Not only is the problem of time profound and fascinating in its own right, it is one whose understanding and resolution is central to other philosophical questions of perennial interest. In the Introduction Quentin Smith will consider just some of the ways in which the debate between the tensed and the tenseless theories of time is related to other philosophical problems.

We have not included a bibliography, since virtually all of the recent work in the areas covered in this book is cited in the introductions and chapters. Nevertheless, we have included a Name Index, which can serve as an aid to identifying reference material. For recent bibliographies on the debate between tensed and tenseless theories of time, as well as closely related issues, see the extensive bibliographies found in D. H. Mellor, *Real Time* (Cambridge: Cambridge University Press, 1981); David Farmer, *Being in Time* (Lanham, Md.: University Press of America, 1990); Robin Le Poidevin, *Change, Cause and Contradiction: A Defence of the Tenseless Theory of Time* (Cambridge: Cambridge University Press, 1992); and L. Nathan Oaklander and Quentin Smith, *Time, Change and Freedom* (New York: Routledge, 1995), as well as the references in Quentin Smith, *Language and Time* (New York: Oxford University Press, 1993), and L. Nathan Oaklander, *Temporal Relations and Temporal Becoming: A Defense of a Russellian Theory of Time* (Lanham, Md.: University Press of America, 1984).

The New Theory of Time

General Introduction: The Implications of the Tensed and Tenseless Theories of Time

QUENTIN SMITH

In *Aspects of Time* George Schlesinger writes about the importance of the issue of temporal becoming. "We shall consider what fairly may be called the profoundest issue in the philosophy of time: the status of temporal becoming. Some philosophers have even regarded it as the profoundest issue in all of philosophy."[1] Although this statement may be an exaggeration, it cannot be denied that the issue of temporal becoming is central to metaphysics, the philosophy of science, the philosophy of mind, the philosophy of religion, the philosophy of logic, and the philosophy of language, as well as other fields of philosophy. The issue of temporal becoming is often formulated differently by different philosophers, but one of the most familiar formulations represents the issue as one about the ontological status of events that occupy different temporal locations. According to this formulation, the issue of temporal becoming is whether events are first future, then become present, and finally become past, or whether events do not "come into being" in this sense but merely exist "without becoming" at their respective temporal locations. If there is temporal becoming, then future events do not yet exist, present events exist, and past events no longer exist; but if there is no temporal becoming, then all events exist equally, regardless of whether they are located in 5,000 B.C., the twentieth century, or the twenty-fourth century.

Philosophers who claim that all successively ordered events have the same ontological status may be called "detensers," or proponents of the tenseless theory of time. They hold that the nature of time can be captured completely by tenseless sentences, such as "The birth of Plato is earlier than the birth of Russell." Philosophers who hold that there is temporal becoming are called "tensers" and are proponents of the tensed theory of time. They believe that tensed sentences, such as "Plato was born a long time ago," are necessary if the complete nature of temporal reality is to be described.

An important issue discussed by tensers and detensers concerns whether temporal properties of a certain sort are exemplified. Some philosophers characterize the debate between tensers and detensers in terms of whether events have monadic temporal properties of presentness, pastness, or futurity (A-properties) or whether the only temporal properties of events are the polyadic properties (relations) of earlier than, later than, and simultaneity (B-relations). For these philosophers, the issue of temporal becoming is whether events possess A-properties in addition to B-relations. Some defenders of the tensed theory of time, such as A. N. Prior and his followers, reject the thesis that there are events with properties of futurity, presentness, and pastness and state the issue between tensers and detensers in terms of whether some propositions change their truth-value with time or whether all propositions have permanent truth-values. What is common to the various versions of the tensed theory of time is the rejection of the claims that the temporal system consists of successively ordered events which are equally real and that tenseless sentences are sufficient to describe the temporal system.

This Introduction focuses upon the implications of the debate about the tensed and tenseless theories for a variety of philosophical subjects. Although the focus of this book is the new tenseless theory of time and the response to it by tensers, in this Introduction I shall indicate the broader implications and importance of this debate by explaining some of its consequences for certain issues in the philosophy of science, the philosophy of mind, the philosophy of religion, the philosophy of logic, and the philosophy of language. The new tenseless theory of time and its relation to the old tenseless theory of time will be explained at the beginning of Part I, but this distinction is not crucial to the ideas discussed in what follows here.

Tensed/Tenseless Theories of Time and the Philosophy of Science

Which of these theories of time is true has important consequences for the interpretation of our best-established scientific laws and theories. As an example, I shall mention the relevance of the tensed/tenseless debate to Einstein's special theory of relativity.

It should be noted at the outset that many thinkers believe the reverse of the above statement; that is, they believe that the result of the philosophical debate about tensed/tenseless theories does not deterine how Einstein's special theory of relativity should be interpreted, but rather, that Einstein's theory has conclusive implications for the debate. Some philosophers, such as Grünbaum, Smart, Quine, and Putnam, as well as virtually all physicists, believe that Einstein's special theory of relativity implies the tenseless theory of time, such that empirical confirmation of the theory of relativity is ipso facto confirmation of the tenseless theory of time. Their reasons for thinking this are not convincing, however, and in the course of explaining why, I will show why my earlier statement is true, namely, that the outcome of the philosophical debate about time will determine how the physical theories are to be interpreted.

The reason why physicists think that Einstein's theory implies the tenseless theory of time is that they adopt Minkowski's 1908 formulation of Einstein's theory and Minkowski presupposes the tenseless theory of time in his interpretation of Einstein's theory. Minkowski writes:

> A point of space at a point of time, that is, a system of values, $x, y, x, t,$ I will call a *world-point*. The multiplicity of all thinkable x, y, x, t systems of values we will christen the *world*. . . . Not to leave a yawning void anywhere, we will imagine that everywhere and everywhen there is something perceptible. . . . We fix our attention on the substantial point which is at the world-point x, y, x, t and imagine that we are able to recognize this substantial point which is at the world-point x, y, x, t and imagine that we are able to recognize this substantial point at any other time. Let the variation dx, dy, dz of the space co-ordinates of this substantial point correspond to a time element dt. Then we obtain, as an image, so to speak, of the everlasting career of the substantial point, a curve in the world, a *world-line,* the points of which can be referred unequivocally to the parameter t from -00 to $+00$. The whole universe is seen to resolve itself into similar world-lines.[2]

This presupposition of the tenseless theory of time, however, is no more implied by Einstein's theory than is the thesis that the tensed theory of time is true. Storrs McCall, D. Dieks, Howard Stein, William Lane Craig, I, and others have shown how Einstein's theory may be interpreted in terms of the tensed theory of time.[3] For example, the Lorentz transformations do not imply that "everywhen there is (tenselessly) something perceptible," since it is consistent to suppose that solutions to these equations correspond to physical events only when these events belong to the origin or past light cone of the relevant reference frames. It may be the case that future light cones do not consist of determinate events but involve what is not yet real. No scientific observation or test can decide between these two interpretations of Einstein's theory; in terms of scientific observations, they are equivalent. (It is arguable, however, that human temporal experience in general confirms one of the two theories. See Part III.) Defenders of the Minkowski interpretation of special relativity sometimes say that Newton's assumption of absolute space is a metaphysical assumption that cannot be verified by scientific observation; but the same could be said of Minkowski's assumption that the tenseless theory of time is true. The tensed/tenseless issue is a matter not of physics but philosophy, and it can only be resolved by the sorts of considerations presented in this book.

However, some thinkers have presented an *argument* that Einstein's theory implies the tenseless theory of time. It is worth seeing why their arguments are unsound. The first argument was put forward by Putnam and Rietdijk and was later restated by Maxwell.[4] It may be stated as a reduction to absurdity of the premise that some events are future, that is, not yet real.

Suppose there is an observer 01 at the space–time position *p*. Some event *E* is in 01's future, so is not yet real to 01 (this is the premise to be reduced to absurdity). Now there is another observer 02 rapidly in motion relatively to 01. 02 passes 01 at the space–time position *p*. While 02 is at the space–time position *p*, 02 is real to 01. However (we may coherently suppose), event *E* is present to 02 while 02 is at *p*; that is, according to 02's reference frame, *E* occurs simultaneously with 02's occupation of the space–time position *p*. Thus, *E* is real to 02 at *p*. Now this shows the tensed theory of time is false, the argument goes, since if 02's being at *p* is real to 01, and *E* is real to 02 at *p*, then *E* is real to 01 at *p*. This contradicts the original assumption that *E* is not yet real to 01 at *p*.

A similar sort of argument can be constructed for any event *E* that is said to be future and not yet real for some observer. Thus it follows that all events are equally real; there are no events that are not yet real. Thus, the tensed theory of time, which implies that some events are not yet real, is false.

However, this attempt to show the special theory of relativity implies the tenseless theory of time fails, since it depends on accepting the dubitable *metaphysical assumption* that the relation of *being real to* is a transitive relation. This assumption is not part of Einstein's special theory of relativity, and there is no good reason to accept it; indeed, there are good reasons for rejecting it, since the special theory of relativity implies that analogous relations are intransitive. For example, *is simultaneous with* is intransitive. Now the defender of the tensed theory of time can coherently adopt the assumption that *being real to* is intransitive. The tenser believes that reality is dependent on an event's position in the A-series, whether it is present, not yet present, or no longer present, and therefore that *being real* is no less relative than *being present*. Thus, it is natural to think that *is real to* is similar to *being present to* in being relative and intransitive. Accordingly, the tenser can coherently maintain that *E* is present and real to 02 at *p*, that 02 at *p* is present and real to 01 at *p*, but that *E* is future and not yet real to 01 at *p*.

Given that this interpretation of the special theory of relativity is possible, it follows that relativity theory does not entail the tenseless theory of time.

A second familiar argument is Grünbaum's general argument that physics makes no mention of physical events possessing the property of presentness, and that if physical events did possess presentness, this would be mentioned in physical theories.[5] Grünbaum regards this as a reason for thinking that physics is committed to the tenseless theory of time. However, his reasoning is arguably based on a confusion, namely, a belief that a subject matter which uniquely pertains to *observational physics* should be represented in *theoretical physics* if it is recognized in physics. Which event or year is present must be decided by observation, just as the location of the earth in the Milky Way galaxy must be decided by observation. Certainly, the presentness of some events does not appear in the Lorentz transformations of the special theory of relativity or in the ten field equations of the general theory of relativity; but that no more shows that no event is present than the fact that the location of the earth is not mentioned in these equations shows that the earth is not located anywhere.

In fact, the tenser may well argue that which time is present *is* mentioned in observational physics and cosmology. For example, one of the most important concepts in observational cosmology is what the physicist R. Dickie calls "the present value of *T*," *T* being the Hubble age (how long the universe has been expanding). The present value of T is 15 billion years.[6] Numerous other tensed temporal locutions appear in observational cosmology.[7] Of course, this does not prove that observational cosmology implies that events possess A-properties, for the use of tensed locutions does not by itself show that the theory is committed to A-properties. It is a philosophical question whether tensed expressions ascribe A-properties or merely refer to B-positions. But it does show that the truth of the tensed or tenseless theory of time cannot be read off from our current physics. It is rather the case that the correct interpretation of physical theories remains an open question until the metaphysical issue regarding the tensed versus the tenseless theory is resolved by philosophers. This supports my earlier contention that the philosophical debate about the tensed and tenseless theories has consequences for how our scientific theories should be interpreted.

It should be added that philosophers of time need not be confined to the assumption that temporal determinations are relative, as described in the special theory of relativity. One may argue on metaphysical grounds that temporal properties and relations are possessed absolutely.[8] If simultaneity or presentness is absolute, this is consistent with the observable behavior of light rays, clocks, rigid bodies, and so forth, being relativistic in nature. If time is absolute, then the word "time" and the variable "*t*" in the special and general theories of relativity should be interpreted as signifying the observable states or changes in the relevant physical bodies, rather than time itself. Such an interpretation of physics is observationally equivalent to the interpretation that regards the physicists' t as referring to time. The assumption that time is not absolute but instead consists of the observably relative behavior of physical bodies *because* the variable t in physical equations refers to this behavior is not a scientific thesis but a philosophical one that may be proved correct or incorrect on the basis of philosophical considerations.[9]

Tensed/Tenseless Theories of Time and the Philosophy of Mind

A number of philosophers believe that the debate in the philosophy of mind between defenders of the *substantival* and *temporal parts* theory of personal identity hinges upon whether the tensed or the tenseless theory of time is true. Philosophers such as Delmas Lewis, Robin Le Poidevin, L. Nathan Oaklander, and Ronald Hoy, among others, believe one or both of the following two theses about time and personal identity: (1) that the substantival theory of personal identity is true only if the tensed theory of time is true, and (2) that the temporal parts theory of personal identity is true only if the tenseless theory of time is true.[10]

The substantival theory of personal identity holds that a person is a substance or particular that endures through time and successively acquires or loses various properties that are normally ascribed to a person (waking, sleeping, being angry, being sad, and so on). The temporal parts theory of personal identity implies that there is no continuing particular but rather a succession of particulars, each being a different temporal part that helps to make up the whole person. On the substantival theory, a person Alice exists through various times and acquires or loses various properties; whereas on the temporal parts theory, "Alice" refers to the whole that is made up of the temporal parts, Alice at t_1, Alice at t_2, Alice at t_3, and so on.

Some philosophers argue that the identity through time implied by the substantival theory requires the tensed theory of time. For example, Delmas Lewis and Roderick Chisholm point out that the same person cannot *have* (tenselessly) incompatible properties, but that this is possible on a tensed view. Chisholm writes:

> There aren't two you's, a present one having one set of properties, and a past one having another. It is rather that you *are* now such that you have these properties and lack those, whereas formerly, you *were* such that you had those properties and lacked these. The "former you" *has* the same

properties that the "present you" now has, and the "present you" *had* the same properties that the "former you" then had.[11]

By contrast, other philosophers, such as Ronald Hoy and Oaklander, believe that the temporal parts theory of personal identity implies the tenseless theory of time.[12] If the temporal parts theory were conjoined with the tensed theory, then a certain paradox would arise that was first articulated by J. J. C. Smart, who writes that the

> notion of pure becoming is connected with that of events receding into the past and of events in the future coming back from the future to meet us. This notion seems to me unintelligible. What is the "us" or "me"? It is not the whole person from birth to death, the total space–time entity. Nor is it any particular temporal stage of the person. A temporal stage for which an event *E* is future is a different temporal stage from one for which event *E* is present or past.[13]

If the substantival theory of personal identity is true, then the "me" that future events come to meet is a continuing particular that passes from one present moment to the next. But if the temporal parts theory is true, then there is no candidate for this moving "me" according to Smart, Oaklander, and others. Accordingly, if the temporal parts theory of personal identity is true, the tenseless theory of time is true.

If these arguments are correct, then one of the central issues in the philosophy of mind—whether the self is a substance or a series of temporal parts—cannot be resolved until we know which of the theories of time is true. For example, if considerations based on McTaggart's paradox (see Part II) show that the tensed theory of time is false, then we will know that the substantival theory of personal identity is false. On the other hand, if arguments from the semantic properties of tensed language (see Part I) and from the psychological properties of the experience of temporal passage (see Part III) show that the tensed theory of time is true, then we will know that the temporal parts theory of personal identity is false.

Of course, the claim that the substantival/temporal parts theories in the philosophy of mind are logically dependent on the tensed/tenseless theories in the philosophy of mind can be challenged. And indeed, I myself have challenged this connection, although Oaklander has countered that my argument against the temporal parts–tenseless time connection is unsound.[14] But, at the very least, it is safe to say that the tense/tenseless issue in the philosophy of time has an intimate and controversial bearing on the substantival/temporal parts issue in the philosophy of mind. Philosophers who work in the field of philosophy of mind can ignore the tense/tenseless debate in the philosophy of time only at their own peril.

Tensed/Tenseless Theories of Time and the Philosophy of Religion

The distinction between the tensed and the tenseless theories of time has relevance for a number of issues in the philosophy of religion, such as whether God should be conceived as existing timelessly or at all times, the nature of divine foreknowledge, and the question of whether God's foreknowledge is compatible with human freedom. The literature on divine foreknowledge has burgeoned since the mid-1960s, and this topic is currently among the most widely discussed issues in the philosophy of religion. Some of the implications of the tensed/tenseless distinction for the issue of the nature of divine foreknowledge may be mentioned.

If the tenseless theory of time is true, then the thesis that God foreknows future events, including future contingents, has a relatively clear formulation. Since all times are equally real and the nature of the universe at each time is fully determinate—note that being determinate is not to be confused with being causally determined—then there is no special difficulty in God knowing what occurs at every time, those later than 1992 as well those earlier than 1992. For example, it is true at every time (or timelessly) *that Jane decides to commit suicide in the year 3002*. This proposition corresponds to reality eternally; that is, it corresponds eternally to the 3002 event of Jane committing suicide, and thus is known by God eternally. "Future events," on this theory, is understood in terms of the tenseless locution "events later than some specified time *t*."

If, on the other hand, the tensed theory of time is true, the theory of divine foreknowledge is considerably more complicated. One reason for this is that there are many versions of the tensed theory of time, ranging from the full future theory to the empty future theory. The full future theory is analogous to the tenseless theory of time in holding that all times later than 1992 are fully determinate; but it differs in implying that the events occupying these later times have the monadic property of futurity and are in the process of becoming present. A half-empty future theory may hold that future times are partly determinate, consisting (say) of events that are effects of deterministic causal chains that extend from the present time but that these future times do not consist of future contingents, which are events that are not effects of deterministic causal chains. For example, it may be said the future time, tomorrow noon, includes the causally determined future event of the sun shining on the earth but that it is indeterminate in respect of John's free decision to read a book during lunch or his free decision not to read a book during lunch (tomorrow noon contains neither a future event of John reading nor a future event of John not reading). The empty future theory holds that (assuming 1992 is present) that there are no determinate events (be they future contingents or causally determined events) later than 1992 and that there is literally nothing to possess a monadic property of futurity. But even these distinctions are not straightforward, and various complications arise in attempting to state them clearly and explain their relevance to divine foreknowledge.

Some of these difficulties can be illustrated in terms of William Lane Craig's recent attempt to combine the tensed theory of time with divine foreknowledge. Craig endeavors to conjoin certain features of the full future theory with certain features of the empty future theory. He adopts both the thesis that God knows propositions about the future free decisions of humans and the thesis that future events do not exist in any sense. He writes of future events that what is "future does not merely not yet exist; it does not exist at all."[15] Yet Craig holds that future-tense propositions are true, where truth is defined in terms of a correspondence theory. Future-tense propositions are propositions expressible by such sentences as "The race will begin." But if future events do not exist in any sense, then to what do future-tense propositions about future free decisions correspond? It would seem natural to suppose that future-tense propositions lack truth-values if future events do not in any sense exist. But Craig rejects this position. He writes that "a denial of Bivalence for any future-tense propositions based on truth as correspondence is a *non sequitur*, for that view only requires that reality *will* correspond to the description given in a future-tense proposition, not that it *does* now correspond."[16]

But Craig's formulation does not seem entirely happy, since it arguably entails a contradiction. When the described event takes place, it is present and then corresponds to a present-tense proposition (a proposition stating the event is present), not a future-tense one. When the race is beginning, the future-tense proposition *that the race will begin* does not correspond to the race's beginning; what corresponds is the present-tense proposition *that the race is beginning*. Indeed, the assumption that future-tense propositions are now true but do not now correspond to anything is an explicitly logical contradiction if truth is correspondence. Consider this statement of Craig's: it is false that "future-tense propositions lack [current] truth-value due to lack of current correspondence."[17] If truth is correspondence, then (omitting "value") this sentence is synonymous with: *it is false that future-tense propositions lack current truth due to lack of current truth.*

One solution to this problem would be to allow that all future events exist in some sense and that future-tense propositions are now true by virtue of corresponding to these existents. This would amount to a full future theory. In fact, this solution is entailed by another position Craig adopts, that "tenseless propositions concerning future contingents would seem to be always true or false."[18] If the tenseless proposition *that the race begins on June 15* is always true, it is true on June 14; given that truth is correspondence to the item described, it follows that it is true on June 14 that there is something to which this proposition corresponds, namely, the race's beginning on June 15. Thus it seems that if divine foreknowledge consists of a knowledge of all future contingents, including all future free decisions of humans, then the full future version of the tensed theory of time is true (if any version of the tensed theory is true).

These reflections show that one must settle one's position on the tensed/tenseless theories of time before one can adopt a position on the nature of divine foreknowledge. Moreover, this latter position must be settled before one can begin discussing whether divine foreknowledge is compatible with human freedom. For these reasons and others, philosophers of religion are dependent on the tensed/tenseless debate no less than are philosophers of science and philosophers of mind.

Tensed/Tenseless Theories of Time and the Philosophy of Logic

It is not possible to decide which logic is adequate until it is first determined whether the tenseless or tensed theory of time is true. The issue of which theory of time is true will determine whether the temporal structure of reality and the semantic content of natural language can be represented by a tenseless logical symbolism—for example, the extensional symbolism of first-order sentential logic—or whether they require a tensed symbolism—for example, Prior's intensional logic with tense operators ("It will be the case that" and "It was the case that"). Moreover, if the tensed theory of time is true, then the particular version of it that is true will have consequences for the symbolism and valid formulae of tense logic.

This tensed/tenseless issue was not addressed at great length by Frege, Russell, Peano, and the other founders of modern predicate and sentential logic. Rather, they tended to build their logics on the basis of the unargued-for philosophical assumption that the tenseless theory of time is true, much as post-Minkowski physicists introduced this philosophical assumption into the interpretation of their equations. This issue in the philosophy of logic did not become widely discussed until Prior and Quine addressed it in the 1950s and 1960s.

Quine aimed to preserve the tenseless extensional symbolism of first-order predicate and propositional logic, which he believed to be adequate to represent the temporal structure of reality and natural language. He defended his belief about the temporal structure of reality by appealing to the fact that physicists, following Minkowski, accepted the tenseless theory of time; but he did not seem to be aware that their acceptance of this theory was a metaphysical assumption.[19] Regarding natural language, Quine believed that it could be paraphrased by a tenseless language that substitutes singular phrases denoting dates for tenses; thus "John was running" as uttered at noon, 1 November 1992, becomes "John runs (tenseless) before noon, 1 November 1992," which is a tenseless locution that can be represented in an extensional logic.[20] However, Quine's belief about the de-tensing of natural language has been challenged frequently, and it runs into many of the difficulties discussed in Part I. Thus, Quine's belief in the adequacy of extensional logic seems to rest on both a faulty understanding of physics and controversial assumptions in the philosophy of language.

Prior adopts a different approach. He believes that the temporal structure of reality and natural language can be represented only in a tense logic. He begins with the first-order sentence calculus but regards the sentences as being in the present tense rather than tenseless. He introduces the tense operators F and P, "It will be the case that" and "It was the case that," so that if *p* is "John is running," P*p* is "It was the case that John is running." Further elements of his tense logic are added to these basics.

Prior is aware that the issue of the advantages of tense logic vis-à-vis tenseless logic turns on controversial philosophical issues about the nature of time and natural language and cannot simply be assumed without argument. He himself advanced some arguments in favor of the tensed theory of time, such as the argument from the experience of passage,[21] but his arguments have been met by counter-arguments—for example, by some of the arguments that are presented in Part III. It seems more difficult to establish the tensed theory of time than Prior's brief arguments for it might lead one to suppose. Moreover, Prior himself advanced only some of the possible versions of the tensed theory of time, and many tensers have rejected his versions. For example, Prior rejected the view that there are events with properties of futurity, presentness, and pastness and never adequately clarified the positive ontological import of his tense operators. For example, if "it was the case that" does not involve the ascription of a property of pastness, then what are its semantic content and ontological import? Mellor interprets Prior as meaning that this operator ascribes the property of pastness to a tensed fact,[22] but Prior himself nowhere says this. Contemporary followers of Prior also adopt his reticent line. For example, Ferrel Christensen says that the information conveyed by tenses or tense operators cannot be represented in "an ontology of individuals and their properties and relations," but he never specifies to which ontological category the relata of tenses or tense operators belong.[23] Christensen and other tensers who follow Prior seem to think that the assumption that events have properties of pastness, presentness, and futurity leads to McTaggart's paradox, but it can be argued that this is not the case (see Part II). If it turns out that events have such properties, then we will want a tense logic that reflects this, and Prior's tense operators may be inadequate for this purpose. If these tense operators are interpreted as ascribing temporal properties, it would seem natural to interpret them as ascribing properties to the truth-values of sentences, propositions, or facts, rather than to events. P*p* would ascribe pastness to *p*'s being true, and we would need a different symbolism to represent the ascription of pastness to the event described by *p*.

These considerations reflect the fact that the question of which logic is adequate to represent the temporal form of reality and the semantic content of natural language is a question of metaphysics and the philosophy of language rather than logic, and requires the sort of complex, detailed metaphysical and linguistic investigations that are undertaken in the book.

Tensed/Tenseless Theories of Time and the Philosophy of Language

The most important development in the philosophy of language in the past twenty or thirty years is the New Theory of Reference of Marcus, Kripke, Kaplan, Putnam, N. Salmon, H. Wettstein, J. Perry, and numerous others. But whether this theory is true depends on whether the tenseless theory of time is true or, if the tensed theory of time is true, on which version of the tensed theory is true.

The New Theory of Reference holds that many expressions, including indexicals such as "now," "here," and "I," refer directly and rigidly to particulars and do not ascribe properties. For example, David Kaplan argues that a use of "now" at noon refers directly to noon and does not ascribe any property. However, if the tensed theory of time is true and uses of "now" ascribe the property of presentness, then the New Theory of Reference is false. Temporal indexicals such as "now" will not be "purely referential," to borrow a phrase from Marcus,[24] but will express a characterizing sense, the property of presentness.

In the introduction to Part I, the relation of the New Theory of Reference to the tensed/tenseless issue will be discussed further, and I will show how the New Theory of Reference was partly instrumental in motivating the new tenseless theory of time. The conflict between the New Theory of Reference and the tensed theory of time will be discussed at length in chapters 11 and 12, by David Kaplan and myself, where the issue of whether the indexical "now" is purely referring will be examined in light of the relevant linguistic data.

In this general introduction to the book, I have hoped to bring out the centrality of the debate between tensers and detensers to a number of issues in various fields of philosophy. I have argued that the outcome of the tensed/tenseless debate is crucial to important theories in the philosophy of science, the philosophy of mind, the philosophy of religion, the philosophy of logic, and the philosophy of language. This debate has crucial import for other philosophical issues as well, for example, the metaphysical question about the nature of existence (for example, on some versions of the tensed theory of time, existence may be identified with presentness), but the few sample topics I have discussed suffice to suggest the wide-ranging ramifications of the debate. We shall now turn to the debate itself, beginning (in Part I) with the linguistic, or semantic, dimension of the argument between the detensers and the tensers.

Notes

1. George Schlesinger, *Aspects of Time* (Indianapolis, 1980), p. 23.
2. H. Minkowski, "Space and Time," in Albert Einstein et al., *The Principle of Relativity* (New York), 1952), p. 76.
3. Storrs McCall, "Objective Time Flow," *Philosophy of Science* 43 (1976): 337–362; D. Dieks, "Special Relativity and the Flow of Time," *Philosophy of Science* 55 (1988): 456–460; Howard Stein, "On Einstein–Minkowski Space-Time," *Journal of Philosophy* 65 (1968): 5–23; idem, "On Relativity Theory and Openness of the Future," *Philosophy of Science* 58 (1991): 147–167; Quentin Smith, *Language and Time* (New York, 1993); William Craig, "God and Real Time," *Religious Studies* 26, no. 3 (1990): 335–347.
4. H. Putnam, "Time and Physical Geometry," *Journal of Philosophy* 64 (1967): 240–247; C. W. Rietdijk, "A Rigorous Proof of Determinism Derived from the Special Theory of Relativity," *Philosophy of Science* 33 (1966): 341–344; N. Maxwell, "Are Probabilism and Special Relativity Compatible?" *Philosophy of Science* 52 (1985): 23–43.
5. Adolf Grünbaum, *Modern Science and Zeno's Paradoxes* (Middletown, Conn.; 1967).
6. R. Dickie, "Dirac's Cosmology and Mach's Principle," *Nature* 192 (1961): 440.
7. See Quentin Smith, "The Mind-Independence of Temporal Becoming," *Philosophical Studies* 47 (1985): 109–119.
8. This is argued in Quentin Smith, *Language and Time.*
9. Ibid.
10. Delmas Lewis, "Persons, Morality, and Tenselessness," *Philosophy and Phenome-*

nological Research 47 (1986): 305–309; Robin Le Poidevin, *Change, Cause, and Contradiction: A Defense of the Tenseless Theory of Time* (New York, 1991); L. Nathan Oaklander, "Persons and Responsibility: A Critique of Delmas Lewis," *Philosophy Research Archives* 13 (1987–88): 181–187; Ronald Hoy, "Becoming and Persons," *Philosophical Studies* 34 (1978): 269–280.

11. R. Chisholm, *Person and Object* (La Salle, Ill., 1976), p. 92.
12. Hoy, "Becoming and Persons"; Oaklander, "Persons and Responsibility."
13. J. J. C. Smart, "Time and Becoming," in P. van Inwagen, ed., *Time and Cause* (Boston, 1981), p. 6.
14. Quentin Smith, "Personal Identity and Time," *Philosophia* 22 (1993): 155–167; L. Nathan Oaklander, "Temporal Passage and Temporal Parts," *Noûs* 26 (1992): 79–84.
15. William Lane Craig, *Divine Foreknowledge and Human Freedom* (Leiden, Brill, 1991), p. 226.
16. Ibid., p. 58.
17. Ibid., p. 59.
18. Ibid., p. 61.
19. W. V. O. Quine, *Word and Object* (Cambridge, Mass., 1960), p. 172.
20. Ibid., sec. 36.
21. A. N. Prior, "Thank Goodness That's Over," *Philosophy* 34 (1959): 12–17.
22. D. H. Mellor, *Real Time* (Cambridge, 1981), p. 95.
23. Ferrel Christensen, "McTaggart's Paradox and the Nature of Time," *Philosophical Quarterly* 24 (1974): 299.
24. Ruth Barcan Marcus, "Modalities and Intensional Languages," *Synthese* 130 (1961): 303–322.

PART I

Time and Tensed Sentences

Introduction: The Old and New Tenseless Theories of Time

QUENTIN SMITH

The difference between the old and the new tenseless theories of time can be explained most simply by taking just one of the many versions of the tensed theory, specifically the version that holds that tensed sentences ascribe monadic properties of futurity, presentness, and pastness to events. The issue of whether the tenseless or the tensed theory of time is true used to be debated in terms of whether tensed sentences could be translated by sentences ascribing relations of simultaneity, earlier than, and later than. "The sun will rise soon" seems to depict the sun's rising as a future event, as something that will become present before too long; "The sun is rising now" seems to depict the event as present, and "The sun arose some time ago" as past. If these tensed depictions are true, this suggests that the sun's rising possessed first the property of futurity, then the property of presentness, and finally that of pastness.

But do tensed sentences depict events as having these temporal properties? If they do not, the fact that many tensed sentences are true cannot count as evidence that events possess these properties. If tensed sentences have the same meaning as tenseless sentences and can be translated into them, then tensed sentences do not ascribe these temporal properties. For example, if "The sun is rising now" (as uttered on 23 April 1992) is translated by "The sun is rising on 23 April 1992," then the "now" as used in the first sentence does not ascribe a property of presentness to the sun's rising; it merely refers to the date on which the sentence is uttered. Likewise, if "The sun will rise soon" as uttered at 3:00 A.M., 22 April, is translated by "The sun rises later than 3:00 A.M., 22 April," then the tensed sentence does not ascribe futurity to the sun's rising but merely states that the sun's rising occurs later than the date on which the sentence is uttered. If these translations succeed, there is no reason to think that temporal becoming is real. Reality then consists of events related to each other by relations of simultaneity, earlier than, or later than; that is, events are temporally characterized only by the characteristics attributed to them in tenseless sentences.

These ideas enable us to define the old tenseless theory of time. This is the theory that tensed sentences or their tokens are translatable into tenseless sentences, and therefore that tensed sentences do not ascribe any temporal determinations not ascribed by tenseless sentences; since tenseless sentences ascribe only relations of earlier than, later than, and simultaneity, and not properties of pastness, presentness, and futurity, it follows that time consists only of these relations. Proponents of the old tenseless theory of time include Bertrand Russell, J. J. C. Smart, Hans Reichenbach, and Nelson Goodman, among others.

The old tenseless theory of time has been abandoned, however, as a result of advancements in the philosophy of language. The New Theory of Reference, first developed by Ruth Barcan Marcus in her 1961 article "Modalities and Intensional Languages" and later expanded upon by Kripke, Kaplan, Perry, Donnellan, Putnam, N. Salmon, and others in the 1970s, became the impetus for rethinking the issue of tensed and tenseless sentences. A key idea in the New Theory of Reference is Marcus's thesis that many expressions, such as proper names, refer directly to objects (and refer to the objects in each possible world in which they exist) rather than refer indirectly by expressing senses that represent the object's properties. For example, "Neptune" refers directly to Neptune rather than expresses some characterizing sense, such as whatever planet is the seventh furthest from the sun.

David Kaplan was one of the first to apply Marcus's theory in a systematic way to indexicals such as "now." Kaplan argued that the rule of use ("character") of "now" is that it refers directly to the time at which it is uttered and does not ascribe any property. This theory supports the tenseless theory of time. If "now" refers directly, then the "now" in "The sun is rising now" refers directly to the date of its utterance. Moreover, this sentence is not translatable into a tenseless sentence. Translation, for Kaplan, requires sameness of meaning as well as of semantic content. The meaning is the rule of use of the sentence. The rule of use of "The sun is rising now" is (in part) that "now" refers to the date on which it is uttered. But the rule of use of "The sun rises at 6 A.M., 23 April 1992," is (in part) that "6 A.M., 23 April 1992," refers to 6 A.M., 23 April 1992, regardless of when this expression is uttered. Since these two sentences have different rules of use (different meanings), it follows that one cannot translate the other. (For the same reason, a token of one of the sentences cannot translate a token of the other.) Nonetheless, it is true that the tensed theory of time is false. This is because the tensed sentence ascribes no temporal determinations to reality that are not ascribed by the tenseless sentence. The rule of use of "now" is that it refers directly to the date of its utterance, not that it ascribes the property of presentness to something.

This idea lies at the basis of the new tenseless theory of time, namely, that tensed sentences (as uttered on some occasion) are untranslatable by tenseless sentences, but that it is nonetheless the case that tensed sentences ascribe no temporal determinations not ascribed by tenseless sentences.

The new tenseless theory of time was first suggested by J. J. C. Smart in his 1980 article "Time and Becoming," but its first systematic development was at the hands of D. H. Mellor and L. Nathan Oaklander.[1] Later contributors include Michelle Beer, Clifford Williams, and others. The idea underlying the new tenseless theory of time is the same for all these philosophers, but the particular formulation varies. Although I explained how this idea may be motivated by Kaplan's theory of indexicals, Smart formulated his theory largely on the basis of Donald Davidson's theory (as explained in Essay 2). The ideas of Mellor and Oaklander were likewise independently motivated; for example, Oaklander uses in part Castañeda's theory of indexicals to formulate his theory. Moreover, some of these philosophers, especially Oaklander, have developed theories of their own about the relation of language to reality, theories which are not drawn from other proponents of the New Theory of Reference although they are nonetheless integral to the new tenseless theory of time. These complexities are discussed in Part I.

Critical reaction to the new tenseless theory of time did not appear until some years after the new tenseless theory appeared in print, making its first published appearance in my 1987 article. "Problems with the New Tenseless Theory of Time."[2] One of the basic criticisms is that the New Theory of Reference, or at least the part of it that is used in the formulations of the new tenseless theory of time, is unsound. This does not amount to a retreat to the traditional, or Fregean, theory of reference (which maintains that reference is achieved indirectly, through the mediation of a sense); it merely entails a modification of the New Theory of Reference to accommodate certain facts about tensed sentences. For example, it is conceded that "now" refers directly to the date on which it is uttered, but it is denied that this fact supports the tenseless theory. It is argued that "now" not only refers directly to the date of its utterance but that it also ascribes the property of presentness to its direct referent. This argument is developed most fully in Essays 8 and 12.

Essays by Mellor, Oaklander, Beer, Williams, Kaplan, and myself comprise Part I. (The important parts of Smart's 1980 article are quoted in Essay 2.) They fall into two groups.

The first (Essays 1–6) is a debate among Mellor, Oaklander, and myself about the adequacy of Mellor's, Smart's, and Oaklander's versions of the new tenseless theory of time. In Essay 1 Mellor sets forth his version of the new tenseless theory of time. In Essays 2–6, which are alternating papers by Oaklander and myself, I offer criticisms of the new tenseless theory and Oaklander defends the new tenseless theory against my criticisms.

In Essay 1 Mellor sets forth his theory that tensed discourse is not translatable by tenseless discourse but that the truth conditions of tensed locutions can be given in a tenseless language and that these tenseless truth-condition sentences are sufficient to describe the nature of time.

In Essay 2 I take issue with Mellor's theory (and Smart's), arguing that tensed sentences have tensed truth conditions that involve the properties of presentness, pastness and futurity.

In Essay 3 Oaklander revises and defends Mellor's and Smart's theories in light of my criticism, presenting some new arguments for the thesis that tensed sentences require only tenseless truth conditions.

In Essay 4 I argue that Oaklander's new arguments are no less faulty than Mellor's and Smart's arguments and that the requirement for tensed truth conditions remains.

Oaklander then introduces, in Essay 5, a novel twist to the debate, by asserting that the detenser is not required to show that ordinary tensed sentences can be explained in terms of tenseless truth-condition sentences. He distinguishes between a "logical language" adequate for analyzing ordinary tensed language and an "ontologically adequate language." The latter is a tenseless language that describes time solely in terms of relations of earlier than, later than, and simultaneity. The analysis of ordinary language is tensed and explains the logical connections among ordinary tensed sentences, but, it is claimed, it does not represent the real nature of time.

I respond to this new line of thought in Essay 6, arguing that Oaklander is mistaken if he attributes this view to Mellor and Smart, since their defense of the new tenseless theory consisted in the claim that ordinary tensed sentences have tenseless truth conditions. But I acknowledge that Oaklander's theory may nevertheless have merits on its own ground as an alternate version of the new tenseless theory of time. The truth or falsity of Oaklander's version of the new tenseless theory can be decided by examining his claim that ontological investigations (as opposed to an analysis of ordinary language) are sufficient to decide the issue in favor of the tenseless theory of time. Oaklander's ontological investigation is based largely on considerations related to McTaggart's paradox. Oaklander maintains that in order to avoid logical paradoxes, one must assume that time consists only of relations of earlier, later, and simultaneity. This ontological debate is reserved for Part II, and I content myself at this juncture with pointing out that if McTaggart's paradox poses no threat to the tensed theory of time (as

I will argue in Part II), then the fact that ordinary tensed sentences have tensed truth conditions is a reason to believe the tensed theory of time.

These remarks essentially conclude the debate among Mellor, Oaklander, and myself about the truth conditions of tensed sentences, and at this point the debate about the new tenseless theory of time shifts to a discussion among Beer, Williams, Kaplan, and myself. This debate largely centers around the semantic content of the temporal indexical "now" and whether the New Theory of Reference shows that "now" does not ascribe a property of presentness. Essays 7–12 constitute the second major debate about the new tenseless theory of time and are outlined as follows.

In Essay 7 Michelle Beer argues that propositions expressed by uses of tensed sentences are different from propositions expressed by uses of tenseless sentences, but that the former propositions nevertheless do not report anything about time that is not reported by the latter. Part of Beer's argument hinges on the thesis that uses of "now" refer to the time at which they are uttered and therefore do not ascribe a property of presentness.

I respond to Beer's argument in Essay 8, arguing that uses of "now" both refer to the time of their use and ascribe to this time the property of presentness.

In Essay 9 Clifford Williams attempts to defend the new tenseless theory by arguing that the fact that "now"-sentences (for example, "It is now noon") have different truth conditions from tenseless sentences does not entail that "now"-sentences ascribe a property of presentness. He points to some analogies among "now," "here," and "I" and argues that if "here" and "I" do not ascribe mind-independent properties of "hereness" or "me-ness," then, by parity of reasoning, we should conclude that "now" does not ascribe a mind-independent property of presentness.

In Essay 10 I argue that Williams's argument about truth conditions is unsound, since it requires the false assumption that "now"-sentences have only tenseless truth conditions. If these sentences have tensed truth conditions, then they do ascribe a property of presentness. I further argue that Williams has not disproved the thesis that "I" and "here" ascribe mind-dependent properties of me-ness and hereness and thus that they are partly analogous to "now" (which ascribes a mind-independent property of presentness).

In Essay 11, which consists of selections from Kaplan's *Demonstratives,* Kaplan argues that the New Theory of Reference implies that uses of "now" do not express a sense but instead refer directly to the time at which they are uttered. He further shows that uses of "now" are rigid designators and have a different semantic character from tensed copulae such as "is" or "am."

I respond to Kaplan in Essay 12 by arguing that the New Theory of

Reference is inconsistent with the semantic facts about indexicals, such as "now," and tensed copulae, such as "is," "was," and "will be," ane needs to be modified to state that indexicals both refer directly to items and express a sense (a property) that characterizes their direct referent. I further argue that if uses of "now" are rigid designators, then the substantival theory of time must be true and the relational theory of time false.

As should be apparent from this outline of the essays in Part I, the issues are complex, and no clear-cut decision is apparent between the tensers and the defenders of the new tenseless theory of time. The debate is ongoing, and it is hoped the reader will use these essays to stimulate further thoughts of his or her own about the respective merits of the tensed theory and the new tenseless theory of time.

Notes

1. Mellor, *Real Time*; L. Nathan Oaklander, *Temporal Relations and Temporal Becoming* (Lanham, Md., 1984), esp. chap. 4, as well as the articles reprinted in this volume.
2. Quentin Smith, "Problems with the New Tenseless Theory of Time," *Philosophical Studies* 52 (1987): 371–391. This is reprinted as Essay 2 of this volume.

ESSAY 1

The Need for Tense

D. H. MELLOR

So much for the canard that tenseless time is spacelike time. I shall have more to say later about how time differs from space, especially in relation to change and causation. But I trust that I may now take at least some difference for granted, and that in particular I will not be charged with spatializing time when in what follows I draw on spatial analogies. I use spatial analogies to help to sell the ensuing argument, because they make it more obvious that something must be wrong with the inferences I mean to discredit, namely, that tenseless sentences could translate, or at least supplant, tensed ones. Not that the falsity of these inferences is more obvious in the spatial case; it is equally obvious that neither "Cambridge is here" nor "It is now 1980" can be either translated or replaced by spatially and temporally tenseless sentences. The point is rather that this does not make anyone adopt a tensed view of space, whereas it does seem to encourage tensed views of time. The spatial inferences must therefore, to start with, be more obviously discreditable. Yet the temporal inferences are valid only if the spatial ones are. So to encourage a proper skepticism about them, I shall start by treating the two cases together.

The Untranslatability of Tense

Why should tenseless space or time be thought to imply tenseless translations for spatially or temporally tensed sentences? The reasoning seems to be as follows. Tenseless facts provide the truth conditions for all tokens of tensed sentences and judgments: tokens of "Cambridge is here" are true if and only if they are in Cambridge; tokens of "It is now 1980" are true if and only if they occur in 1980, and so on. For many, stating a sentence's truth conditions gives its meaning; and if it does, a sentence that states another sentence's truth conditions surely has the same meaning. For example, any English sentence of the form "*X* is half empty" is true if and only if *X* is half full; so ". . . is half full" means the same in English as ". . . is half empty."

Now apply this train of thought to the tenseless truth conditions of tensed sentences, allowing for the fact that the truth conditions of tokens of "Cambridge is here" and "It is now 1980" vary respectively from place to place and from time to time. Nonetheless, although these truth conditions vary, they do so in peculiar and characteristic ways; and a statement of how they vary does seem to give the meanings of those sentence-types. Anyone who knows that for any place *X,* tokens of "*X* is here" are true if and only if they are at *X,* and that for any date *T,* tokens of "It is now *T*" are true if and only if they occur at *T,* surely knows what ". . . is here" and "It is now . . ." mean in English. So the tenseless sentences I have just used to give those meanings should themselves mean the same thing and hence provide tenseless translations of these tensed sentences.

But they don't. Let *X* = Cambridge and *T* = 1980, and let *R* be any token of "Cambridge is here" and *S* be any token of "It is now 1980." (*R* and *S* must of course not themselves be token-reflexive names or descriptions.) Then *R* is true if and only if it occurs in Cambridge, and *S* is true if and only if it occurs in 1980. If a sentence giving another's truth conditions means what it does, *R* should mean the same as "*R* occurs in Cambridge" and *S* should mean the same as "*S* occurs in 1980." But these sentences have different truth conditions. In particular, if true at all, they are true everywhere and at all times. If *R* does occur in Cambridge, that is a fact all over the world, and if *S* occurs in 1980, that is a fact at all times. You need not be in Cambridge in 1980 to meet true tokens of "*R* occurs in Cambridge" and "*S* occurs in 1980." But you do need to be in Cambridge in 1980 to meet the true tokens, *R* and *S;* for only there and then can *R* and *S* themselves be true. At all other places and times those tensed sentences would be false, whereas their alleged translations are true everywhere and always.

Now it may be contentious whether meanings are in general given by truth conditions and whether sameness of truth conditions guarantees sameness of meaning. But few will deny that same meaning means same truth conditions. Two sentences can hardly mean the same if, as here, they are true in quite different circumstances. And for all those reluctant to relate meaning to truth conditions at all, there is the following consideration. Because these tenseless sentences, if true at all, are nowhere and never false, the truth of any token of them says nothing about where or when it occurs. To be told, of some token *R,* "*R* occurs in Cambridge," is to be told nothing at all about where you are; and "*S* occurs in 1980" is similarly unenlightening about what time it is. Tokens of these tenseless sentences are quite useless for telling people where they are or what the time is. But those are the chief uses of the tensed sentences they purport to translate. Now, however meaning may relate in general to truth conditions, it is an undisputed canon of modern philosophy to relate it to usage. And however loosely that canon is applied, no sentence could possibly mean the same as another when, as here, it cannot be used at all as the other one standardly is. No theory of meaning, therefore, could make *R* mean "*R* occurs in Cambridge" or *S* mean "*S* occurs in 1980."

Obviously, nothing tenseless will translate tokens *R* and *S* if "*R* occurs in Cambridge" and "*S* occurs in 1980" do not. And if simple tensed sentences such as these have no tenseless translation, then no tensed sentence does. The fact, I think, needs arguing no further, least of all to opponents of tenseless time. The only question is what it implies for tenseless time. It is tempting to infer from it that tensed sentences mean more than their token-reflexive truth conditions reveal and that tensed facts may after all be needed to say what. How is the temptation to be resisted?

First, by resisting a bad analogy. Sometimes, indeed, there must be more to meaning than truth conditions: in mathematics, for example. If mathematical truths are all necessarily true, then each is true in the same conditions, namely, in all conditions. Thus "2 + 2 = 4" is true if and only if there is no greatest prime number. But "2 + 2 = 4" does not mean that there is no greatest prime number. Someone who already understands "2 + 2 = 4" will not learn what "There is no greatest prime number" means by learning that they have the same truth conditions.

But the contingent tensed sentences that we are concerned with are not like that. They may not *have* the same meaning as the tenseless sentences that give their truth conditions, but those truth conditions surely *give* their meaning. As I have already remarked, anyone who knows that, for all dates *T*, "It is now *T*" is true during and only during *T* knows what "It is now . . ." means. Given just this knowledge, he can use and understand tokens of any such present-tense sentence and distinguish it from all past- and future-tense sentences of which he also knows the token-reflexive truth conditions. (Granted, he also has to recognize what I have elsewhere called the "presence of experience," in order to tell that he is *now* hearing someone say "It is now *T*"; but I have already given a tenseless account of that phenomenon.) Here, unlike mathematics, correct usage *is* explained by people knowing how the truth of what they say depends on when and where they say it; in particular, the *different* meanings of different sentences are differentiated, as they are not in mathematics, by their different truth conditions. "Cambridge is here" and "Cambridge is ten miles away" are used differently because they are known to be true in different places; and similarly in time for "It is now 1980" and "1980 is two years ago."

Truth conditions give meanings less problematically here than in mathematics and elsewhere for another reason too. For a sentence's truth conditions to give its meaning, its being true in them must be more than a coincidence. Otherwise, so far as truth conditions go, the English sentence "Snow is white" could just as well mean that grass is green, since "Snow is white" *is* true and grass *is* green. "Snow is white" is indeed true if and only if grass is green. But that, of course, is just a coincidence. Even if grass were not green, "Snow is white" would still be true—provided snow was still white. "Snow is white" is true not only in the real world, it would also be true in any other world in which snow was white and false in any world in which it wasn't. That is really why the sentence means in English what it does, rather than meaning that grass is green. To give meanings, therefore, truth conditions generally have to include imaginary conditions as well as real ones. But it is a very moot point how enlightening about meanings reference to truth in imaginary conditions is. I could not, for example, teach anyone that "Snow is white" doesn't mean that grass is green by taking him to a world in which it isn't and showing him there that the English still take "Snow is white" to be true, because there is in reality no such world. On the contrary, we only know that "Snow is white" would be true in some such imaginary world because we already know what it means and assume for that very reason that snow could be white even if grass were not green.

This is a serious objection to using truth conditions to give meanings, but it does not apply when the conditions are token-reflexive. I *can* teach someone what "Cambridge is here" means by showing him, or describing to him in tenseless terms the real places where it is true and thus distinguishing them from the real places where it is false. And likewise for "It is now 1980." When truth conditions are being given for token sentences, they need not include imaginary conditions in any problematic way. They must admittedly apply to imaginary as well as to real tokens. All the tokens of "Cambridge is here" in Cambridge could by chance have occurred in King's College, making tokens of that sentence true as a matter of fact if and only if they occur in King's. But that again would be only a coincidence and would not give the real meaning of "Cambridge is here." So we must consider all the actual places where tokens of this English sentence *would* be true, even if none is actually here. But although we have therefore to consider imaginary tokens, we have only to consider them in actual places.

To give the (spatial or temporal) meanings of essentially tensed contingent sentences, we need only say when and where in the real world they would actually be true. In this case truth conditions seem to me both innocuous and effective: I find it hard to see what aspect of tensed meaning they fail to accommodate. And certainly they fail to do nothing that tensed facts could do better. For the only aspect of a sentence's meaning which tensed facts could supply is its truth conditions. This, after all, is what they are defined to do. That Cambridge is here and that it is now 1980 make "Cambridge is here" and "It is now 1980" objectively true if they do anything at all. They do nothing to make these sentences important or memorable or short or poetic: merely true. But that job has already been done by the tenseless facts adduced earlier. If doing that job does not suffice to give the meaning of tensed sentences, the lacuna will not be filled by tensed facts.

But if their tenseless truth conditions give the meaning of tensed sentences, why have they no tenseless translations? Actually, the reason is quite simple. To translate a sentence is to find another sentence with the same meaning: in particular, therefore, one with the same truth conditions. Now the truth conditions of tokens of spatially tensed sentence-types vary, as we have seen, with their spatial position, and those of temporally tensed types with their temporal position. So, therefore, must the truth conditions of tokens of their translations. But what makes sentence-types tenseless is that the truth conditions of their tokens do *not* vary in this way. Spatially tenseless tokens' truth conditions are the same at all places, and temporally tenseless ones' are the same at all times. So far as space and time are concerned, the truth conditions of tenseless sentences are not token-reflexive. No tenseless sentence, therefore, can have tokens whose truth conditions are everywhere and always the same as those of a tensed sentence, because, by definition, the latter vary from place to place or time to time, and the former do not. That is why no tenseless sentence can mean the same as a tensed one does; tenseless token-reflexive and non-token-reflexive truth conditions are bound to differ. Far from the tenseless translatability of tensed sentences following from the tenseless views of space and time, it follows that they are not thus translatable. So that obvious fact is no objection to a tenseless view of either space or time. On the contrary, only the tenseless view explains it. For if tensed sentence-types had tensed and non-token-reflexive truth conditions, that is, stated (if true) tensed facts which did not vary from token to token, it would so far as I can see, just be an unexplained axiom of tense that no such fact is ever identical with what any tenseless sentence states.

The Indispensability of Tense

But if tensed sentences are not translatable, why are they indispensable? If there are no tensed facts, why must we think and speak as if there were? Here, after all, we are only concerned with sentences expressing judgments, that is, stating what people take to be facts; not with expressing emotions, giving commands, and so on. Someone saying "Cambridge is here" or "It is now 1980" is stating what he takes to be a fact. But if all facts are tenseless, why not state them tenselessly? Why on the tenseless view must we make tensed judgments at all?

Yet it is as evident that we must do so as it is that tenseless sentences will not express them. We must obviously know more than the dates and grid references of things and events. To govern our interactions with them and with other people, we must also know which we are faced with at any time, that is, which are spatially and temporally present. But does this not mean that tenseless fact, like patriotism, is not enough, that we need tensed sentences because they state other, tensed facts that we also need to know?

Not at all. We do need to make and express tensed judgments, but not because we need to know and state tensed facts. The fact is that we need to make and express judgments whose truth conditions are token-reflexive. That is the real reason why tensed judgements are indispensable to our thought and speech, despite there being no tensed facts. But why do we need to make token-reflexive judgments? To answer that question, I must first say more clearly what judgments are.

I have so far left the idea of judgment deliberately inexplicit. I introduced it (in chapter 2 of *Real Time*) by remarking that we can think things past, present, or future without saying or writing anything; that is, there can be mental as well as physical tokens of tensed sentences. I said no more about them then, except that they have dates, because that was all I needed to give them tenseless token-reflexive truth conditions. What else judgments are like was immaterial. But now that I have to explain why we need to make judgments with these truth conditions, I shall have to go into more detail.

Basically, to judge that something is so is to believe it. So far, I have given only cases of conscious belief, and only then, I dare say, do we talk of "making a judgment." But belief, including tensed belief, can exist without being conscious. I keep left while driving in Britain, for example, because I believe that is still the rule, but I do not have the rule consciously in mind all the time I am following it. Again, I suppose a dog can believe it will soon be fed, without it being capable of making conscious judgments at all. Of course, not everyone accepts animal or other unconscious beliefs, but since it costs me nothing to admit them, I might as well sidestep that controversy by doing so. So from now on I shall talk of tensed beliefs rather than tensed judgments.

The idiom of belief also helps me to apply to mental tokens an important distinction drawn in chapter 2 of my book. I have so far called judgments events, namely, event tokens of whatever types of sentence would express them. Having a belief, however, is not an event. Acquiring or losing one is, since it is a real change in a believer's state of mind; and so is making a judgment, because it involves the conscious acquisition or reinforcement of a belief. But once acquired, beliefs persist to guide action even when the believer's conscious attention has turned elsewhere (as my beliefs do about which side of the road to drive on).

A person with a belief is what I have elsewhere called a "thing token," not an event token, of a sentence type that would express the belief. That is, a believer is like a printed sentence rather than a spoken one. The difference is that, unlike event tokens, thing tokens have a truth-value at each B-series instant and can therefore change in truth-value from one instant to another. And the truth-values of tensed beliefs do change from time to time as the things and events they are about change their A-series positions. So to keep them true, one must every so often change their tenses, from tokens of future- to tokens of present- to tokens of past-tense types. These changes are the psychological reality behind the myth of passing time, as will emerge in more detail (chapter 7 of *Real Time*). Here their importance is that we need them to undertake timely action. That is why beliefs themselves need to be changeable properties of believers, rather than events.

This way of distinguishing judgments from beliefs is unusual, but it suits my purposes and will not, I trust, be as contentious as the role of consciousness in judgment is. Consciousness, of course, is not the only contentious aspect of belief, but fortunately I can afford to be agnostic about most of the others as well. All I shall need to assume about belief is that it is a state of mind aimed exclusively at truth, is what perception produces, and combines with desire to generate action. Whether belief needs language, whether it is a relation believers have to propositions, whether it comes by degrees, how it manages to be about particular people and things, how it is embodied in the brain—all these hard and important questions I can evade, because different answers to them make no difference to the ensuing argument.

I should, however, enlarge somewhat on what I am assuming. It is nothing very contentious and needs, I think, no more explanation and defense than plausible illustration can provide. But that much it does need, since there is more to it than the last paragraph might suggest.

To start with, belief's peculiarly intimate connection with truth is not easy to pin down. It is not just that believing a sentence or proposition *P* is believing it to be true. Hoping that *P* is likewise hoping that *P* is true, and similarly for intending, regretting, and all the other so-called propositional attitudes. What for my purposes is special about belief can best be illustrated by the paradox pointed out by G. E. Moore.[1] It would, he said, be absurd for anyone to say "*P* is true but I don't believe it"; whereas there is generally nothing absurd in saying "*P* is true but I don't regret it" or "*P* is true but I don't intend it." Now this is not because "*P* is true but I don't believe it" could not be true; there are many truths we don't believe. But even if for example, the mail has come without my realizing it, "It is true that the mail has come but I don't believe it" would still, though true, be a most paradoxical thing for me to say. The reason is that I cannot say such a thing sincerely without turning it into a falsehood, because I cannot think something to be true without consciously believing it. Other propositional attitudes produce no such paradoxes, because settling for oneself a proposition's truth-value does not automatically settle one's other attitudes towards it. But it does settle whether one believes it. For me to believe something just *is* for me to have settled (for myself and for the time being) that it is true. This is what I mean by saying that belief is a state of mind aimed exclusively at truth in a way that other propositional attitudes are not.

This means in particular that, in order to dispose of tensed facts, we need only account tenselessly for the truth of tensed beliefs. Whenever other attitudes seem to imply tensed facts, they do so only by implying either a correspondingly tensed belief or the truth of such a belief. Regretting that something happened in the past, for example, implies believing that it did; and all there is to the *temporal* rightness of that regret is the truth of the tensed belief it incorporates. (Whether the happening was regrettable in other respects is not my business.) Similarly for other propositional attitudes—even knowledge, which differs from the others in implying truth and not merely belief. Some philosophers admittedly deny that knowledge implies belief; but even they will admit that I can know something only if, had I believed it instead, that belief would have been true. So in this case, too, accounting tenselessly for the objective truth or falsehood of a tensed belief will also account, so far as tense goes, for the objective rightness or wrongness of the other tensed attitude to the same proposition. In short, if tensed facts are not needed to make tensed beliefs true or false, they will not be needed at all. In discussing what the indispensability of tensed attitudes implies for tenseless time, therefore, we need not run the whole gamut of propositional attitudes. We need only consider tensed belief.

What makes tense indispensable, therefore, is that we cannot help having tensed beliefs. Partly we cannot help it because belief, including tensed belief, is what perceptions give us. To see or hear something is in the first place always to acquire or reinforce a belief about it: perception is nothing if not the acquisition or reinforcement of belief. Whatever has perceptions, therefore, must also have beliefs.

And whatever *acts* on the basis of its perceptions must have tensed beliefs.

Action is what really makes tensed belief indispensable. That action is, as I have said, generated by combinations of belief and desire is obvious and not seriously disputed. But that—and why—actions need *tensed* beliefs is less obvious. The fact that they do, however, and especially the token-reflexive reason for the fact, is both the crux of my case and the reason for the persistent and fatal attraction of the idea that tenses are real. The fact, therefore, and the reason for it need developing in some detail, largely as before by illustration.

Suppose I push a switch to turn on the radio. To motivate this action, I need some beliefs and desires: a desire to hear something on the radio, presumably, and a belief that pushing the switch will turn it on. But whatever the desire and the belief, desire and belief are both needed, since neither attitude engenders action on its own. I will not act unless I desire some outcome of my action; but nor will desiring an outcome make me act unless I believe it is an outcome. It is not enough for the switch actually to turn the radio on; I will still not push it unless I believe that it does.

(One must allow in general for comparative degrees of belief and desire, since people sometimes act to avoid a relative evil rather than to get anything they positively want, and sometimes they gamble on what they believe are very unlikely outcomes. But the greater complexity of other cases is not really germane. Any action needs some degree of belief in its efficacy as a means to some end, and it also needs some degree of tensed belief. The need for tense does not depend on the degree of belief involved; so we may as well stick to a simple case. The ensuing discussion will, I think, make it evident enough that tense, if needed there, will be needed everywhere.)

Now suppose more specifically that I want to hear the one o'clock news; so I push the switch at one o'clock. Why did I do that at one o'clock and not some minutes or hours earlier or later? Well, obviously, because I wanted to hear the one o'clock news. But that on its own is not enough, even given that I know that pushing the switch turns the radio on. I could have been wanting to hear the one o'clock news for hours, and I could have mastered the radio years ago. Something more than these two steady states of desire and belief is needed actually to propel my finger to the switch at a specific time. Obviously, what I also need to believe is that it is *now* one o'clock. Until I acquire that present-tense belief, I shall do nothing, however much I want to hear the news and however strong my tenseless belief in the efficacy of the switch.

What action needs, in short, is the belief, not the fact. Even if there were tensed facts such as one o'clock being now, they would not make me turn the radio on or do anything else. I turn the radio on when I *believe* it is now one o'clock, whether it actually is or not. My beliefs are what make me act, not the facts that make them true or false. But of course, when I act on false beliefs, tensed or tenseless, my actions are apt to fail, that is, not get me what I want. If it is not in fact one o'clock when I push the switch, or if the switch does not in fact turn on the radio, I am likely to miss the news I want to hear. So although the truth-value of my beliefs is irrelevant to explaining my actions, it is by no means irrelevant to explaining their success or failure. And since actions are meant to be successful, the truth conditions of beliefs are highly relevant to determining how actions depend on them. Truth being the sole aim of belief, it is all belief can contribute to successful action. In general, therefore, action so depends upon the truth conditions of the agent's beliefs that their being satisfied will suffice for its success (waiving again the extra complexities of action based merely on degrees of belief).

In particular, action will be timely if it satisfies the token-reflexive truth conditions of the tensed beliefs it depends on. To hear the one o'clock news, I need to turn the radio on at one o'clock. I turn it on when I believe it is *now* one o'clock, because the truth conditions of that belief are satisfied, and only satisfied, at one o'clock. It is a belief such that, if I act on it when it is true, my action will succeed. And to have that desirable property, it must be a tensed belief, because only tensed beliefs have truth conditions that include the time at which the belief is held. Tenseless beliefs, if true, are always true; so acting when they are true does nothing to secure the success of an action which needs to be done at a particular time.

This dependence of timely action on the token-reflexive truth conditions of tensed beliefs is no coincidence. Nor is it merely the result of acting on a prudential maxim ("If you want to succeed in life, act when your beliefs are true") that feckless people might ignore. That might be so if believing something were a state of mind independent of how one acts as a result of being in it—but it isn't. Suppose it was: suppose, for example, that people's beliefs were what they consciously thought them to be, so that self-deception about belief was not merely rare but impossible. Now imagine a husband who is always finding implausible pretexts to return home unexpectedly, that is, is acting in a way that makes sense only as being designed to succeed in showing his wife to be unfaithful if the belief that she is is true. Yet the husband is quite sincere in the flimsy pretexts he gives; he is consciously quite unaware of believing his wife to be unfaithful. We should have to say, then, that he did not really believe it after all, because he did not think he did, even though he acts in a way that then makes no sense. But we would say no such thing. He really is jealous, as his actions show; only he is at the same time deceiving himself into thinking that he isn't.

If a man wants above all else to do something at a particular time, knows how to do it, and is able to, it makes no sense to suppose he believes that time to be present but still does nothing because he is too feckless to act on his belief. If in those circumstances he does not act on that present-tense belief, he does not have it. To have it just *is* to be disposed to act in ways that will succeed if it and the other relevant beliefs are true. Explaining people's actions by their beliefs and desires means crediting them, willy-nilly, with beliefs whose truth would ensure the actions' success. And, in particular, explaining the timeliness of people's actions means crediting them, willy-nilly, with temporally token-reflexive and therefore tensed beliefs.

It follows from all this that a tensed belief, conscious or unconscious, must be a real psychological property of whoever has it. By this I mean that a change in the belief has to be a real event in a believer's mental life, that is, an event which has effects. This must be so because that is how tensed beliefs explain actions, namely, causally. If I already want to hear the one o'clock news and know how to work the radio, a change from future to present in the tense of my belief about one o'clock is what *causes* me to turn the radio on. Unless my states of tensed belief were real properties of mine, changes in them could not be, as usually they are, the immediate causes of my actions.

A proper discussion of change and of the distinction between real and other properties of people and things must wait for later. For the time being, an illustration will have to serve to indicate the significance of the distinction. Fame, for example, is not a real property of a famous man, since it resides in the acclaim of others, not in him. That is, although his hearing of his fame can immediately affect him, his merely becoming or ceasing to be famous cannot. Fluctuations in his fame occur at a distance from him, and causes do not produce their immediate effects at a distance. Conversely, what does affect him immediately, for example, by being an immediate cause of his actions, is not remote from him. That is the crucial difference between his beliefs and his fame, the difference that puts them among his real psychological properties—that is, among those which materialists will have to suppose to be embodied somehow in definite and distinct states of his brain.

Changes in tensed beliefs, being real events, will normally have causes as well as effects. The change from my thinking one o'clock future to thinking it present does not occur automatically at one o'clock; it takes a cause to set it off. The cause could come from the "internal clock" that any agent, human or animal, needs to have to change its tensed beliefs from time to time to try and keep them true. Or it could be external, for example my seeing a real clock change its display from "12:59" to "1:00." This perception, of course, needs tenseless beliefs as well to make it work—for example, that the clock is accurate and that my perceptions of its changes occur virtually as soon as they do. But given that I have those beliefs, the change in the clock will at once cause me to believe that it is now one o'clock and hence to turn on the radio to hear the news.

As this example illustrates, tensed beliefs about B-series times like one o'clock are really only intermediaries between perceiving some events and acting to affect others. The primary subjects of our tensed beliefs are events, not dates. We have beliefs about dates only because, being fixed attributes of many events, they conveniently sum up what we need to know about them. To know that one o'clock is present is to know that any event is present which has that time. But ultimately, coming to act on what we see and hear is a process of acquiring past- or present-tense beliefs about what we perceive, inferring present- or future-tense beliefs about events we wish to effect and then acting on them.

These inferences, from what we have perceived to what we wish to effect, need mediating by tenseless beliefs—for example, that the radio news starts as the clock strikes one or that a ball I am trying to catch will reach me a second or so after I see it. But mediation is no use without something to mediate; the timing of our action depends in the first place on the tense of beliefs our senses give us, about clocks striking, things to be caught or avoided, and so forth.

This, more precisely, is what I meant earlier by action based on perception needing tensed belief. The main function of perception, indeed, is to give us the tensed beliefs we must have to act on at the right time. This must be done for all action whose desired outcome has to occur at a particular time—or at the same time as a particular event—and that means almost all action. My radio example is no isolated case. To catch a train or a bus, as well as a ball, I must be in the right place at the right time; even to talk to someone, I must get him and the sound waves I emit somewhere simultaneously. There may be actions to whose success timing is completely immaterial, but not many. And for the rest, we need perceptions to give us tensed beliefs, because we need to derive from them beliefs with the right token-reflexive truth conditions to make us act in time.

Conversation is an important special case of an activity that calls for tensed belief, which it does for two main reasons. First, conversation conveys information whose value, as we have seen, often depends upon its being tensed. People can say what time it is, for example, as well as clocks can, and they use explicitly tensed forms of words, for example, "It is now one o'clock," to do so. These spoken tokens are meant to do what a clock does, namely, give their hearer corresponding mental tokens just when they are true, that is, in time to be acted on successfully. They may not do that in fact, because people can lie or be mistaken just as clocks can go wrong, but that is their function; and knowing that is all it takes to use—or misuse—these tensed expressions. Similarly, of course, for tensed information about particular events: "The bus is due in half an hour" tells you how soon and how fast to set off to meet it at the bus stop. Here again the importance of tense lies in the way the sentence being true only at certain times makes for the success of actions caused by coming to believe it at those times.

But even if I know all this and want to tell the truth, I produce tensed sentences when I believe them to be true, not necessarily when they actually are true. There is no more magic in its now being one o'clock to make me say "It is now one o'clock" than there was to make me turn the radio on. It is the beliefs that tensed sentences express, not the tensed facts that allegedly make them true, that cause us to use them when we do. Now many philosophers take the content of a belief to be whatever we need to know to use and understand sentences expressing it; and although the moral to be drawn from this claim is in general debatable, it seems to me plain enough here. We have seen that using tensed sentences demands nothing more than knowledge of when they are true and when false, that is, of their tenseless token-reflexive truth conditions; so that, I suggest, is all there is to the tensed beliefs they express. The idea that tensed sentences also express non-token-reflexive beliefs in tensed facts, while not yet positively disproved, is a gratuitous and idle supposition. It adds nothing to the token-reflexive explanation of why and how uses of these sentences and reactions to them depend on how action needs to be timely in order to be successful.

The token-reflexive truth conditions of tensed sentences incidentally explain not only how we usually use them but also how our usage varies when their tokens take time to arrive. So far, I have taken it that you hear what I say almost as I say it, so that what you hear will be true if what I say is true. When that is so, as it usually is, all I need do, to try and tell the truth, is to say what I believe. But when it takes a substantial or unpredictable time for my message to reach you, this may not work for tensed sentences. I cannot, for instance, expect to tell you the time by writing "It is now one o'clock" in a letter, since I have no idea at what o'clock you will read it. That being understood, the rule is that if I do write such a thing, you should ignore its truth-value when you read it and infer instead that it was true when I wrote it, that is, that I wrote it at one o'clock. On the other hand, if I want to tell or celebrate a time or an event in a tensed way at a temporal distance, I am allowed by convention to do so by dispatching a message that is false when I send it, so long as I intend it to be true when it arrives. That, for example, is why signing and sending birthday and Christmas greetings in advance, when what they say is still false, is not a case of lying.

These variant uses of tensed sentences are commoner still in the case of spatial tenses, since we more often communicate at significant spatial distances. In reading my letters, no one dismisses my tales of life "here" just because they are not true there. The reader infers instead that "here" refers to wherever I wrote the letter. And it is, of course, mainly to facilitate such inferences, both spatial and temporal, that correspondence customarily includes the date and place of its composition. Where that information is lacking, the tensed contents of correspondence become correspondingly unintelligible. "I wish you were here now" says a postcard, undated, unstamped, unsigned, and unaddressed. What, in tensed terms, could be more explicit? Yet the postcard clearly tells us almost nothing: not who, not where, not when. The token-reflexive account tells us why not.

The second reason why conversation needs tensed beliefs stems from its being a means of communication. Even if I am telling you something tenseless, I still need tensed beliefs in order to tell you it. The reason lies in a psychological mechanism of communication described by H. P. Grice as part of a theory of linguistic meaning.[2] Whether or not the whole theory is right, the part that demands tensed beliefs is undeniable and runs roughly as follows. When I tell you something, I may or may not persuade you that it is so. But I should at least persuade you that I believe it, so that insofar as you accept my authority, you will incline for that reason to believe it too. This, at any rate, is the normal situation, on which the exceptions are parasitic. In lying, for instance, I try to exploit the normal assumption, that people believe what they say, to mislead you about my beliefs and thereby about the facts. Either way, therefore, I mean you to form a belief about my beliefs, correct if I am speaking sincerely, incorrect if I am lying. And either way, the beliefs I intend to make you believe something about are the beliefs I have as I am speaking. But I cannot intend something without knowing what it is, which in this case means having a belief (as I speak) about what my own beliefs (as I speak) are, so that I know what I am trying to get you to believe I believe. Which means having a belief about what other beliefs I have at the same time, that is, a belief with the token-reflexive truth conditions of the present tense. So even if the belief I am trying to communicate is wholly tenseless, I still need a present-tense belief about what it is in order to communicate it.

In short, tensed beliefs are indispensable, not only for timely action in general and timely conversation in particular, but also for the conscious communication, sincere or insincere, of anything at all, tensed or tenseless. And the reason in both cases is that no agent, and above all no user of language, can do without token-reflexive beliefs. So, far from the tenseless view of time, with its token-reflexive analysis of tensed belief, implying that tensed beliefs are dispensable, it alone explains exactly why they are not.[3]

Notes

1. G. E. Moore, "A Reply to My Critics," in P. A. Schlipp, ed., *The Philosophy of G. E. Moore* (La Salle, Ill., 1942), pp. 535–677. The quoted passage occurs on p. 543.
2. H. P. Grice, "Meaning," *Philosophical Review* 66 (1957): 377–388.
3. This essay was originally published in Mellor, *Real Time,* as chap. 5.

ESSAY 2

Problems with the New Tenseless Theory of Time

QUENTIN SMITH

Since the early decades of this century the central debate in the philosophy of time has been between those who espouse a tenseless theory of time and those who hold a tensed theory. The proponents of the tenseless theory maintain that events are not really future, present, or past; they merely sustain unchanging relations to each other of simultaneity, earlier than, or later than, such that the obtaining of these relations is describable in a tenseless language. The upholders of the tensed theory claim that events really do possess transient properties of futurity, presentness, and pastness, and consequently that tensed descriptions are essential to any complete theory of time. From 1903, the date of publication of Russell's *Principles of Mathematics,*[1] up until the early 1980s, this issue was debated primarily in terms of the translatability of contingent tensed sentences (or their tokens) into tenseless sentences about permanently related events. The defenders of the tenseless theory argued that contingent tensed sentences or tokens *are* translatable into tenseless sentences and therefore do not refer to events with transient temporal properties. This position was argued by Russell, Smart, Goodman, Fisk, and many others.[2] The proponents of the tensed theory, on the other hand, maintained that contingent tensed sentences and their tokens are *not* translatable into tenseless sentences and therefore do refer to future, present, or past events. Philosophers like Broad, Prior, Taylor, and Gale espoused this viewpoint.[3]

But recently this debate has taken a new direction, largely due to problems that began to appear in the late 1960s and the 1970s with the assumptions underlying the traditional tenseless theory of time. For the most part, these problems did not arise from within the philosophy of time but from a different quarter, the philosophy of language. Philosophers of language working in the area of demonstratives or indexicals—Castañeda, Davidson, Kaplan, Perry, Wettstein, Lewis, and others—began to amass evidence and arguments to the effect that sentences containing indexicals are untranslatable into sentences not containing them.[4] Moreover, several of these philosophers, such as Castañeda, went further and argued that the basic kinds of indexicals are irreducible to each other, so that tenses and "now," for example, cannot be systematically translated by indexical-containing expressions like "this time" or "simultaneously with this utterance." Castañeda recently summarized his position by explaining that he has argued for "the untranslatability of *now*-statements into non-indexical statements: I have also argued that such statements cannot be translated into other indexical statements not containing 'now.'"[5] These linguistic investigations did not have an immediate impact upon the proponents of the tenseless theory of time, but beginning with the publication of Smart's "Time and Becoming" in 1980, proponents of this theory began in increasing numbers to abandon the old, or traditional, tenseless theory and adopt a new one. Although the theories of demonstratives developed by all the above-mentioned philosophers were influential in creating the climate for the new theory, the new theory had its specific impetus in some remarks about demonstratives in Davidson's "Truth and Meaning":

> Corresponding to each expression with a demonstrative element there must be in the theory [of meaning] a phrase that relates the truth conditions of sentences in which the expression occurs to changing times and speakers. Thus the theory will entail sentences like the following:
>
> "I am tired" is true as (potentially) spoken by *p* at *t* if and only if *p* is tired at *t*.
>
> "That book was stolen" is true as (potentially) spoken by *p* at *t* if and only if the book demonstrated by *p* at *t* is stolen prior to *t*.
>
> Plainly, this course does not show how to eliminate demonstratives: for example, there is no suggestion that "the book demonstrated by the speaker" can be substituted ubiquitously for "that book" *salva veritate*.[6]

Two points are of interest in this passage. First, Davidson is claiming that demonstratives are not eliminable in favor of, or translatable by, nondemonstrative constructions. Second, Davidson is affirming that sentences with demonstratives have truth conditions that are specifiable by demonstration-free clauses.

Smart first realized the implications of this theory for the tenseless theory of time.[7] He offers this response to one of the defenders of the tensed theory of time: "Taylor's criticisms of 'the attempts to expurgate becoming' seem to me to turn on the impossibility of translating indexical expressions, such as tenses, into non-indexical ones. I agree on the impossibility, but I challenge its metaphysical significance, since the semantics of indexical expressions can be expressed in a tenseless metalanguage."[8] Smart offers some examples of these semantics: "When a person *P* utters at a time *t* the sentence 'Event *E* is present,' his assertion is true if and only if *E* is at *t*. More trivially, when *P* says at *t* 'time *t* is now,' his assertion is true if and only if *t* is at *t*, so that if *P* says at *t* '*t* is now' his assertion is thereby true."[9] The idea behind these passages is that sentences with temporal indexicals are untranslatable into sentences without them, but this does not entail that tensed sentences refer to events with properties of futurity, presentness, or pastness. For the fact that tensed sentences have truth conditions statable in a tenseless metalanguage, truth conditions that involve only unchanging relations of simultaneity, earlier than, and later than, suffices to explain the conditions under which tensed sentences are true. The assumption that transient temporal properties are needed to make tensed sentences true is accordingly unjustified and gratuitous.

By this means the tenseless theory of time can be saved from the dangers posed by the theories of demonstratives developed in the late 1960s and the 1970s. The advocates of the tenseless theory can give the defenders of the tensed theory their thesis of untranslatability but at the same time claim that all temporal facts or determinations are describable in a tenseless theory, inasmuch as the metaphysical implications of tensed sentences (their implications about what time really is) are statable in the tenseless language that describes the truth conditions of tensed sentences.

Besides Smart's contributions to the new tenseless theory of time, versions of it have been developed or endorsed by David Mellor, Murray MacBeath, Paul Fitzgerald, and others.[10] But while the development of this new tenseless theory has been proceeding at a fast pace, there has been no response whatsoever to it (at least that I am aware of) from the defenders of the tensed theory of time. It has not even been mentioned by the tensers, let alone criticized.

It is this lacuna that I intend to fill in this essay. I shall argue that the new tenseless theory of time is faced with insurmountable problems and that it ought to be abandoned in favor of the tensed theory. I will examine the two extant versions of the new theory, the version worked out by Mellor and endorsed by Fitzgerald (in the next section), and the version suggested by some of the ideas of Smart and MacBeath (in the following section).

Mellor's Version of the New Tenseless Theory of Time

Mellor's *Real Time* (1981) was received with enthusiasm by some detensers. Paul Fitzgerald in particular believed that the new tenseless theory of time it contained solved once and for all the problem of whether events possess transient temporal properties:

> D. H. Mellor in *Real Time* deftly unties once and for all some knots that had been loosened by prior work. He gives us, I think, the final word on the supposed A-*determinations* (pastness, presentness, and futurity), and on what *tenses* do and do not indicate about the world and our thought. . . . He concedes to Gale and other *objectivist* A-theorists (believers in the consciousness-independent reality of A-determinations) their claim that A-statements cannot be translated into non-A-statements . . . [but he shows that] A-determinations or tensed facts are not needed to give an account of the truth-conditions of tensed statements. So A-determinations are otiose.[11]

I believe this appraisal of Mellor's theory is incorrect, since the theory is self-contradictory in a crucial respect. Mellor inconsistently holds all five of these positions: (1) tensed sentences have different truth conditions from tenseless sentences and thus are untranslatable by them; (2) tensed sentences have tenseless truth conditions, namely, tenseless facts; (3) these tenseless facts are the only facts needed to make tensed sentences true; (4) tensed sentences state the facts that are their truth conditions; and (5) tensed sentences state the same facts that are stated by the tenseless sentences that state the former sentences' truth conditions. After explaining these five positions, I will show that (1) is incompatible with (5).

Point 1 expresses Mellor's conviction that sameness of truth conditions is a necessary condition of one sentence translating another: "To translate a sentence is to find another sentence with the same meaning: in particular, therefore, one with the same truth conditions" (p. 27 above). The untranslatability of tensed by tenseless sentences (which for Mellor implies the untranslatability of any given token of a tensed sentence by any given tenseless sentence-token) is due to the difference in the temporal conditions in which they are true. "No tenseless sentence . . . can have tokens whose truth conditions are . . . always the same as those of a tensed sentence, because, by definition, the latter vary from . . . time to time, and the former do not" (p. 27). For example, any token *S* of "It is now 1980" is true *iff* [if and only if] *S* occurs in 1980, and any token *T* of "It is now 1981" is true *iff T* occurs in 1981. But truth conditions of tokens of tenseless sentences are not temporally variable in this way. Consider any token *U* of the tenseless sentence "*S* occurs in 1980." *U* need not occur in 1980 in order to be true: if it is the case that *S* occurs in 1980, then *U* is true regardless of when *U* occurs. "You need not be in . . . 1980 to meet true tokens of . . . '*S* occurs in 1980'" (p. 24); true tokens of this sentence can be met in 1990 or 1970 no less than in 1980.

Point 2 is implicit in the first point: tokens of tensed sentences have tenseless truth conditions, which are facts about the unchanging relations of simultaneity, earlier than, or later than that the tensed sentence-tokens bear to other phenomena. For instance, any token *S* of "It is now 1980" "is true if and only if it occurs in 1980." The clause following the biconditional states a tenseless fact, one that holds at all times: "if *S* occurs in 1980, that is a fact at all times" (p. 24).

Mellor's third point is that tenseless facts are all that are needed to make tokens of tensed sentences true. Any token *S* of "It is now 1980" is made true by the fact that *S* occurs in 1980, and no additional tensed fact, such as that it is now 1980, needs to be postulated: "The truth of a tensed statement depends only on how much earlier or later it is made than whatever it is about. Whether its subject matter is also past, present or future is irrelevant to its truth; so such statements can quite well be objectively true or false even though nothing in reality is past, present or future at all."[12] Point 4 is connected to the third point; tensed sentences do not state tensed facts (indeed, there *are* no tensed facts), but tenseless facts, particularly the tenseless facts that are their truth conditions. The fact stated by any token *S* of "It is now 1980" is the fact that *S* occurs in 1980. Mellor maintains that tensed sentences express tensed beliefs and that the content of these beliefs—what is believed to be a fact—is the truth conditions of the sentences: "We have seen that using tensed sentences demands nothing more than knowledge of when they are true and when false, that is, of their tenseless token-reflexive truth conditions; so that, I suggest, is all there is to the tensed beliefs they express. The idea that tensed sentences also express non-token-reflexive beliefs in tensed facts . . . is a gratuitous and idle supposition" (p. 35). Point 5 follows from the others, the fifth point being that a tensed sentence-token states the same fact that is stated by the tenseless sentence stating the tensed token's truth conditions. A token *S* of "It is now 1980" states the same fact that is stated by any token *U* of "*S* occurs in 1980."

To show that points 1 and 5 are incompatible, certain more or less implicit assumptions that Mellor makes must be made fully explicit. Mellor's crucial assumptions concern facts, but he never defines "fact"; nor does he defend his assumptions against those who reject facts or reduce facts to other items (for example, propositions). Since, however, I am interested in developing an *internal* critique of Mellor's theory, I shall also assume facts and shall use the word "fact" in the way Mellor does. Mellor's usage of this term appears to commit him to several theses, the first being that (a) facts correspond to true tokens of sentences, but not to false sentence-tokens. Accordingly, if we talk of a sentence "stating a fact" and do not wish to prejudge the sentence to be true, this phrase is to be understood as elliptical for some phrase like "stating what is taken to be a fact by the sentence-user." Mellor himself usually uses the shorter phrase but sometimes uses a phrase like the longer one, for example, in his remarks that he is only concerned with declarative sentences: "we are only concerned with sentences expressing judgments, that is, stating what people take to be facts" (p. 28). Mellor's second assumption has already been discussed in the foregoing, that (b) truth conditions, conditions that are necessary and sufficient to make sentences true, are facts. A third assumption that is implicit in Mellor's theory follows more or less directly from the first two assumptions, the third assumption being that (c) if a sentence as tokened on some occasion states a fact F_1, then the sentence as tokened on that occasion is true *iff* F_1 and every fact implied by F_1 exists. A fact F_1 implies another fact F_2 *iff* F_1 cannot exist unless F_2 also exists; for example, there cannot be the fact that the sun is in motion unless there is also the fact that the sun is extended.

Now assumptions (a), (b), and (c) entail the *principle of the identity of truth conditions* (as I choose to call it):

PITC If two tokens of the same sentence or two tokens of different sentences state the same fact, F_1, they have the same truth conditions; that is, are true *iff* F_1 and every fact implied by F_1 exist.

Mellor's theory is in contradiction with PITC and thus with its own assumptions, for the conjunction of points 1 and 5 contradicts PITC.

To see this more clearly and to see how this contradiction might be resolved, let us examine in detail the main example of tensed and tenseless sentences that Mellor offers:

> Let *R* be any token of "Cambridge is here" and *S* be any token of "It is now 1980." (*R* and *S* must of course not themselves be token-reflexive names or descriptions.) Then *R* is true if and only if it occurs in Cambridge, and *S* is true if and only if it occurs in 1980. If a sentence giving another's truth conditions means what it does, *R* should mean the same as "*R* occurs in Cambridge" and *S* should mean the same as "*S* occurs in 1980." But these sentences have different truth conditions. In particular, if true at all, they are true everywhere and at all times. If *R* does occur in Cambridge, that is a fact all over the world, and if *S* occurs in 1980, that is a fact at all times. You need not be in Cambridge in 1980 to meet true tokens of "*R* occurs in Cambridge" and "*S* occurs in 1980." But you do need to be in Cambridge in 1980 to meet the true tokens, *R* and *S;* for only there and then can *R* and *S* themselves be true. (p. 24)

Four things should be noted concerning the sample sentences "It is now 1980" and "*S* occurs in 1980." (1) Any token of the tenseless sentence "*S* occurs in 1980" states the fact that *S* occurs in 1980. (2) This fact is the tenseless truth condition of any token *S* of "It is now 1980." Moreover (3), this tenseless fact is the only fact stated by *S,* for *S* is a token of a tensed sentence, and tensed sentences express beliefs in "their tenseless token-reflexive truth conditions. . . . The idea that tensed sentences also express non-token-reflexive beliefs in tensed facts . . . is a gratuitous and idle supposition" (p. 35). And finally (4), any token of "*S* occurs in 1980" has different truth conditions from any token *S* of "It is now 1980," for *S* is true *iff* it occurs in 1980 and "*S* occurs in 1980," if true at all, is true "at all times" it is tokened. The conjunction of (1), (2), and (3) contradicts (4), for, by principle PITC, if two tokens of different sentences state the same fact, they have the same truth conditions.

This contradiction can be resolved if we reject (4), the claim that tokens of "It is now 1980" and "*S* occurs in 1980" have different truth conditions. The rejection of (4) seems *prima facie* plausible, inasmuch as *S* and any token *U* of "*S* occurs in 1980" both appear to be made true by one and the same fact, that *S occurs in 1980*. Call this fact F_1; since *S* and *U* both state F_1, they are both true *iff* F_1 and every fact implied by F_1 exist.

But why does Mellor not see this truth-conditional similarity between *S* and *U*? Mellor seems to think that *S* and *U* have different truth conditions, in that it is necessary and sufficient for *S* to be true that it occur in 1980, whereas it is neither necessary nor sufficient for *U* to be true that it occur in 1980. But Mellor is mistaken on this point, for this difference is not a difference in the *truth conditions* of *S* and *U,* the facts that make *S* and *U* true. It is a difference in what these facts are *about*. The fact that *S* occurs in 1980 is a fact about *S,* not *U,* and because of this only the occurrence of *S* is truth-conditionally restricted to 1980. Furthermore, the fact that *the fact that S occurs in 1980* is about *S* and not *U* is not a further fact uniquely among the truth conditions of *S,* for this further fact is also among the truth conditions of *U*. Certainly *U* cannot be true unless the fact that *S* occurs in 1980 is about *S* and not *U,* for if it is not about *S* and is about *U,* then this fact would not make any token of "*S* occurs in 1980" true.

If the contradiction in Mellor's theory is resolved in this way, a large price must be paid; for this resolution reduces Mellor's "new tenseless theory of time" to the *old* tenseless theory of time. Mellor's only grounds for holding that tokens of tensed sentences cannot be translated by tokens of tenseless sentences are that these tokens have different truth conditions, and once these truth conditions are seen to be the same, Mellor is deprived of his reasons for subscribing to the thesis of the new theory that tensed tokens are untranslatable. Mellor's theory becomes just another version of the old theory: tensed tokens are translatable by tenseless tokens and therefore do not imply that events have transient temporal properties.

This reduction of Mellor's theory to the old theory will not overly trouble detensers, for their primary goal is to show that there are no transient temporal properties, and whether this goal be achieved by the old or the new theory is of secondary importance. The tensers will be vindicated only if it can be shown that both versions of Mellor's theory, the new and the old, are unviable. This could be done by uncovering a contradiction in the old version of his theory as well, a contradiction that can be resolved only by adopting a tensed theory of time.

I believe that there is such a contradiction, and that the following considerations bring it to light. The sentence

(1) It is now 1980

entails the sentence

(2) 1980 is present.

In the language of facts, this means that there cannot be a fact statable by any token *S* of (1) unless there is a fact statable by any token *V* of (2). In other words, a fact statable by *S* implies a fact statable by *V*, and consequently a fact statable by *V* is among the truth conditions of *S*. But these considerations are incompatible with the claim that *S* has the same truth conditions as any token *U* of "*S* occurs in 1980," for a fact statable by *V* is not among the truth conditions of *U*. The tenseless fact stated by *V*, that *V* occurs in 1980, is not implied by the fact that *S* occurs in 1980, for "It is now 1980" could be tokened in 1980 even if "1980 is present" is not. And supposing—contrary to Mellor's assumptions—that there is also a tensed fact statable by *V*, that fact surely is not among the truth conditions of *U*, for *U* is a token of a tenseless sentence. Therefore, since a fact statable by *V* is a truth condition of *S* but not of *U*, it follows that *S* and *U* have different truth conditions and fail to translate each other.

The explanation of the difference in the truth conditions of *S* and *U* is at the same time an explanation of the entailment of (2) by (1). (1) cannot entail (2) by virtue of the *tenseless* facts stated by tokens of (1) and (2), since (to repeat) *that S occurs in 1980* does not imply *that V occurs in 1980*. There must be some other facts statable by *S* and *V* that explain this entailment, namely, *tensed facts*. The tensed fact statable by *S* is that *it is now 1980*, and the tensed fact statable by *V* is *1980 is present*, and these two facts imply each other. Alternatively, one could argue that these two facts are really one and the same fact and that (1) and (2) entail each other because the same tensed fact is statable by tokens of each.

These reflections enable the difference in truth conditions between *S* and *U* to be explained. *S* cannot have the same truth conditions as *U,* since in addition to *S*'s tenseless truth conditions there are its tensed truth conditions, and *U* does not have these tensed truth conditions. *S* and *U* share the same tenseless truth conditions, namely, the fact that *S occurs in 1980* and every fact implied by this fact, but *S* has tensed truth conditions not possessed by *U.*

Notice that I am describing the tenseless facts as *stated* by and the tensed facts as *statable* by *S* and *V;* this is because the tenseless facts need to be stated by *S* and *V* in order to exist, whereas the tensed facts need not be. The reason for this difference is that the tenseless facts are that *S* and *V* occur in 1980 and consequently require that *S* and *V* occur then, which is tantamount to saying that they require themselves to be stated by *S* and *V.* The tensed facts, on the other hand, are not about these tokens and hence do not require their occurrence. "1980 is present" need not be tokened in order for there to be the fact that 1980 is present, for this fact is not about *V* or any other token and implies no facts about *V* or any other token. The tenseless facts, then, have the necessary property of being stated by *S* and *V,* and the tensed facts have instead the dispositional property of being statable by *S* and *V.* It is this difference that helps to explain why (1) entails (2) even though the tokening of (1) does not require the tokening of (2).

Detensers will have a seemingly obvious objection to my introduction of these tensed facts. If the tenseless fact that *S* occurs in 1980 and the facts implied by this fact are necessary and sufficient to make *S* true, then, by definition, other facts—including tensed facts—are neither necessary nor sufficient for *S*'s truth and therefore cannot be truth conditions of *S.*

My response is that the tenseless facts are necessary and sufficient for the tenseless truth of *S* but are not sufficient for its tensed truth. To explain this response, let me first clarify "tenseless truth" and "tensed truth." The tenseless truth of the sentence-token *S* is expressed in such locutions as "*S* is (tenselessly) true at t_1" where "t_1" refers to the date of *S*'s occurrence. If *S* occurs only at t_1, then *S* is not true at t_0 or t_2 (since *S* is not located at these times), but it is true to say at t_0 and t_2 that "*S* is (tenselessly) true at t_1." The tensed truth of the sentence-token *S* is expressed in such locutions as "*S* is (in the present tensed sense) true" or "*S* is now true." *S* possesses tensed truth only when it is present.[13] If *S* was uttered in 1980 and is not now occurring, *S* is now neither true nor false. However, it is now true that *S* was true, for in 1980 when *S* was present it was true to say "*S* is (in the present tensed sense) true."

Given this account of tenseless and tensed truth, my response that "tenseless facts are insufficient for the tensed truth of *S*" can be explained. The tenseless fact that *S* occurs in 1980 and the facts implied by this fact are necessary and sufficient for *S* to possess the value of true on the date of its occurrence, namely, some time in 1980; but they are not sufficient for *S* to now possess the value of true. And this is not simply due to the fact that it is not now 1980. For suppose it is 1980 now; the fact that *S* occurs in 1980 and the facts implied by this fact still would not suffice to make *S* true now, for the fact that *S* occurs in 1980 does not imply the fact that it is now 1980. The tenseless fact and its implications cannot make *S* true now unless they are conjoined with the tensed fact that it is now 1980. But even this conjunction is not sufficient; an additional tensed fact is also required. Suppose *S* occurs in March 1980 and it is now September 1980; in that case

(3) *S* occurs in 1980

and

(1) It is now 1980

do not entail

(4) *S* is now true,

for *S* is not now occurring and thus now possesses no truth-value. In order for (4) to follow from (3) and (1), an additional premise is needed, that

(5) *S* is now occurring.

The fact that *S* is now occurring is not implied by the fact that it is now 1980 and must be regarded as an additional tensed fact *stated* (rather than *statable,* since this fact requires the occurrence of *S* for its existence) by *S*[14]

These considerations suggest that there is room for truth conditions of *S* other than the tenseless ones, namely, the tensed facts about the presentness of *S* and 1980 that are necessary conditions for the tensed truth of *S*. But are these tensed facts also *sufficient* for the tensed truth of *S*? If "being sufficient for the tensed truth of *S*" means that these facts and their implications suffice to make *S* true, then the answer is affirmative. Surely nothing else is needed for the tensed truth of *S* than the facts that it is now 1980 and *S* is now occurring and all the facts implied by these two facts (which include all the facts implied by the *conjunction* of these two facts). However, if "being sufficient for the tensed truth of *S*" means that these tensed facts could make *S* true now even if there were no tenseless fact that *S* occurs in 1980, then the answer is negative, for if it were 1980 now, it would be impossible for *S* to be true now unless it were a fact that *S* occurs in 1980. These two answers are of course consistent, for the conjunction of the two tensed facts *implies* the tenseless fact, inasmuch as

(1) It is now 1980

and

(5) *S* is now occurring

entail

(3) *S* occurs in 1980

A defender of Mellor's theory might respond that my introduction of a tensed truth-value of *S* is *ad hoc* and is merely designed to counter the objection that no other facts than the tenseless ones are necessary or sufficient to make *S* true. But I would reply that my introduction of a tensed truth-value is by no means unwarranted but is required by my explanation of the entailment of (2) "1980 is present" by (1) "It is now 1980," an entailment that is inexplicable given only Mellor's tenseless truth conditions and values. (1) entails (2) because the same tensed fact is (or logically equivalent tensed facts are) statable by each, namely, that it is now 1980 or that 1980 is present; and if this tensed fact is a necessary condition for the truth of *S* and *V,* then, by definition, it is a necessary condition for their tensed rather than tenseless truth. (In case this is not obvious, it follows from the true assumption that the tensed fact is not a necessary condition for the tenseless truth of *S*; by definition, the only other truth of *S* for which it could be a necessary condition is a tensed one.)

The detenser might object that if tensed facts are unnecessary for *S*'s tenseless truth, then *S* could be true without being true, that is, it could be (tenselessly) true without being (tensedly) true, which is paradoxical.

The response is that this is not paradoxical. Suppose *S* was truly uttered in 1980. In that case *S* possesses the value of true on the date of its occurrence. But *S* does not now possess the value of true, since it is not now 1980. So it is perfectly intelligible that *S* can be true (tenselessly) without being true (tensedly). The only fact about *S*'s tensed truth that *S*'s tenseless truth requires is that *S* be tensedly true *at some time,* that is, that *S* either now be true or will be true or has been true.

The detenser might then put forward a seemingly stronger objection: the idea that *S* has both a tenseless truth-value and a tensed truth-value entails the absurd notion that *S* has *two* truth-values. But by definition any sentence-token has only one truth-value.

My response is that the expression "*S* has a tensed truth and a tenseless truth" has misled the detenser if it suggested to him that there are two truth-values. In fact, there is only one truth-value possessed by *S,* such that *S*'s possession of this truth-value has both intransient temporal relations to other phenomena, relations describable in a tenseless language, and transient temporal properties, which are describable in tensed terms. If it is now 1980 and *S* is now true, the value of "true" it now possesses is not a different value than the one it possesses on the date of its occurrence in 1980. *S* possesses only one truth-value, but *S*'s *possession* of this value is considered on the one hand in respect of its presentness and on the other hand in respect of its simultaneity with the events of 1980. To say that *S* can be true (tenselessly) without being true (tensedly) is not to say that *S* can possess one truth-value (the tenseless one) without possessing the other (the tensed one), but that *S*'s possession of its one and only truth-value can have a simultaneity relation to the events of 1980 at a time when it does not have the property of presentness. It is now, in 1985, say, one of these times, for at this time *S*'s possession of its truth-value is both past and simultaneous with the events of 1980.

There are many other detenser objections and tenser responses concerning the issue of the tensed truth-values and conditions of tensed sentence-tokens, but my goal here is not so much to develop a full-blown tenser theory as it is to present an internal critique of the new tenseless theory of time. Accordingly, I shall forgo stating further responses to further objections and shall turn to a critical examination of the second of the two main versions of the new tenseless theory of time.

Smart's and MacBeath's Version of the New Tenseless Theory of Time

Mellor's version of the new tenseless theory of time is not the only possible version, and it is conceivable that another version succeeds where Mellor's fails. We have seen that Mellor's version fails because it can neither prove the basic premise of the new theory to be true nor prove the basic argument of the theory to be valid. The argument, summarily stated, is:

Premise: Tokens of tensed sentences have tenseless truth conditions, even though they are untranslatable by tokens of the tenseless sentences stating these truth conditions.

Conclusion: Therefore, the only real temporal facts are those stated by tokens of tenseless sentences.

Mellor failed to prove the premise to be true, since his explication of this premise entailed something that contradicted the premise, namely, that tokens of tensed sentences *are* translatable by tokens of the tenseless sentences stating the tensed tokens' truth conditions. Moreover, Mellor could not validate the inference to the conclusion, since he could not establish that the tenseless facts are the *only* truth conditions of tensed sentence-tokens; tensed facts need to be assumed to account for the entailment relations between tensed sentences for which Mellor's tenseless facts could not account.

However, ideas developed by other proponents of the new tenseless theory, namely, Smart and MacBeath, might be able to remedy these problems. Let us begin with the problem of validity. We saw that Mellor's tenseless truth conditions could not explain the logical equivalence of "It is now 1980" and "1980 is present," since *S occurs in 1980* neither implies nor is implied by *V occurs in 1980*. But perhaps Mellor is mistaken in thinking that the tenseless truth conditions of tensed sentence-tokens are token-reflexive, that is, that they are facts about the temporal relations of the subject matters of the sentence-tokens to the *sentence-tokens themselves*. Perhaps the real truth conditions of tensed sentence-tokens are facts about the temporal relations of the subject matters of the sentence-tokens to the *dates* on which the sentence-tokens occur. This is the view adopted by Murray MacBeath and Smart. Smart, for example, writes that "when *P* says at *t* 'time *t* is now,' his assertion is true if and only if *t* is at *t,* so that if *P* says at *t* '*t* is now,' his assertion is thereby true."[15] Applying this to our sample sentences, we get

(1) When *P* says in 1980, "It is now 1980," his assertion is true *iff* 1980 is at 1980.

(2) When *P* says in 1980, "1980 is present," his assertion is true *iff* 1980 is at 1980.

On this view, the same tenseless fact, that 1980 is at 1980, is the truth condition both of the 1980 tokens of "It is now 1980" and of the 1980 tokens of "1980 is present," and consequently the logical equivalence of these two sentences has a purely tenseless explanation.

But this won't do, for the alleged truth-condition sentences 1 and 2 are in reality no more truth-condition sentences than is

(3) When *P* says in the presence of white snow, "This snow is white," his assertion is true *iff* the white snow is in the same place as the white snow.

(1), (2), and (3) are not truth-condition sentences, since the real truth conditions of the sentence-tokens are stated *before* the biconditional. The sentence-tokens are made true by their occurrence in 1980 or in the presence of white snow, and the conditions specified after the biconditional are therefore redundant, being but some of the many tautologies implied by the real truth conditions (other of these tautologies being that the 1980 tokens occur when they occur and that the tokens occurring near the snow occur where they occur).

A second consideration showing that (1) and (2) are not real truth-condition sentences is that the 1980 tokens of "It is now 1980" and "1980 is present" are tokens of contingent sentences, and "1980 is at 1980" is a tautology, and no tautology can be both necessary and sufficient to make tokens of a contingent sentence true. Sentences like (1) and (2) can at best create the (misleading) appearance that this is the case by building the necessary and sufficient conditions of the truth of the contingent sentence-tokens into the clause before the biconditional.

These reflections should make it apparent that the fact that 1980 is at 1980 cannot explain the logical equivalence of "It is now 1980" and "1980 is present" or of the 1980 tokens of these sentences. For no tautological fact can make two logically contingent sentences or tokens true in all and only the same circumstances. A logically contingent fact is required, such as the fact that *it is now 1980*.

Readers of recent literature on indexicals and rigid designators might object that the indexical "now" rigidly designates the time of its tokening and that, consequently, a 1980 token of "It is now 1980" is tautological. I have elsewhere argued that, despite appearances to the contrary, tokens of "now" are not rigid designators;[16] but instead of repeating these arguments here, I shall show that even if tokens of "now" are rigid designators, it is false that 1980 tokens of "It is now 1980" are tautologically true. Let us name an arbitrarily selected 1980 token of this sentence, "*S*." Now for any tautologically true sentence-token, the truth of the token is entailed by premises stating the relevant tautological fact and that the token occurs. But

(1) 1980 is at 1980

and

(2) *S* occurs

do not entail

(3) *S* is true.

To derive (3), we need the additional premise that

(4) *S* occurs in 1980.

Note that (4) is not entailed by (2), since "*S*" is a proper name of the sentence-token and (if the Marcus–Kripke theory is correct) thereby directly refers to *S* without imparting any information about it, such as that it has the property of occurring in 1980.

This argument does not show that the 1980 token of "now" is not a rigid designator of 1980, for it is consistent with the view that 1980 is rigidly designated by this token. What it shows is that rigidly designating 1980 is not the only semantic property of this token, that this token in addition imparts some information about 1980. In tenseless terms, the imparted information is stated by the additional premise (4), but, as we have seen in the previous section, that cannot be the only information imparted. It must also impart the tensed information that 1980 is present. The 1980 token of "now" rigidly designates 1980 and ascribes to it the property of presentness. This does not of course imply that the token of "now" refers to 1980 *via* the property of presentness, which would be impossible, since the property of presentness does not refer to anything. Rather the token directly refers to 1980 and ascribes to its direct referent the property of presentness. If we use "*A*" as a directly referential name of 1980, we may say that the tensed semantic content of the "now" token may be represented as the ordered pair:

<*A*, the property of presentness>.

A second argument for the logical contingency of the 1980 token of "It is now 1980" is based on the difference between the semantic content of the tokens of "now" and "1980." The year 1980 is the twelve-month period that is 1979 years later than the birth of Christ. "1980" in its normal use expresses the sense that is also expressible by an attributive use of the definite description "the twelve-month period that is 1979 years later than the birth of Christ." The token of "now," assuming it is a rigid designator, does not refer to 1980 by expressing this sense, but directly; it directly refers to the set of all and only those events that, in fact, possesses the property of being the twelve-month period that is 1979 years later than the birth of Christ. If we call this set "*A*," we can say that the 1980 token of "It is now 1980" directly refers to *A* and asserts the identity of *A* with the twelve-month period that is 1979 years later than the birth of Christ. But this identity is contingent! For there are possible worlds in which *A* ***is not*** the twelve-month period that is 1979 years later than the birth of Christ. In world W_1 *A* exists but is not preceded by any set that includes the event of Christ's birth. In W_1, Christ was not born (even though people mistakenly believe otherwise), and consequently *A* does not possess the property of being 1979 years later than Christ's birth. In this world, everything happens as if Christ was born (and everything else is otherwise identical with the actual world), but in fact there is no birth to which subsequent sets of events can stand in the relation of *later than*. And in another world, W_2, Christ was born, but twenty years later than when he was actually born; in this world people mistakenly believe that Christ was born twenty years earlier than he was born. In W_2, the set *A* possesses the property of being 1959 years later than Christ's birth. Thus, to return to the actual world, the 1980 token of "It is now 1980" is only contingently true; the set *A* is identical with whatever twelve-month period is 1979 years later than Christ's birth in some but not all of the worlds in which *A* exists.

The problem with the "date version" of the new tenseless theory of time does not concern only its treatment of allegedly trivial sentence-tokens such as the 1980 tokens of "It is now 1980." It also concerns its treatment of sentence-tokens that it recognizes to be logically contingent. Consider Smart's truth-condition analysis of "*E* is present":

> When a person *P* utters at a time *t* the sentence "Event *E* is present," his assertion is true if and only if *E* is at *t*.[17]

(By "assertion," Smart means utterance.) Applying this to a concrete example, we get:

> When John Doe utters in 1814 the sentence "The Battle of Waterloo is present," his utterance is true if and only if the Battle of Waterloo occurs in 1814.[18]

Let us call John's utterance "*U*." Is a necessary condition of *U*'s truth that the Battle of Waterloo occur in 1814? The year 1814 is the set of all and only the events that occur during the twelve-month period that possesses the property of being 1813 years later than the birth of Christ. Let us call the set that actually possesses this property "*B*." Now *B* exists in many possible worlds; but in some of these worlds it is preceded by some year-long sets different from the ones that it is actually preceded by. For example, in world W_1 *B* is not preceded by any year-long set of events that includes the birth of Christ, and in this world *B* does not possess the property of being the year 1814, the year-long set of events that is 1813 years later than the birth of Christ. And yet *U* is true in W_1, for in this world it occurs when the Battle of Waterloo occurs. Surely John Doe utters something true in W_1 if he utters "The Battle of Waterloo is present" simultaneously with the Battle of Waterloo. Consequently, the occurrence of the Battle of Waterloo in 1814 is not a necessary condition of *U*'s truth.

It might be objected that the date-expressions in the truth conditions of *U* should be read *de re* rather than *de dicto*. That is, "1814" should not be given its normal reading in which it abbreviates the definite description "the twelve-month period that is 1813 years later than the birth of Christ" but should be read as directly referring to the set of events that is, in fact, 1813 years later than the birth of Christ. This is the set of events *B*. It follows from this reading that in W_1 *U* is true, since in this world the Battle of Waterloo *is* a member of *B*.

But this *de re* reading of "1814" does not suffice to save the date version of the new tenseless theory of time, for there are other possible worlds in which *U* is true but in which the Battle of Waterloo does not belong to *B*. There is a world W_3 in which John Doe utters *U* while the Battle of Waterloo is occurring but in which a leaf that actually fell simultaneously with this battle does not fall. That is, in W_3 *U* and the Battle of Waterloo belong to a set that is identical with *B* in every respect but for the falling of the leaf. Since sets with different members are different, the set in W_3 is not the set *B* but some different set, *C*. This shows that *U* is true in W_3 even though the Battle of Waterloo is not a member of *B*. Therefore the battle's occurrence in 1814 is not a necessary condition of *U*'s truth even if "1814" is read *de re* as referring directly to *B*.

What these examples reveal is that the occurrence of an event on a certain date is not necessary and sufficient for the truth of a tensed sentence-token. The tenseless truth conditions of such a token concern its temporal relation to the event it is about (for example, *U* is tenselessly true if and only if it is simultaneous with the Battle of Waterloo). The tensed truth conditions of such a token involve the temporal property the event possesses (for example, *U* is tensedly true if and only if the Battle of Waterloo and *U* are present).[19]

I conclude that the date version of the new tenseless theory of time is no more successful than the token-reflexive version developed by Mellor and is in some respects less successful, since it fails to account even for the tenseless truth conditions of tensed sentence-tokens. This suggests that the challenge to the tenseless theory of time posed by theories of the untranslatability of indexicals developed in the 1960s and 1970s has not been met by the new tenseless theory of time. The tenseless theory must be either abandoned or radically reworked.[20]

Notes

1. Bertrand Russell, *The Principles of Mathematics* (New York, 1903), chap. 54.
2. See, e.g., ibid.; J. J. C. Smart, *Philosophy and Scientific Realism* (New York, 1964), pp. 133 ff.; Nelson Goodman, *The Structure of Appearance* (Cambridge, Mass., 1951), pp. 287–298; Milton Fisk, "A Pragmatic Account of Tenses," *American Philosophical Quarterly* 8, no. 1 (1971): 93–98.
3. See, e.g., C. D. Broad, *Examination of McTaggart's Philosophy*, vol. 2, pt. 1 (New York, 1976); A. N. Prior, *Past, Present and Future* (Oxford, 1967); Richard

Taylor, *Metaphysics* (Englewood Cliffs, N.J., 1974); Richard Gale, *The Language of Time* (New York, 1968).

4. See, e.g., Hector-Neri Castañeda, "Indicators and Quasi-Indicators," *American Philosophical Quarterly* 4, no. 2 (1967): 85–100; Donald Davidson, "Truth and Meaning," in Jay F. Rosenberg and Charles Travis, eds., *Readings in the Philosophy of Language* (Englewood Cliffs, N.J., 1971); David Kaplan, "Dthat," in Peter A. French, Theodore E. Uehling, Jr., and Howard K. Wettstein, eds., *Contemporary Perspectives in the Philosophy of Language* (Minneapolis, 1979); John Perry, "The Problem of the Essential Indexical," *Noûs* 13 (1979): 3–21; Howard K. Wettstein, "Indexical Reference and Propositional Content," *Philosophical Studies* 36 (1979): 91–100; David Lewis, "Attitudes *De Dicto* and *De Se*," *Philosophical Review* 88 (1979): 513–543.
5. Castañeda, letter to Quentin Smith, 8 Apr. 1985, p. 2.
6. Davidson, "Truth and Meaning," p. 464.
7. Smart, "Time and Becoming"; see nn. 9 and 24.
8. Ibid., p. 11.
9. Ibid., p. 5.
10. Essay 28; Mellor, *Real Time;* idem, "McTaggart, Fixity and Coming True," in R. Healy, ed., *Reduction, Time and Reality* (Cambridge, 1981); Essay 26, Essay 27; Paul Fitzgerald, Critical Notice of Mellor's *Real Time, Philosophy and Phenomenological Research* 45, no. 2 (1984): 281–286. L. Nathan Oaklander has developed some ideas pertinent to the new tenseless theory of time in his *Temporal Relations,* chap. 4, but he subscribes to a version of the old tenseless theory of time: "Surely, any expressions in which the indexical word 'now' occurs could be replaced by one in which 'this moment' or 'this utterance' occurs" (p. 127).
11. Fitzgerald, Critical Notice, p. 281.
12. Mellor, *Real Time,* p. 5.
13. It is possible to distinguish two senses of "tensed truth"; one may use this phrase to mean *present-tensed truth* (as I am using it) or *future-, present-, or past-tensed truth* (as I am not using it). In the latter sense, *S* now possesses tensed truth if and only if *S* will be true, is true, or was true.
14. It should be noted that not all the facts stated or statable by *S* need to be explicitly stated or statable; presumably only the fact that it is now 1980 is explicitly statable, and the others are tacitly stated. This seems to be confirmed by the relevant data about communication, e.g., when I hear a true token of "It is now 1980," I explicitly grasp the fact that it is now 1980 and only tacitly comprehend that *S* is now occurring and that *S* is occurring in 1980.

    This distinction enables a problem about the conditions of translation to be solved. If sameness of truth conditions is a necessary condition of translation, as Mellor suggests, then "It is now 1980" does not translate "1980 is present," which is counter-intuitive. I suggest that this condition be dropped and that we assume instead that *sameness of facts explicitly stated or statable* is a necessary condition of translation, where "sameness" means logical equivalence.
15. Smart, "Time and Becoming," p. 5.
16. Quentin Smith, "Tenses, Temporal Indexicals and Time" (mimeograph, 1986).
17. Smart, "Time and Becoming," p. 5.

18. "1814" may be understood as short for "18 June 1814," which, strictly speaking, is the date of the battle.
19. The argument that *U* requires tensed truth conditions in addition to its tenseless token-reflexive truth conditions proceeds in the same way as the argument regarding the tokens of "It is now 1980" and "1980 is present." The token *U* of "The Battle of Waterloo is present" states a fact that is logically equivalent to the fact stated by a token *Υ* of "The Battle of Waterloo is taking place" (where the "is" is present-tensed). However, the fact that *U* is simultaneous with the Battle of Waterloo is not logically equivalent to the fact that *Υ* is simultaneous with the Battle of Waterloo. A tensed fact is needed to explain the equivalence, viz., the fact that the Battle of Waterloo is present.

    It should be added that this argument from logical equivalence is not the only argument that establishes the existence of tensed facts. Other semantic arguments are presented in my "Tenses, Temporal Indexicals and Time," and some phenomenological arguments are presented in my book *The Felt Meanings of the World: A Metaphysics of Feeling* (West Lafayette, Ind., 1986), chaps. 4 and 6.
20. Proponents of the new tenseless theory of time propound two arguments for the tenselessness of time that I have not addressed in this essay. First, they argue on the basis of McTaggart's paradox that tensed facts are self-contradictory. I have countered this argument elsewhere (Essay 15), by showing that McTaggart's paradox warrants no such conclusion. Second, they argue that the physical sciences represent the world tenselessly. I have rebutted this argument in "Mind-Independence of Temporal Becoming."

    The old tenseless theory of time has been critically discussed in my "Sentences about Time," *Philosophical Quarterly* 37, no. 146 (1987): 37–53, and "The Impossibility of Token-Reflexive Analyses," *Dialogue* 25 (1986): 757–760.

ESSAY 3

A Defense of the New Tenseless Theory of Time

L. NATHAN OAKLANDER

As we ordinarily think and talk about time, it is a truism that time passes. Dates, like the events that occur at those dates, are once in the future, then become present, and then recede into the more and more distant past with the passage of time. To think of time as passing and events as changing with respect to the characteristics of pastness, presentness, and futurity is to conceive of the transient aspect of time, or temporal becoming. One central issue in the philosophy of time concerns the metaphysical nature of temporal becoming. Do events exemplify the nonrelational properties of *pastness, presentness,* and *futurity,* as the tensed theory maintains, or are they intrinsically tenseless, exemplifying only the unchanging relations of *simultaneity, earlier than,* and *later than,* as the tenseless view believes? Although the issue is metaphysical, the dispute between the tensed and the tenseless views has, until quite recently, centered around temporal language. Defenders of the tenseless view have often argued that since tensed discourse could be eliminated or translated without loss of meaning into tenseless discourse, an adequate account of the nature of time need not countenance any special kind of tensed fact or tensed properties. In other words, the old tenseless theory of time assumed that a logical analysis of ordinary language that eliminates tensed discourse supported an ontological analysis of time that rejects transient temporal properties. The tenser shared that assumption but argued that, since no tenseless translations were successful, temporal becoming in some form or another (for example, as the acquiring and shedding of transitory temporal properties, or as the moving NOW) is necessary in any adequate account of time. Tensers claim, in other words, that because tensed discourse is ineliminable, the detenser is mistaken, and tensed properties and facts must exist.

For a variety of reasons, some having to do with arguments in the philosophy of time and some having to do with arguments in the philosophy of language, recent defenders of the tenseless view have come to embrace the thesis that tensed sentences cannot be translated by tenseless ones without loss of meaning.[1] Nevertheless, recent detensers have denied that the ineliminability of tensed language and thought entails the reality of temporal properties. According to the new tenseless theory of time, our need to think and talk in tensed terms is perfectly consistent with its being the case that time is tenseless. Tensed discourse is indeed necessary for timely action, but tensed facts are not, since the truth conditions of tensed sentences can be expressed in a tenseless metalanguage that describes unchanging temporal relations between and among events.

In the previous essay (originally published in 1987), Quentin Smith offers a provocative response to the new tenseless theory of time. He argues that since the new tenseless theory is faced with insurmountable problems, it must be either radically reworked or abandoned in favor of the tensed theory. Although he offers numerous arguments in support of these contentions, his central arguments purport to show that the detenser gives a logically inadequate analysis of ordinary temporal discourse and a metaphysically inadequate (because incomplete) account of the truth conditions of tensed sentences. I argue here that the tenseless theory of time need not be abandoned or radically reworked (although it perhaps needs to be clarified), since the difficulties Smith raises are indeed surmountable.

Smith launches his attack on the new tenseless theory by criticizing D. H. Mellor's token-reflexive version of it. On Mellor's view the world is intrinsically tenseless, in that events and things are not in themselves past, present, or future. Of course, we do make judgments (and have beliefs) about the tense of things, and such judgments (or beliefs) are sometimes true, but the truth conditions of a tensed sentence- or judgment-token can be given in terms of a tenseless and not a tensed fact. On the token-reflexive account that Mellor propounds, the temporal relation between the date at which one says, thinks, or writes down a tensed sentence and the event or thing that it is about provides an objective basis for the truth-value of any tensed sentence. A present-tense sentence-token is true if and only if it occurs (exists tenselessly) at (roughly) the same time as the event it is about; a past-tense token is true if and only if it occurs at a time later than the event it refers to, and so on. Thus, on the token-reflexive account, the truth conditions of tensed sentence- and judgment-tokens are tenseless facts.[2]

Mellor argues that we should not be misled into thinking that tensed discourse is eliminable, translatable, or has the same meaning as tenseless discourse. For a necessary condition of one sentence being the translation of another is that they both have the same truth conditions, but tensed sentences *have different truth conditions* from tenseless ones. At this point, through a judicious selection of quotes, Smith argues that Mellor contradicts himself because he also maintains that tensed sentences *have the same truth conditions* as the tenseless sentences that state their truth conditions and thus *are* translatable in terms of the tenseless ones. Consequently, Smith claims that Mellor's tenseless account of time is internally inconsistent, since he maintains that tensed sentences both do and do not have the same truth conditions as tenseless sentences and that tensed sentences both are and are not translatable by tenseless ones.

At the outset we may as well admit that Mellor is not always as clear as he might be and that, therefore, there is some basis in the text for attributing an internal inconsistency to him. Nevertheless, the token-reflexive version of the new tenseless theory of time can avoid the contradiction that Smith appears to uncover by distinguishing between sentence-types and sentence-tokens. To see how this distinction helps, let us begin by clarifying it. A sentence-token is a particular object that exists at a definite time and a definite place. A sentence-type is either the sum of all the tokens of that type or the geometrical property (the shape) that is common to all tokens of that type. As Mellor puts it:

> For me, the important feature of tokens as opposed to types is that a token is a particular object, in this case an arrangement of ink on the particular piece of paper you are looking at, that is, a thing which is in a definite place at every moment of its existence. The sentence type, by contrast, is a much more widespread object than any of its tokens, if indeed it is an object at all. The sentence type you are now reading a token of, for instance, is scattered across the world as widely as—I hope—copies of this book are. Sentence types are in fact not so much objects as properties of objects, namely of all the objects that are their tokens.[3]

Now, consider the sentence-type

(1) It is now 1980

and call it "*S*." What are the truth conditions of *S*? Insofar as *S* is construed as a tensed sentence-type, it does not strictly speaking have truth conditions; only its tokens do. As a consequence, we should also say that, strictly speaking, tensed sentence-types have no truth-value. Nevertheless, we can speak of the "truth conditions" of *S* in a Pickwickian sense, in which case they will vary from time to time; that is, they will depend on when a token of *S* is thought or uttered or written down. A sentence-token "is uttered" when the words, either written or spoken, of which it is composed, are produced on a given occasion. And, in the case of token-reflexive sentences like *S*, "is true" and "is false" apply to sentence-tokens. Thus, a token of *S* uttered in 1980, call it "*S* (1980)," is true because it occurs (exists tenselessly) in 1980, whereas another token of *S* uttered in 1981, call it "*S* (1981)," is false because it does not occur in 1980. Clearly, then, "some tokens of the same tensed type will differ in truth-value depending on their date" (p. 173). On the other hand, none of the tokens of the same tenseless type will vary in truth-value from time to time. For example, all tokens of "*S* (1980) occurs in 1980" have the same truth-value regardless of the date.[4] Thus the truth conditions of (the different tokens of) the sentence-types "*S*" and "*S* (1980) occurs in 1980" are different, because a token of the tenseless sentence may be true at a time when a token of the tensed sentence is false.

However, if we are considering "*S*" and "*S* occurs in 1980" as sentence-tokens, then their truth conditions are the same: they are true if and only if *S* occurs in 1980. Smith, on the other hand, interprets Mellor to be saying that "any token of '*S* occurs in 1980' has different truth conditions from any token *S* of 'It is now 1980,' because *S* is true *iff* it occurs in 1980 and "*S* occurs in 1980," if true at all, is true 'at all times' it is tokened" (p. 44). But this way of putting the point is misleading. If we are talking about a token of "*S* occurs in 1980," then it is nonsense to speak of it as being true at all times it is tokened. Tenseless sentence-types are "true" (or "false") at all times they are tokened (that is, tokens of a tenseless sentence-type are either all true or all false), but tokens are not themselves tokened at different times. Furthermore, to maintain that *S* is not true "at all times" is ambiguous and masks a confusion. If "*S*" stands for a sentence-type, then any token of it is true if it is produced in 1980. But if "*S*" stands for a 1980 token of "It is now 1980," then, like the tenseless sentence that states its truth conditions, *S is* true at all times. To quote Mellor once again: "The whole point of the type/token distinction is that tensed tokens, as opposed to types, have definite and temporally unqualified truth-values. . . . A saying or writing of '*E* is past' which occurs before *E* always was and always will be just plain false" (p. 172). Thus it does not follow that tensed sentence-tokens have different truth conditions from the tenseless sentence-tokens which state these truth conditions. Smith could only think they were different by confusing sentence-tokens with sentence-types. More important, Mellor's version of the token-reflexive theory can be modified to avoid an alleged internal inconsistency. For there is no inconsistency in claiming that tensed and tenseless sentence-*types* have tokens with different truth conditions, while also claiming that tensed and tenseless sentence-*tokens* themselves have the same truth conditions.

From Smith's point of view, this way out of the contradiction will provide Mellor with little solace. For if tensed tokens have the same truth conditions as tenseless ones, then Mellor's view reduces to the old tenseless theory of time which he explicitly denies. As Smith argues: "Mellor's *only* grounds for holding that tokens of tensed sentences cannot be translated by tokens of tenseless sentences is that these tokens have different truth conditions, and once these truth conditions are seen to be the same, Mellor is deprived of his reasons for subscribing to the thesis of the new theory that tensed tokens are untranslatable" (p. 44, my emphasis).

Smith's reasoning here is not very convincing. In the first place, for Mellor, having the same truth conditions is a necessary but not a sufficient condition for translatability. Thus, even if a tensed and tenseless sentence-token have the same truth conditions, it does not follow that the former can be translated by the latter. Furthermore, it is simply not true that Mellor's *only* ground for denying the translatability thesis is that tensed and tenseless sentences have different truth conditions. Mellor gives other reasons for denying the translatability thesis. He claims that in order for two sentences to have the same meaning, they must have the same use. Now, one of the chief uses of tensed sentences is to tell people what time it is. For example, it is perfectly correct to answer the question "When are we going to the movies?" by the retort "We are going to the movies now." On the other hand, we cannot use the tenseless sentence "Our going to the movies is simultaneous with the token 'We are going to the movies now'" to inform a questioner when we are going to the movies. Hence Mellor concludes that tensed and tenseless sentence-tokens have different meanings, and that is one reason, other than their having different truth conditions, why tokens of tensed sentences cannot be translated by tokens of tenseless sentences (see Essay 1).

There are other reasons for denying the translatability thesis, but we need not pursue them, for on that point recent detensers and tensers both agree. Moreover, the crucial question is not whether Mellor's view is internally consistent but, rather, whether the token-reflexive version of the new tenseless theory of time is true. And that turns on the following question: "Do tokens of tensed sentences have only those tenseless truth conditions stated by tokens of tenseless sentences?" Smith argues that they do not, since tensed facts must also be introduced, but I shall argue that his arguments fail.

Smith begins his argument against the token-reflexive account of the truth conditions of tensed sentences by noting that:

(1) It is now 1980

entails the sentence

(2) 1980 is present.

He writes: "In the language of facts, this means that there cannot be a fact statable by any token *S* of (1) unless there is a fact statable by any token *V* of (2). In other words, a fact statable by *S* implies a fact statable by *V*, and consequently a fact statable by *V* is among the truth conditions of *S*" (p. 45). He then claims that the tenseless truth conditions (or the fact statable by any token *S*) of (1), namely, *S* occurs in 1980, does not entail the tenseless truth conditions (or the fact statable by any token *V*) of (2), namely, *V* occurs in 1980 (for (1) could be true although no token *V* of (2) is uttered). He concludes that tenseless truth conditions are not sufficient to explain the logical relations between (1) and (2); tensed truth conditions must be introduced for that purpose. Thus, Mellor's token-reflexive version of the tenseless theory fails because "he could not establish that the tenseless facts are the *only* truth conditions of tensed sentence-tokens; tensed facts need to be assumed to account for the entailment relations between tensed sentences for which Mellor's tenseless facts could not account. . . . Mellor's tenseless truth conditions could not explain the logical equivalence of 'It is now 1980' and '1980 is present,' since *S occurs in 1980* neither implies nor is implied by *V occurs in 1980*" (p. 50).

There does indeed *appear* to be a difficulty here for the token-reflexive analysis. For if one sentence logically implies a second, then we should be able to justify the inference on the basis of truth conditions; we should be able to show that what makes the first true must make the second true. If we cannot do this, there would seem to be grounds for concluding either that we are mistaken about the putative entailment relations or that we have not got the right truth conditions for the sentences in question.

Although Mellor does not directly consider this objection, his most recent pronouncements on time and tense suggest a way out of the difficulty Smith raises.[5] It involves employing Kaplan's views on demonstratives and indexicals, and arguing that one can thereby account for the logical equivalence of (1) and (2) in terms of tenseless truth conditions.[6] According to Kaplan (and Mellor), the meaning of an indexical sentence-type (and all its tokens) is a semantic function (rule) from facts about tokens of that type (their context of utterance) to their tenseless truth conditions. In particular, the meaning of (1) and (2) is a semantic function from the context of utterance, namely, the time at which their tokens are produced, to their tenseless truth conditions. Since the context of utterance varies, so do the truth conditions of their tokens, but in each case the truth conditions are tenseless. Thus, any token of (1) is true with respect to the context in which it is produced (namely, the time at which it is uttered), if and only if the year of that context is 1980, and the same may be said of any token of (2). Consequently, since the truth conditions of (tokens of) (1) and (2) are the same, the difficulty of getting (1) and (2) to be logically equivalent vanishes.[7]

In the second part of his essay Smith critically examines the date version of the new tenseless theory of time. Like the token-reflexive account, the date version of the tenseless theory is not new. Earlier proponents of the tenseless theory such as Russell, Goodman, and Quine adopted it.[8] In its old form the date version of the tenseless view maintained that a sentence in which the word "now" or its equivalent is used can be translated through the use of a second sentence formed by replacing the "now" in the first sentence with any date-expression used to refer to the time at which the first sentence was uttered. Consider, for example, Quine's statement of this view: "Logical analysis is facilitated by requiring rather that each *statement* be true once and for all or false once and for all, independently of time. This can be effected by rendering verbs tenseless and then resorting to explicit chronological descriptions when need arises for distinctions of time. . . . The sentence (1) 'Henry Jones of Lee St., Tulsa, is ill' uttered as a tensed sentence on July 28, 1940, corresponds to the statement 'Henry Jones of Lee St., Tulsa, is [tenseless] ill on July 28, 1940'."[9]

The new date version denies the thesis of linguistic reducibility and claims instead that corresponding to every tensed sentence-token is a tenseless sentence that gives its truth conditions. For example, Smart claims that "the notion of becoming present seems a pretty empty notion, and this is even more obvious when we recognize the indexical nature of words like 'present,' 'past,' and 'future.' When a person *P* utters at a time *t* the sentence 'Event *E* is present,' his assertion is true if, and only if, *E* is at *t*. More trivially, when *P* says at *t* 'time *t* is now,' his assertion is true if, and only if, *t* is at *t*, so that if *P* says at *t* '*t* is now' his assertion is thereby true.[10]

This view has recently been modified, renamed the "co-reporting thesis," and defended by Richard Gale and Michelle Beer.[11] Again, the heart of the co-reporting thesis is that temporal indexicals like "now," "this time," and "the present," as used on a given occasion, are referring terms which denote a time. On this view, if a temporal indexical sentence such as "Event *E* is now occurring" is uttered at t_1, then it reports an event that is identical with the event reported at any time by the use of the nonindexical sentence "Event *E* is occurring at t_1." On this view, indexicals and proper names such as dates are rigid designators. Thus a tensed sentence like "It is now 1980," uttered in 1980, reports the same fact as the necessary truth reported by "It is 1980 in 1980" or "1980 is at 1980." It does not follow, and it is not part of the co-reporting thesis to maintain, that "It is now 1980" and "1980 is at 1980" express the same proposition or have the same meaning. On this version of the new tenseless theory of time, as on the other, two sentences can have different meanings, while still having the same truth conditions or corresponding to the same fact.

Smith's main arguments against the date version of the new tenseless theory of time purport to demonstrate that the truth conditions of tensed sentences are not what the tenseless theory claims them to be. He does this by arguing (Essay 2) that (1) a 1980 token of "It is now 1980" is logically contingent, and (2) that Smart's truth-conditions analysis of "*E* is present" is mistaken. His first argument in support of (1) is stated as follows:

> Even if tokens of "now" are rigid designators, it is false that 1980 tokens of "It is now 1980" are tautologically true. . . . Now for any tautologically true sentence-token, the truth of the token is entailed by premises stating the relevant tautological fact and that the token occurs. But
>
> (1) 1980 is at 1980
>
> and
>
> (2) *S* occurs
>
> do not entail
>
> (3) *S* is true.

It is not clear to me why (2) alone does not entail (3). Since "*S*" is the name of a 1980 token of "It is now 1980," premise 2 could also be read as

(2′) A 1980 token of "It is now 1980" occurs

and (2′) does entail (3). Smith might object that (2) cannot be replaced by (2′) because "*S* is a proper name of the sentence-token and (if the Marcus–Kripke theory is correct) thereby directly refers to *S* without imparting any information about it, such that it has the property of occurring in 1980" (pp. 51–52). I would reply that since "*S*" names a sentence that contains a 1980 token of "now," (2) can be replaced by (2′), or at least (2) entails (2′), which in turn entails (3).

But suppose that Smith is correct and that (3) is not entailed by (1) and (2), so that a 1980 token of "It is now 1980" is logically contingent. What metaphysical significance does this have? Plenty, according to Smith, for "it shows . . . that rigidly designating 1980 is not the only semantic property of ['now'], that this token in addition imparts some information about 1980. . . . It must also impart the tensed information that 1980 is present. The

1980 token of 'now' rigidly designates 1980 and ascribes to it the property of presentness" (p. 52). Clearly Smith is drawing ontological conclusions from logical investigations, but I do not think that what he says follows merely from the fact that a 1980 token of "It is now 1980" is contingent. Before we can see why this is so, let us consider one more reason Smith gives for supposing that a 1980 token of "now" ascribes the property of presentness to 1980.

Smith argues that if "It is now 1980" and "1980 is present" are contingent sentences, then "the fact that 1980 is at 1980 cannot explain the logical equivalence of 'It is now 1980' and '1980 is present' or of the 1980 tokens of these sentences. For no tautological fact can make two logically contingent sentences or tokens true in all and only the same circumstances. A logically contingent fact is required, such as the fact that *it is now 1980*" (p. 51). On the one hand, Smith's argument raises an irrelevant objection, for he does not seem to realize that the date version analysis of the truthmakers of tensed sentences need not also explain the logical equivalence of (1) and (2); the synonymity of "present" and "now" in ordinary usage is sufficient to do that. On the other hand, Smith does raise at least a *prima facie* problem for the date analysis. He claims that any token "It is now 1980," uttered in 1980, is only contingently true; the same token might have occurred in 1990 and so have been false. However, the truth conditions ascribed to it by the date analysis—namely, *1980 is at 1980*— could not have failed to obtain. Consequently, on the date analysis a 1980 token of "It is now 1980" is necessarily, not contingently, true. However, if a token is genuinely only contingently true, then no sentence which is necessarily true can state that token's truth conditions. Thus Smith concludes that the date analysis must be mistaken and that the truth conditions of a 1980 token of "It is now 1980" must include the fact that *it is now 1980*.

In reply, a defender of the date analysis may concede that a 1980 token of "It is now 1980" is contingent but allow its truth conditions to vary from context to context (in this case, the contexts are possible worlds). In the actual world, its truth conditions are that *1980 is at 1980,* so the token is actually true. In another world, where the token occurs at 1990, its truth conditions are that *1990 is at 1980,* and so it is false in that world. Thus, even though the token in question is contingent, the proper analysis of its truth conditions does not yield any tensed fact or tensed property.

Smith's final criticism of the new tenseless theory concerns the date version analysis of the truth conditions of contingent sentence-tokens. On this analysis, when, for example, John Doe utters in 1814 the sentence "The Battle of Waterloo is present," his utterance is true if and only if the Battle of Waterloo occurs in 1814. Smith writes: "Let us call John's utterance '*U*.' Is a necessary condition of *U*'s truth that the Battle of Waterloo occur in 1814? The year 1814 is the set of all and only the events that occur during the twelve-month period that possesses the property of being 1813 years later than the birth of Christ. Let us call the set that actually possesses this property '*B*'" (p. 53). Smith then considers a possible world, W_1, in which *U* is true (because it occurs when the Battle of Waterloo occurs) but in which *B* is not 1813 years later than the birth of Christ, because Christ was not born. On the basis of W_1 he concludes that "the occurrence of the Battle of Waterloo in 1814 is not a necessary condition of *U*'s truth" I shall argue, however, that whether we understand "1814" attributively or referentially, Smith's argument is unsuccessful.

In ordinary discourse the date-expression "1814" is used to attribute to an event or set of events the relational property of being 1813 years later than the birth of Christ. In metaphysical discourse, however, "1814" is used to refer to the moment or period of time denoted by that date. We shall assume, as Smith apparently does, that the referent of "1814" is a certain set *B* of successive and overlapping events. Now, admittedly, there is some possible world, W_1, in which *B* does not have the property ordinarily associated with the use of "1814." Nevertheless, the possibility of W_1 does not demonstrate that the occurrence of the Battle of Waterloo in 1814 is not a necessary condition of *U*'s being true. For if we are using "1814" to attribute a certain property to the events in set *B,* then W_1 is a world in which neither the Battle of Waterloo nor *U* occur in 1814. But if *U* does not occur in 1814, then *U* cannot be a true 1814 token of "The Battle of Waterloo is present." Thus, if "1814" is being used attributively, then Smith is mistaken in maintaining that *U*'s truth is compatible with the battle's not occurring in 1814. On the other hand, if we give "1814" a *de re* reading and claim that "1814" refers directly to *B,* then the existence of the possible world Smith conjures up cannot prove his point. For even in W_1 both the Battle of Waterloo and *U* belong to set *B*—that is, the time denoted by "1814"—and so *U* is true if and only if the Battle of Waterloo occurs in 1814.

To his credit, Smith considers giving a *de re* reading of "1814" but does not think that this avoids the problem. There is, he claims, a possible world, W_3 in which "*U* is true, but . . . the Battle of Waterloo does not belong to *B*" (p. 54). W_3 is a world in which John Doe utters *U* while the Battle of Waterloo is occurring, but in which a leaf that did fall in the actual world does not fall. Smith continues: "Since sets with different members are different, the set in W_3 is not the set *B* but some different set, *C.* This shows that *U* is true in W_3 even though the Battle of Waterloo is not a member of *B*. Therefore the battle's occurrence in 1814 is not a necessary condition of *U*'s truth even if '1814' is read *de re* as referring directly to *B*" (p. 54). I disagree. What Smith's argument shows is that it is possible for *U* to be true even if the Battle of Waterloo is not a member of the set denoted by a *de re* reading of "1814." But why would that undermine the date-version analysis of the truth conditions of *U*? Insofar as *U* is a token, it has a date. Thus, even if there is a possible world, W_3, in which *U* and the Battle of Waterloo do not belong to *B*, and so do not exist in 1814, they still exist at the same time. In W_3 a different date, say "1814*," will refer directly to set *C*, and then a token of "The Battle of Waterloo is present" uttered in 1814* will be true if the Battle of Waterloo occurs in 1814*. To put the point otherwise, "1814" may not denote a time that contains *U* and the Battle of Waterloo, but then there is another possible world, W_3, where some other time *t* contains *U* and the Battle of Waterloo, and in that world *U* is true at time *t* if the Battle is at time *t*. Furthermore, if in W_3 the Battle of Waterloo is not a member of *B* and *U* is simultaneous with the battle, then *U* does not occur in *B* either. And if *U* does not occur in *B*, then *U* cannot be a true 1814 token of "The Battle of Waterloo is present." Thus, whether we interpret "1814" attributively or referentially, Smith cannot claim that (an 1814 token) *U* is true even if the Battle of Waterloo does not occur in 1814.

Smith has not shown that the (tenseless) occurrence of an event *E* at a certain time *t* is not a necessary condition for the truth of "Event *E* is present" uttered at time *t*. We have also seen that his earlier argument against the token-reflexive account failed to establish that it is internally inconsistent or that tenseless truth conditions are not sufficient to account for the truth of tensed sentences. I conclude, therefore, that contrary to what Smith maintains,[12] the new tenseless theory of time need not be abandoned or radically reworked.[13]

Notes

1. See, e.g., Essays 1, 3, 27; also Oaklander, *Temporal Becoming,* chap. 4; Jeremy Butterfield, "Indexicals and Tense," in I. Hacking, ed., *Exercises in Analysis* (Cambridge, 1985), pp. 69–87; and Keith Seddon, *Time: A Philosophical Treatment* (New York, 1987), chap. 13.
2. Mellor, *Real Time,* chap. 2.
3. Ibid., p. 35.
4. I wrote the sentence-type as "S(1980) occurs in 1980" because, to be precise, "S occurs in 1980" does not have all true (or false) tokens. Since "S" is the name of a sentence-type, when tokens of S do not occur in 1980, tokens of "S occurs in 1980" will be false; but when tokens of S do occur in 1980, tokens of "S occurs in 1980" will be true. In other words, if "S" stands for a sentence-type, then the truth-value of (the tokens of) *both* "S" and "S occurs in 1980" will vary from time to time depending on their date.

5. D. H. Mellor, "I and Now," *Proceedings of the Aristotelian Society* 89, pt. 2 (1988/89): 79–94.
6. D. Kaplan, "On the Logic of Demonstratives," *Journal of Philosophical Logic* 8 (1978): 81–98; rpt. in N. Salmon and S. Soames, eds., *Propositions and Attitudes* (Oxford, 1988), pp. 66–82.
7. In Essay 5 I propose a different way out of the problem Smith raises for the token-reflexive account.
8. Bertrand Russell, "Review of MacColl's *Symbolic Logic and its Applications*," *Mind* 15 (1906): 255–260; Goodman, *Structure of Appearance;* W. V. O. Quine, *Elementary Logic* (New York, 1941).
9. Quine, *Elementary Logic,* p. 6.
10. Smart, "Time and Becoming," p. 5.
11. Michelle Beer and Richard Gale, "An Identity Theory of the A- and B-Series," unpublished; and Essay 7.
12. Smith furthers the debate between the new tenseless and the tensed theories in Essay 8. However, a consideration of his discussion lies beyond the scope of the present essay.
13. I wish to thank Neil Cooper, Hugh Mellor, Quentin Smith, and anonymous referees for *Philosophical Quarterly,* in which this essay was first published, for their useful comments on earlier versions. I also wish to thank the University of Michigan–Flint for an award that partially funded the underlying research.

ESSAY 4

The Truth Conditions of Tensed Sentences

QUENTIN SMITH

In Essay 2 I argued that Mellor's and Smart's versions of the new tenseless theory of time are unsuccessful. I shall here respond to Mellor's, Smart's, and Oaklander's defense of these versions of the new tenseless theory against the criticisms offered there.

In the earlier essay I argued that Mellor's theory is both self-contradictory and fails to explain the entailment relations among tensed sentences. I shall first address the issue of whether the self-contradiction can be avoided.

We recall that Mellor holds a token-reflexive version of the new tenseless theory of time. He shares with the old token-reflexive tenseless theory of time the idea that "the tenseless facts that fix the truth-values of tokens of simple tensed sentences and judgments . . . include facts about the tokens themselves—their relative whereabouts in the B-series." "This account of what makes them true is called a 'token-reflexive' account."[1] But Mellor denies the translatability thesis. He is concerned with showing that A-sentences and their tokens are untranslatable by a particular sort of tenseless sentence or token, namely, one that states the truth conditions of the A-sentence and its tokens. The truth conditions of any token *S* of "It is now 1980" are that *S* is true if and only if it occurs in 1980. But the sentence stating this truth condition, the sentence "*S* occurs in 1980," is such that neither it nor any of its tokens translates *S* or the sentence-type "It is now 1980." The reason, Mellor argues, is that the tenseless sentence and tokens have different truth conditions from the tensed ones. The sentence-token *S* is true if and only if it occurs in 1980 (and "It is now 1980" is true if and only if it is tokened in 1980), but these are not the truth conditions of "*S* occurs in 1980" or any of its tokens. Rather, the tenseless sentence or token is true regardless of when it is tokened, in 1980 or 1990, and requires merely that *S* occurs in 1980.

It seems to me that this theory is self-contradictory and that the removal of one of its contradictory theses results in its reduction to the old token-reflexive theory of A-sentences. Let us consider *S* and any token *T* of "*S* occurs in 1980." According to Mellor, *T* does not translate *S*, for *S* and *T* have different truth conditions. But this contradicts Mellor's account of *S* and *T*, for Mellor holds that *S* is true if and only if *S occurs in 1980* and *T* is true if and only if *S occurs in 1980* (p. 24). They have the same truth conditions! Mellor was led to this self-contradictory position through fallaciously inferring from

(1) It is true of *S* but not of *T* that *it is true if and only if it occurs in 1980*

that

(2) *S* and *T* have different truth conditions.

This is mistaken, since (1) does not state a difference in the truth conditions of *S* and *T* but a difference in the relation of their common truth conditions, that *S* occurs in 1980, to *S* and *T*. (1) states that these truth conditions are about *S* and are not about *T*.

Mellor responded to an earlier statement of this criticism by saying that it is wrongheaded to talk of truth conditions being "about" sentence-tokens or other phenomena.[2] But the language of aboutness is precisely the language used by Mellor throughout *Real Time* to describe the relation of truth conditions (which Mellor identifies with facts) to sentence-tokens or other phenomena. For example, he uses such expressions as "facts *about* them" and writes that "the tenseless facts that fix the truth-values of tokens of simple tensed sentences and judgments . . . include facts *about* the tokens themselves."[3]

But suppose we adopt Mellor's later position that facts or truth conditions are not "about" phenomena. This would not remove the contradiction that *S* and *T* both do and do not have the same truth conditions or the fallacious inference from (1) to (2). One reason—indeed, the only reason I can think of—for denying that truth conditions are about sentence-tokens or other phenomena is that they include these phenomena. (Mellor is not forthcoming on this matter.) But if we substitute "include" for "about" (or some other relevant relation *R* for "about"), the criticism of Mellor remains sound. We would then say that (1) does not state that *S* and *T* have different truth conditions but merely that their common truth conditions, that *S occurs in 1980,* includes *S* but not *T* (or stand in *R* to *S* but not *T*).

Oaklander responded (Essay 3) to my earlier statement of this criticism of Mellor by saying that Mellor holds merely that the relevant sentence-types are untranslatable and allows that their tokens, such as *S* and *T,* have the same truth conditions. It follows, according to Oaklander, that "there is no inconsistency in claiming that tensed and tenseless sentence-*types* have tokens with different truth conditions, while also claiming that tensed and tenseless sentence-*tokens* themselves have the same truth conditions." But Oaklander misunderstands Mellor's theory, which is that the sentence-tokens also have different truth conditions. For example, Mellor writes:

> Now the truth conditions of *tokens* of spatially tensed sentence-types vary, as we have seen, with their spatial position, and those of temporally tensed types vary with their temporal position. So, therefore, must the truth conditions of *tokens* of their translations. But what makes sentence-types tenseless . . . is that the truth conditions of their *tokens* do not vary in this way. . . . No tenseless sentence, therefore, can have *tokens* whose truth conditions are everywhere and always the same as [the truth conditions of *tokens*] of a tensed sentence, because by definition the latter vary from place to place or time to time and the former do not. (p. 27, emphasis of "tokens" added)

The type–token distinction, consequently, cannot be used to demonstrate the consistency of Mellor's theory. But Oaklander makes a second attempt to show that it is consistent. He adds that even if the tensed and tenseless tokens have the same truth conditions, it still follows from Mellor's own principles that they do not translate one another, since the tenseless and tensed tokens have a *different usage*. Oaklander alleges that Mellor's theory is that differences in meaning correspond to differences in usage, not just to differences in truth conditions. But this also is a misunderstanding of Mellor's theory, for Mellor defines differences in usage and meaning of the relevant sentence-tokens in terms of differences in their truth conditions. In the case of temporally and spatially tensed sentence-tokens, "correct usage *is* explained by people knowing how the truth of what they say depends on when and where they say it; in particular, the *different* meanings of different sentences are differentiated, as they are not in mathematics, by their different truth conditions" (p. 25). In the case of the logically contingent sentences or sentence-tokens, if they have the same truth conditions, then they have the same meaning and usage.

I believe that the contradiction is indeed present in Mellor's theory and can be removed only by altering his theory.

Oaklander wishes to emphasize (p. 61) that "the crucial question is not whether Mellor's view is internally consistent but, rather, whether the token-reflexive version of the new tenseless theory of time is true." Oaklander is of course correct in stressing this, and it is worth considering whether his revision of Mellor's theory has merits on its own grounds, apart from whether it succeeds in defending Mellor against the criticisms offered in Essay 2.

Oaklander addresses my argument that

(1) It is now 1980

entails the sentence

(2) 1980 is present

and therefore that a tensed fact involving the *presentness of 1980* is among the truth conditions of both sentences. He believes that Kaplan's theory of indexical sentences can explain this entailment without any need to appeal to tensed facts. He employs Kaplan's theory as follows:

> The meaning of an indexical sentence-type (and all its tokens) is a semantic function (rule) from facts about tokens of that type (their context of utterance) to their tenseless truth conditions. In particular, the meaning of (1) and (2) is a semantic function from the context of utterance, namely, the time at which their tokens are produced, to their tenseless truth conditions. Since the context of utterance varies, so do the truth conditions of their tokens, but in each case the truth conditions are tenseless. Thus, any token of (1) is true with respect to the context in which it is produced (namely, the time at which it is uttered), if and only if the year of that context is 1980, and the same may be said of any token of (2). Consequently, since the truth conditions of (tokens of) (1) and (2) are the same, the difficulty of getting (1) and (2) to be logically equivalent vanishes. (pp. 62–63)

However, this does not solve the problem I pointed out in Essay 2; it merely reproduces it in a different guise. The tenseless truth conditions of tokens of (1) and (2) are not the same, and Oaklander can create the appearance of sameness only by equivocating upon "it." Any token *S* of (1) is true with respect to the context of its utterance if and only if the year of its context of utterance is 1980, and any token *V* of (2) is true with respect to the context of *its* utterance if and only if the year of *its* context of utterance is

1980. But once we replace the occurrences of "it" by names of the relevant tokens, this appearance of similar truth conditions vanishes. The tenseless truth conditions are these:

(3) Any token *S* of (1) is true with respect to the context of *S*'s utterance if and only if the year of *S*'s context of utterance is 1980.
(4) Any token *V* of (2) is true with respect to the context of *V*'s utterance if and only if the year of *V*'s context of utterance is 1980.

Oaklander states in Essay 3 that, strictly speaking, sentence-types do not have truth-values or truth conditions; rather, only their tokens do. So let us consider two 1980 tokens of (1) and (2) and state their truth conditions. Let us call these two tokens "*S*" and "*V*" respectively. We have

(5) *S* is true with respect to the context of *S*'s utterance if and only if the year of *S*'s context of utterance is 1980.
(6) *V* is true with respect to the context of *V*'s utterance if and only if the year of *V*'s context of utterance is 1980.

The two tenseless facts mentioned after the biconditionals in these truth-condition sentences are *S occurs in 1980* and *V occurs in 1980*. We are now back in the situation I described in Essay 2. These two tenseless facts do not entail each other. *S* could occur in 1980 even if *V* does not occur at all, and vice versa. Consequently, these facts are insufficient to explain the logical equivalence of *S* and *V*. The logical equivalence is explained, however, if we assume that the tensed fact, *1980 has presentness,* belongs to both their truth conditions and is asserted by both of them. Thus, it appears that we still have reason to think that the tensed theory of time is better suited to the facts about language and time than the token-reflexive version of the new tenseless theory of time.

The date version of the new tenseless theory of time also fails to specify adequate truth conditions for tensed sentence-types or tokens. Smart claims that the tenseless truth conditions of tensed-sentence utterances involve dates. For example, "When a person *P* utters at a time *t* the sentence 'Event *E* is present,' his assertion [utterance] is true if and only if *E* is at *t*."[4] Let us say that *P*'s utterance of this sentence at time *t* is the utterance *U*. I argued that this truth-condition sentence is false, since there is some possible world *W* in which *U* and *E* occur simultaneously (and thus in which *U* is true) but at some time *t'* different from *t*. (See the discussion of the Battle of Waterloo example in Essay 2.) In other words, *U*'s occurrence at time *t* is not a necessary condition of its truth.

Oaklander believes (p. 67) that Smart can be defended by claiming that in *W*, *U*'s truth conditions are that *U* is true in *W* if and only if *E* is at *t*′. But this fails as a defense of Smart, since Smart's theory aims to give the truth conditions of the utterance *U*, not world-indexed truth conditions of *U*. Note that Smart's truth-condition sentence says:

(1) When a person *P* utters at a time *t* the sentence "Event *E* is present," his assertion [utterance] is true if and only if *E* is at *t*.

It does not say:

(2) When a person *P* utters at a time *t* in the world *W* the sentence "Event *E* is present," his assertion [utterance] is true in *W* if and only if *E* is at *t* in *W*.

Thus Oaklander's introduction of the different truth-condition sentence (2) cannot show that my criticism of Smart's truth-condition sentence (1) is false.

Of course, the more important issue is whether the date version of the new tenseless theory of time can be salvaged if it is altered so that all truth-condition sentences become world-indexed. I believe it cannot, since world-indexed truth conditions are insufficient to *give the meaning (semantic content) of tensed sentence-tokens.* I do not mean by this that clauses stating these truth conditions fail to translate the tensed sentence-tokens, since the failure of translation is something the proponents of the new tenseless theory insist on in any case. I mean, rather, that the world-indexed truth conditions do not give the meaning in the following sense. World-indexed truth-condition sentences do not give the semantic content of utterances, since the criterion of truth for these sentences is that the clause on the right-hand side of the biconditional that has the world-index phrase appended to it has the same truth-value in the mentioned world as the utterance denoted by the clause on the left-hand side of the biconditional. This allows any clause with the same truth-value in *W* as "*E* is at *t*" to be substituted for "*E* is at *t*," and this prevents (2) from giving the meaning of the utterance. Consider a concrete example:

(3) "Henry is ill" as spoken by John on 28 July 1940 in *W* is true in *W* if and only if Henry (is) ill on 28 July 1940 in *W*.

Suppose that *W* is the actual world and that the utterance mentioned is true. Then we may substitute any true clause for "Henry (is) ill on 28 July 1940," and (3) will remain true. Thus, it is true that

(4) "Henry is ill" as spoken by John on 28 July 1940 in *W* is true in *W* if and only if the sun is 93 million miles from the earth in *W*.

If (3) gives the meaning of the utterance "Henry is ill" by stating with truth world-indexed truth conditions of this utterance, then (4) would give the meaning of this utterance for the same reason. But, obviously, "Henry is ill" as used by John does not mean that the sun is 93 million miles from the earth. A truth-condition sentence that gives this semantic content, or at least gives it up to logical equivalence, must state conditions that obtain in all and only the worlds in which this utterance is true, and this can be done only in terms of truth conditions that are not world-indexed.

The world-indexing of the truth conditions runs into the same problem as does an extensional reading of the biconditional in truth-condition sentences that are not world-indexed. It is worthwhile pointing this out, since Smart writes that his preferred response to my criticism is to make the biconditional extensional rather than to read it as intensional and world-index the truth conditions.[5] But if we make extensional the biconditional in

(5) "Henry is ill" is true as spoken by John on 28 July 1940 if and only if Henry (is) ill on 28 July 1940,

then (5) no longer gives the meaning of the utterance "Henry is ill." If (5), as read extensionally, did give the meaning of the utterance "Henry is ill," this would be due to the fact that this utterance has the same actual truth-value as "Henry (is) ill on 28 July 1940," since the criterion of truth for such extensional truth-condition sentences is that the clause after the biconditional has the same actual truth-value as the utterance mentioned before the biconditional. But since the utterance "Henry is ill" also has the same actual truth-value as "The sun is 93 million miles from the earth," the conclusion that this utterance meant that the sun is 93 million miles from the earth would be equally warranted. Only a stricter criterion of truth can make truth-condition sentences meaning-giving, namely, the criterion that the clause mentioned after the biconditional is true in all and only the possible worlds in which the mentioned utterance is true. Mellor makes this point in an intuitively clear manner in *Real Time:*

> For a sentence's truth conditions to give its meaning, its being true in them must be more than a coincidence. Otherwise, so far as truth conditions go, the English sentence "Snow is white" could just as well mean that grass is green, since "Snow is white" *is* true and grass *is* green. "Snow is white" is indeed true if and only if grass is green [if the biconditional is read extensionally]. But that, of course, is just a coincidence. Even if grass were not green, "Snow is white" would still be true—provided snow was still white. "Snow is white" is true not only in the real world, it would also be true in any other world in which snow was white and false in any

> world in which it wasn't. That is really why the sentence means in English what it does, rather than meaning that grass is green. To give meanings, therefore, truth conditions generally have to include imaginary conditions as well as real ones. (p. 26)

Accordingly, I do not think that making the truth-condition sentences extensional or world-indexed can salvage the date version of the new tenseless theory of time. Since the token-reflexive version also fails the test of semantic adequacy, we should conclude that the new tenseless theory of time is false. Only by assuming tensed facts can we explain the semantic content of tensed-sentence utterances.

Notes

1. Mellor, *Real Time,* p. 42.
2. Mellor, private communication of 11 Dec. 1988.
3. Mellor, *Real Time,* pp. 30, 31, 42; my italics.
4. Smart, "Time and Becoming," p. 5.
5. Smart, private communication of 5 Jan. 1988.

ESSAY 5

The New Tenseless Theory of Time: A Reply to Smith

L. NATHAN OAKLANDER

In Essay 2 Quentin Smith argued that the two extant versions of the new tenseless theory of time (the token-reflexive version and the date version) are open to insurmountable difficulties and so must be either radically reworked or abandoned in favor of the tensed theory.[1] The purpose of this essay is to defend the new tenseless theory against Smith's objections. I shall argue that Smith's central arguments raise irrelevant objections, because they rest upon assumptions that are accepted by the old tenseless theory of time but rejected by the new tenseless theory.

Recent defenders of the tenseless view have come to embrace the thesis that tensed sentences cannot be translated by tenseless ones without loss of meaning. Nevertheless, they have denied that the ineliminability of tensed language and thought entails the reality of temporal properties. According to the new tenseless theory of time, tensed discourse is indeed necessary for timely action, but tensed facts are not, since the truth conditions of tensed sentences can be expressed in a tenseless metalanguage that describes unchanging temporal relations between and among events.

On the token-reflexive version of the new tenseless theory, the temporal relation between a tensed sentence-token and the event or date that such a judgment is about provides an objective basis for the truth-value of any tensed sentence. For example, any token *S* of "It is now 1980" is true if and only if *S* occurs in 1980; any token *R* of "It was 1980" is true if and only if *R* is later than 1980; and so on. Thus, on the token-reflexive account, the truth conditions of tensed sentence and judgment-tokens are tenseless facts.

Smith begins his argument against the token-reflexive account of the truth conditions of tensed sentences by noting that

(1) It is now 1980

> entails the sentence
>
> (2) 1980 is present.
>
> In the language of facts, this means that there cannot be a fact statable by any token *S* of (1) unless there is a fact statable by any token *V* of (2). In other words, a fact statable by *S* implies a fact statable by *V*, and consequently a fact statable by *V* is among the truth conditions of *S*. (p. 45)

As this passage makes apparent, Smith assumes that a logical entailment among sentences in ordinary language must be represented by a "logical entailment" among the facts that make those sentences true. That is, he assumes that since (1) entails (2), the truth conditions of (1) must entail the truth conditions of (2). He then argues that since the tenseless truth conditions (or the fact statable by any token *S*) of (1), namely, *S* occurs in 1980, does not entail the tenseless truth conditions (or the fact statable by any token *V*) of (2), namely, *V* occurs in 1980, then, in addition to tenseless truth conditions, (1) and (2) must also have tensed truth conditions. As Smith puts it, the token-reflexive version of the new tenseless theory fails to establish that "tenseless facts are the *only* truth conditions of tensed sentence-tokens; tensed facts need to be assumed to account for the entailment relations between tensed sentences. . . . [Token-reflexive] tenseless truth conditions could not explain the logical equivalence of 'It is now 1980' and '1980 is present,' since *S occurs in 1980* neither implies nor is implied by *V occurs in 1980*" (p. 50). We may agree that tenseless truth conditions cannot explain the logical equivalence of (1) and (2), but that constitutes an objection to the tenseless view only if we presuppose a conception of analysis that is shared by proponents of the old tenseless theory of time but rejected by proponents of the new theory.[2]

To begin to see what is involved in this last point, note that the early defenders of the tenseless view believed that a complete description or analysis of time could be symbolically represented in a nonindexical tenseless language. To give a complete description or analysis involves constructing a single language that performs two functions. First, in its "logical" function, this perspicuous or ideal language (IL) is a symbolic device for representing or transcribing the logic of sentences contained in ordinary language. For example, in ordinary language, arguments are given that involve the entailment of one sentence by another, and in its logical function the IL represents the correct logical form that all sentences and all entailments in a natural language can take. The second function of the IL, call it the "ontological" function, is to provide a representation of the kinds of entities that there are, as well as the facts that exist. One might conceive of the IL in its ontological function as containing expressions that are neither true nor false, but are ontological explanations for (some) true sentences in ordinary language or "stand-ins" for the facts represented by them. By assuming that both these functions could be performed by a single IL, the old tenseless theory drew ontological conclusions from logical considerations. Specifically, they argued that since the logic of ordinary temporal discourse could be represented in a tenseless language, the ontological nature of time consisted of unchanging temporal relations between terms that did not have tensed properties.

Given the assumptions concerning analysis implicit in the old tenseless theory, Smith's argument against the token-reflexive account is very strong. For in order to perform its logical function, the analysis of tensed discourse must be able to explain the inference from (1) "It is now 1980" to (2) "1980 is present." However, in order to perform its ontological function, the analysis of tensed discourse must represent those sentences as tenselessly expressing temporal relations between a sentence-token and the time at which it occurs. The problem, then, is that the ontological description expressible in a tenseless language cannot explain the logical entailment of (2) by (1). Thus, on the tenseless view, the logical representation of tensed discourse is inadequate, and given that the logical and ontological representations are to be performed by a single language, it follows that the ontological representation is also inadequate and that, therefore, there must be temporal properties and tensed facts.

Thus, Smith does indeed have a point. He has shown that the token-reflexive account of tense cannot be a complete description or analysis of time insofar as it purports to represent, within a single language, both the logical form of ordinary temporal discourse and the metaphysical nature of time. But that is not an argument against the new tenseless theory of time, because, in rejecting the criterion of translatability as a method for determining the metaphysical nature of time, proponents of the new tenseless theory are, or should be, rejecting the conception of analysis upon which Smith's argument rests.

The new tenseless theory accepts the tenser's claim that tensed discourse and thought are ineliminable. It therefore agrees that any logically adequate representation of temporal language—that is, any language capable of representing the meaning and logical implications of our ordinary talk about time—must be tensed. The detenser denies, however, that from an ontological point of view, a perspective that attempts to represent the nature of time, tense is ineliminable. Smith understands very well that recent detensers have maintained that tensed sentences cannot be replaced by tenseless ones without loss of meaning. What he fails to appreciate is that in accepting the irreducible nature of tensed discourse, the new tenseless theory is abandoning the analytic ideal of arriving at a single language that is adequate for both ontological and logical investigations. Once these two functions of language are separated and kept distinct, it is open to the defender of the tenseless view to maintain that logical connections among sentences in ordinary language do not represent ontological connections between facts in the world. Thus, although (1) and (2) mean the same thing and entail each other, it does not follow that there must be a necessary connection between the facts that provide the basis for their truth. Nor does it follow that tensed facts must be introduced to explain their logical equivalence. According to the new token-reflexive account of time, not only can two sentences, such as "It is now 1980" and "*S* occurs in 1980," with different meanings correspond to the same fact, but two sentences, such as (1) and (2), with the same meaning can correspond to different facts. These are the consequences of rejecting the conception of analysis upon which Smith's criticism is based. By failing to acknowledge them, Smith's argument, while applicable to the old token-reflexive version of the tenseless theory, is inapplicable to the new theory.

Smith's main argument against the date version of the new tenseless theory of time also raises an irrelevant objection, and for the same reason. On the date analysis, temporal indexicals like "now," "this time," and "the present," as used on a given occasion, and proper names such as dates are referring terms which rigidly designate a time. Thus, the truth conditions of the tensed sentence "It is now 1980," uttered in 1980, are expressible by the use of the necessary truth "1980 is at 1980," uttered at any time. And this is just the point. The metaphysical implications of tensed discourse are nil. An event or time being "now" is nothing more than its occurring at the time at which it occurs.

Smith attempts to avoid this conclusion by arguing that a 1980 token of "It is now 1980" is logically contingent and, for that reason, must impart the tensed information that 1980 has the property of presentness. He begins his argument by claiming that the date "'1980' in its normal use expresses the sense that is also expressible by an attributive use of the definite description 'the twelve-month period that is 1979 years later than the birth of Christ.'" He then argues that a 1980 "token of 'now' . . . refers to the set of all and only those events that, in fact, possesses the property of being the twelve-month period that is 1979 years later than the birth of Christ. If we call this set '*A*,' we can say that the 1980 token of 'It is now 1980' directly refers to *A* and asserts the identity of *A* with the twelve-month period that is 1979 years later than the birth of Christ. But this identity is contingent!

For there are possible worlds in which *A is not* the twelve-month period that is 1979 years later than the birth of Christ" (p. 52). Smith's reasoning is valid, but his conclusion, that "the 1980 token of 'It is now 1980' is only contingently true," constitutes an objection to the new tenseless theory only if he confounds the two functions of language that the new theory insists must be distinguished.

To see why this is so, consider that in ordinary language the date-expression "1980" has the same meaning as "the twelve-month period that is 1979 years later than the birth of Christ." Thus, if the representation of "It is now 1980" is to preserve its informational content and capture its meaning, then we cannot transcribe it as the necessary truth "1980 is at 1980." In other words, in a logically adequate language—a language that represents the meanings and entailments of sentences in a natural way—"1980" cannot be a rigid designator of the time referred to by a 1980 use of "now," since, if it was, the transcription would not convey the information that we ordinarily associate with a tensed sentence like "It is now 1980." On the other hand, in an ontologically adequate language—a language used to represent the metaphysical nature of time—"1980" cannot be replaced by the description that captures its meaning, since, if it was, the transcription would no longer be a perspicuous representation of the tenseless theory of time. If, however, we keep the logical and ontological functions of language distinct, then detensers can agree that in a language constructed to represent the logical form of sentences in ordinary language, "It is now 1980" is contingent, while also maintaining that in a language constructed to represent the metaphysical nature of time, it is a trivial truth perspicuously represented as "1980 is at 1980."

Smith's central arguments against both versions of the new tenseless theory of time result from a fusing of logical and ontological considerations. He thereby presupposes a methodological framework or conception of analysis that is shared by proponents of the old tenseless theory but rejected by defenders of the new tenseless theory. Although Smith is not alone in assuming this framework, I believe that recent detensers are correct in abandoning it. Since it is beyond the scope of this essay to give a general argument in support of that point, I shall conclude with the more modest claim to have defended the new tenseless theory of time against Smith's central objections.

Notes

1. See notes to Essay 2 for references to proponents of the new and old tenseless theories of time. Contrary to what Smith says in his essay, I do subscribe to a version of the new tenseless theory in my book *Temporal Relations and Temporal Becoming*.
2. The conception of analysis that I shall discuss has its roots in the philosophy of logical atomism. Those roots are examined in L. N. Oaklander and S. Miracchi, "Russell, Negative Facts, and Ontology," *Philosophy of Science* 47 (1980): 434–455. See also E. B. Allaire, "Relations and Recreational Remarks," *Philosophical Studies* 34 (1978): 81–90.

ESSAY 6

Smart and Mellor's New Tenseless Theory of Time: A Reply to Oaklander

QUENTIN SMITH

In Essay 5 Oaklander charges me with misunderstanding the new tenseless theory of time of Smart and Mellor. I shall show here that Oaklander misunderstands the theories of Smart and Mellor and confuses their version of the new tenseless theory with his own.

Oaklander writes: "Smith assumes that a logical entailment among sentences in ordinary language must be represented by a 'logical entailment' among the facts that make those sentences true. That is, he assumes that since ['It is now 1980'] entails ['1980 is present'], the truth conditions of ['It is now 1980'] must entail the truth conditions of ['1980 is present']" (p. 78). I go on to argue, as Oaklander notes, that the entailment relation between the truth conditions can be explained only if we assume that tensed facts are among the truth conditions of these sentences.

Now Oaklander concedes in this essay that "we may agree that tenseless truth conditions cannot explain the logical equivalence of ['It is now 1980'] and ['1980 is present']" (p. 78). However, he claims that this does not count as an objection to the new tenseless theory of time, since the new theory does not aim to provide truth conditions of ordinary tensed sentences. The new theory, according to Oaklander, aims not to explain the meaning of ordinary sentences but to develop a semantics that is adequate to the metaphysical nature of time. In his terminology, the goal of the new tenseless theory is to construct not a language that has the "logical" function of representing the meaning of ordinary language but merely one that has the "ontological" function of representing the metaphysical nature of time:

> The new tenseless theory accepts the tenser's claim that tensed discourse and thought are ineliminable. It therefore agrees that any logically adequate representation of temporal language—that is, any language capable

> of representing the meaning and logical implications of our ordinary talk about time—must be tensed. The detenser denies, however, that from an ontological point of view, a perspective that attempts to represent the nature of time, tense is ineliminable. Smith understands very well that recent detensers have maintained that tensed sentences cannot be replaced by tenseless ones without loss of meaning. What he fails to appreciate is that in accepting the irreducible nature of tensed discourse, the new tenseless theory is abandoning the analytic ideal of arriving at a single language that is adequate for both ontological and logical investigations. Once these two functions of language are separated and kept distinct, it is open to the defender of the tenseless view to maintain that logical connections among sentences in ordinary language do not represent ontological connections between facts in the world. (pp. 79–80)

It seems to me, however, that it is Oaklander who misunderstands the new tenseless theory of time. The new theory, as described in my article and as espoused by Smart, Mellor, MacBeath, and others, is based on the thesis that the tensed theory of time is false, *since the truth conditions of ordinary A-sentences and their tokens can be stated in a tenseless metalanguage*. That is, it is the theory that tenseless truth-condition sentences provide a "logically adequate representation of ordinary temporal language" and therefore that the tenseless theory of time is true. This is precisely the point of Smart's claim that "the semantics of indexical expressions can be expressed in a tenseless metalanguage."[1] Smart is here saying that the tenseless metalanguage adequately expresses the meaning of (in the sense of "gives the truth conditions of") ordinary indexical expressions such as "*E* is present," and Smart infers from this the thesis, of "metaphysical significance," that the tenseless theory of time is true. *Contra* Oaklander, Smart is concerned with the logical structure of ordinary language and is interested in how the meaning of ordinary expressions should be understood or represented in theories of meaning for ordinary language. Thus he makes such remarks as "I think that ordinary adverbs should be understood in terms of predicates of events, as has been suggested by Donald Davidson. Tenses should be handled differently, by means of a tenseless metalanguage."[2] Clearly, "tenses" here means "ordinary tenses," and "handled" means "treated in a way that perspicuously reveals their semantic content."

This is even more obviously the case in Mellor's theory, which is the most developed version of the new tenseless theory of time. Mellor explicitly and repeatedly says that his representation of the truth conditions of ordinary A-sentences and their tokens captures the semantic content or meaning of these tokens; for example, when he avers that "Anyone who knows that for any place *X*, tokens of '*X* is here' are true if and only if they are at *X*, and that for any date *T*, tokens of 'It is now *T*' are true if and only if they occur at *T*, surely knows what '. . . is here' and 'It is now . . .' mean in English" (p. 24), a comment Mellor reproduces almost verbatim in another context on the next page. For Mellor, tensed sentences "may not *have* the same meaning as the tenseless sentences that give their truth conditions, but those truth conditions surely *give* their meanings." Mellor's truth-condition theory aims to explain correct ordinary usage of tensed sentences: "correct usage *is* explained by people knowing how the truth of what they say depends on when and where they say it; in particular, the *different* meanings of different sentences are differentiated, as they are not in mathematics, by their different truth conditions" (p. 25). And again, "the token-reflexive truth conditions of tensed sentences incidentally explain not only how we usually use them but also how our usage varies when their tokens take time to arrive" (p. 35).

Thus Oaklander is wrong when (in Essay 5) he describes the new tenseless theory of time of Smart and Mellor as not having the aim of constructing tenseless truth-condition sentences constitutive of "a logically adequate language—a language that represents [in truth-condition sentences] the meanings and entailments of sentences in a natural language." But this fact does not eliminate the relevance of Oaklander's essay, since it may be understood as espousing an alternative version of the new tenseless theory of time. Oaklander's version of the new tenseless theory shares with Smart's and Mellor's version the claim that the *translatability* of tensed by tenseless discourse is not required by the tenseless theory; but Oaklander goes further and argues that *providing tenseless truth conditions of tensed sentences* is not required by the tenseless theory. Oaklander in effect regards the semantic content of ordinary tensed discourse as not relevant, or at least not crucial, to the truth or falsity of the tenseless theory of time. There is a distinction between the "ontological language," which represents the real nature of time, and "ordinary language," which does not. The justification for this distinction, according to Oaklander, lies in considerations based on McTaggart's paradox and the paradox of "the rate at which time flows." Oaklander's argument is that the tensed theory of time is implicitly self-contradictory and therefore cannot be accepted in our ontology (although it may be useful as an analysis of ordinary tensed discourse). I shall address Oaklander's argument based on McTaggart's paradox in Part II, where I will contend that the tensed theory of time is not self-contradictory and hence that there is no reason to think that ordinary tensed discourse does not reflect the ontological nature of time. If ordinary tensed sentences (such as

"It is now noon") that we all accept as true on many occasions have tensed truth conditions, then that is a reason to think that the tensed theory of time is true.

Notes

1. Smart, "Time and Becoming," p. 11.
2. Ibid., p. 15.

ESSAY 7

Temporal Indexicals and the Passage of Time

MICHELLE BEER

We commonly think of events as being past, present, and future and of present events becoming past as future events become present. This process of temporal becoming is often conceived, moreover, as involving the "shift" or "passage" of a mysterious, transcendent present, or NOW, to ever later moments or events. Philosophers, however, have frequently objected to this notion of temporal passage, on the grounds that the motion of the present, or now, along the temporal series cannot be accounted for in any intelligible way.[1] Motion in space involves a rate of change of spatial position with respect to temporal position. What, then, could be meant by the rate at which the present shifts along the temporal series? It seems that this question places us squarely on the horns of a dilemma. For if we postulate a second- and higher-order temporal series with respect to which the rate of the moving present can be determined, then we would presumably be committed to the existence of a second moving present, which, in turn, would lead us to postulate a third temporal series, and so on, ad infinitum. And if we do not, then the only available alternative is to give the tautological reply that the present shifts tediously at the rate of one time unit per time unit, as if this made the notion of the moving present any less perplexing.

To dispel the philosophical perplexities which surround the notion of temporal passage, some philosophers have claimed that events are past, present, or future only in relation to a perceiver or subject of consciousness, thereby relegating the passage of time to the status of mere mind-dependency. Others have attempted to show that the A-series of events running from the past to the present to the future is reducible to, and thus is nothing more than, the B-series of events running from earlier to later. That the A-series can be reduced to the B-series—hereafter called "the B-reduction thesis"—has usually been based on a linguistic reduction in which a proposition describing an event as past, present, or future is claimed to be identical with a proposition describing an event as earlier than, simultaneous with, or later than another event or moment of time. A proposition of the first sort, to be called an "A-proposition," is expressible by the use of a temporal indexical or tensed sentence, which is not freely repeatable, in that its use on successive occasions with the same meaning can express propositions differing in truth-value; a proposition of the second sort, to be called a "B-proposition," is expressible by the use of a freely repeatable or tenseless sentence which describes a temporal relation between two events and/or moments.[2] Thus, it is claimed, if it is now time t_7, the A-proposition expressed now by the use of, say, "Jones is now (was, will be) running" is identical with the B-proposition expressed at any time by the use of "Jones is (tenselessly) running simultaneous with (earlier than, later than)t_7" or, for short, "Jones is (tenselessly) running at (before, after) t_7."

In what follows, the thesis that every A-proposition is identical with a B-proposition will, for various reasons, be found to be wanting. I shall therefore put forth a new version of the B-reduction thesis which does not make any claim for the linguistic reducibility of A-propositions but claims instead that for every A-proposition there is a B-proposition which reports one and the same event. As will be seen, this new version, to be called "the co-reporting thesis," can serve to establish an ontological reduction of the A-series to the B-series and therefore has the same therapeutic value as does the older version of the B-reduction thesis.

In particular, the co-reporting thesis will be defended on the grounds that temporal indexical terms, as they are used on a particular occasion, are referring terms which denote times. The temporal indexical "now," like "this time," refers to a time, just as the spatial indexical "here" refers to a place. This is shown by the fact that "now" can be used to answer such temporal questions as "At what time is Jones running?" and "When does the race begin?" The sentences "Now is t_7" and "Now is the time at which the race begins," therefore, are used to assert an identity between times, just as "Here is 34 Biscayne Boulevard" is used to assert an identity between places. That "now" is a referring term that denotes a time shows, moreover, the absurdity of one way of characterizing the passage of time, according to which the present shifts to ever later times. For this notion of temporal passage would require that this very time—the present, or now—become a different time at ever later times. This way of conceiving the passage of time is clearly absurd; for how is it possible that this time should cease to be identical with itself?

There are at least two objections which can be addressed to the thesis that every A-proposition is identical with a B-proposition. One objection concerns the role that propositions play as objects of propositional attitudes. For, according to the linguistic reducibility claim in question, if it is now t_7, the A-proposition expressed now by the use of

(1) Jones is running now

is identical with the B-proposition expressed at any time by the use of

(2) Jones is (tenselessly) running at t_7.

But if so, how can it be possible for someone who understands both propositions to have a propositional attitude toward one of them without having the same propositional attitude toward the other? And yet it does seem possible for someone who understands both propositions to believe that Jones is running now without believing that Jones is (tenselessly) running at t_7. Moreover, it is not unreasonable to suppose that, given any proposition P and any proposition Q, a necessary condition for P being identical with Q is that the sentences expressing them be interchangeable in all contexts *salva veritate*. From this it follows that the proposition expressed by now using (1) is not identical with the proposition expressed by using (2), because it could be the case that it is true, say, that Smith believes that Jones is running now and false that Smith believes that Jones is (tenselessly) running at t_7.

A second objection to the thesis that every A-proposition is identical with a B-proposition is that, in certain cases, it violates Leibniz's principle of the indiscernibility of identicals. Consider, for example, the temporal indexical sentence

(3) It is t_7 now.

If it is indeed the case that every A-proposition is identical with some B-proposition, then the proposition expressed by the use of (3), if true, is the very same proposition which is expressed by the use of

(4) It is t_7 at t_7.

The difficulty here is that although we can know a priori that it is t_7 at t_7, we cannot know a priori that it is t_7 now. Thus, the proposition expressed by the use of (4) is knowable a priori, while that expressed now by the use of (3) is not knowable a priori. Therefore, by Leibniz's principle, the propositions in question are not identical, since they do not share all the same properties.[3]

The co-reporting thesis, which will be defended here, escapes the objections cited above, since it does not claim that every A-proposition is identical with a B-proposition but rather that every A-proposition reports one and the same event as does some B-proposition. By the event *reported* by a proposition, we mean the event referred to by the participial nominalization of the sentence expressing the proposition. Thus, the event reported on a given occasion by the use of the temporal indexical sentence "Jones is running now" is the event referred to on that occasion by "Jones's running now," while the event reported at any time by the use of the nonindexical sentence "Jones is (tenselessly) running at t_7" is the event referred to by "Jones's running at t_7."

As has previously been noted, temporal indexical terms are used referentially in the same way that spatial indexicals are used. The sentence "Now is t_7" expresses an identity proposition which is true just in case the time at which it is used is time t_7. Thus, the indexical "now," as used on a given occasion, is a referring term which denotes a time. To justify the co-reporting thesis, let us then suppose that it is now t_7. A sufficient condition for an event E_1 being identical with an event E_2 is that E_1 and E_2 involve the same subject instantiating the same property at the same time. From this it follows that the event reported by the use of

(1) Jones is running now

is identical with the event reported at any time by the use of

(2) Jones is (tenselessly) running at t_7.

For, given that "now" refers to a time, it follows that if now is t_7, the use of "now" in (1) denotes time t_7. As a result, the event of Jones's running now, which is the event referred to by the participial nominalization of (1), involves the same subject instantiating the same property at the same time as does the event of Jones running at t_7. Therefore, the A-proposition expressed by now using (1) reports the same event as does the B-proposition expressed by using (2).

Since the co-reporting thesis does not claim that an A-proposition is identical with a B-proposition, it does not in any way attempt to establish a linguistic reduction of A-propositions to B-propositions. However, by holding that an A-proposition does not report anything over and above what is reported by some B-proposition, the co-reporting thesis shows that an event's having an A-determination—its being past, present, or future—is identical with that event's bearing a temporal relation to some moment of time. For example, if it is now t_7, then what is reported by the proposition expressed now by "Jones's running is present," which is synonymous with

"Jones's running is now," is nothing above and beyond what is reported by "Jones's running is (tenselessly) simultaneous with t_7." Therefore, given that it is in fact now t_7, it follows that Jones's running being present or now is identical with Jones's running being simultaneous with the moment t_7. And, since "past" and "future" are respectively synonymous with "earlier than now" and "later than now," the same considerations would show that Jones's running being past (future) is identical with Jones's running being earlier than (later than) t_7.

It might be objected that moments of time are themselves subjects of A-determinations. Can our account, then, be extended to show that, say, t_7's being present or now is identical with t_7's being simultaneous with t_7? One might be tempted to reply that it cannot, in view of the fact that the proposition expressed by now using "t_7 is now" informs us that t_7 is present or now, whereas that expressed by "t_7 is (tenselessly) simultaneous with t_7" does not. But is it indeed the case that these propositions cannot be co-reporting if one of them is informative and the other is not? The answer to this is clearly in the negative. The reason is that since "now" denotes a time, "t_7's being now" is synonymous with "t_7's being simultaneous with this time." And, given that "this time," like "now," denotes the moment t_7, it follows that the referring expressions "t_7's being simultaneous with this time" and "t_7's being simultaneous with t_7," though they differ in sense, are co-referential in that they refer to the same subjects or relata and ascribe the same relation between them. Therefore the propositions expressed now by "t_7 is now" and by "t_7 is tenselessly) simultaneous with t_7" are co-reporting despite the fact that they differ in their informative content.

The co-reporting thesis thus establishes the identity of the A-series and the B-series, for it shows that an event or moment having an A-determination is identical with that event or moment being earlier than, simultaneous with, or later than some moment of time. But we can draw an even stronger conclusion, namely, that the A-series is reducible to, and thus is nothing more than, the B-series. We are not, of course, considering an elimination of the A-series, for we are not saying that events or moments do not have A-determinations. Rather, the reduction being considered here is, in some respects, comparable to such noneliminative scientific reduction claims as, for example, that water is nothing but H_2O and that lightning is nothing more than a flow of ionized particles. The idea underlying such reduction claims is that scientific terms can be used to provide a better description and understanding of the world than can co-referring terms contained in ordinary language. Accordingly, given that water is in fact identical with H_2O, what justifies the assignment of a direction to the claim that water is nothing but

H_2O is that, of the two referring terms "water" and "H_2O," the latter is preferable since it is contained in the language of an accepted scientific theory. Hence the important thing to note in establishing our reduction of the A-series to the B-series is that indexical terms, including temporal indexical ones such as "now," "this time," and "the present," do not in any way enter into the formulation of scientific laws and theories. The reason for this is that, in its formulation of laws and theories, science seeks to abstract from whatever is peculiar to the situation of the observer, such as the observer's particular biological makeup and spatiotemporal location. Scientific laws and theories, therefore, are to be formulated only through the use of freely repeatable sentences, so that they hold true for any place and for any time. Consequently, what justifies the assignment of a direction to our reduction is that nonindexical expressions designating temporal relations are employed to obtain a better description of the world—namely, one which accords with the way in which science describes the world—than that given by ordinary language.

By thereby establishing an ontological reduction of the A-series to the B-series, the co-reporting thesis enables us to achieve the purpose of demystifying the puzzling and troublesome notion of the passage of time. For, given that an event's having an A-determination is reducible to that event's being earlier than, simultaneous with, or later than some moment of time, it follows that the temporal becoming of an event—its being future and subsequently becoming present and then past—turns out to be nothing more than that event's bearing different temporal relations of succession and simultaneity to different times. For example, if it is now t_7, the temporal becoming of, say, Jones's running now consists of nothing more than that event's being later than some time earlier than t_7, its being simultaneous with t_7, and its being earlier than some time later than t_7. Moreover, since the results of the co-reporting thesis hold just as well for moments of time, the same considerations would obviously hold for the becoming of a moment of time. Thus, the co-reporting thesis shows that the passage of time does not involve anything over and above the B-series which runs from earlier to later. And this enables us to clear the way for a better understanding of the nature of time. The co-reporting thesis, of course, does not settle the issue of what constitutes the nature of time. For, while it commits us to the existence of moments of time, it does not commit us to any particular view about the nature of a moment of time. We may, for example, conceive of a moment of time either as an instant of absolute time or as a set of overlapping events or as a set of simultaneous events. But while we are now left with the problem of analyzing the nature of a moment of time, we can make far greater progress in clarifying the nature of a moment than we can in analyzing the nature of some mysterious, transcendent entity or property which shifts relentlessly to ever later times.

Notes

1. See, e.g., D. C. Williams "The Myth of Passage," *Journal of Philosophy* 48 (1951): 463–464; and Smart, "Time and Becoming," p. 4.
2. For a more elaborate discussion of A- and B-propositions, see Gale, *Language of Time,* pp. 37–52.
3. Some philosophers object to the thesis that every A-proposition is identical with a B-proposition on the grounds that the proposition expressed by the use of (4) is necessarily true, whereas that expressed by the use of (3) is merely contingently true if true at all. To this it might be replied, however, that both "t_7" and "now," as used in a given context, are rigid designators, in that they denote the same time with respect to every possible world in which that time exists, and that, as a result, the proposition expressed by the use of (3) is necessarily true if it is in fact true.

ESSAY 8

The Co-reporting Theory of Tensed and Tenseless Sentences

QUENTIN SMITH

In the previous essay Michelle Beer presented the so-called co-reporting theory of tenseless and tensed sentences, a theory originated by Beer and subsequently expounded by Richard Gale[1] and endorsed by L. Nathan Oaklander (Essay 3). I believe the central arguments of this theory to be unsound, as I shall endeavor to show in what follows.

Unlike proponents of the old tenseless theory of time,[2] Beer and Gale do not claim that tensed and tenseless sentence-tokens express the same proposition. They express different propositions but nonetheless "report" that the same event occurs at the same time, say t_7. This is a version of the new tenseless theory of time, originated by Smart in 1980 and given its most extensive development to date by Mellor.[3] The new tenseless theory of time accepts the semantic irreducibility of tensed sentence-tokens but argues that this is consistent with them ascribing only B-relations (earlier than, later than, and simultaneity) and no A-properties (monadic properties of pastness, presentness, and futurity). I believe, however, that this cannot be consistently maintained. I argued this in Essay 2 with regard to Smart's, Mellor's, and MacBeath's versions of the new tenseless theory of time,[4] and I will show here that an analogous problem infects Beer's and Gale's version of this theory.

Beer writes that an utterance at t_7 of "It is t_7 now" expresses a different proposition from an utterance of "It is t_7 at t_7," such that the two utterances "differ in sense" (p. 91). Beer further claims that the tensed utterance and the tenseless utterance are "co-reporting"; that is, that "they refer to the same subjects or relata [namely, t_7 and t_7] and ascribe the same relation between them [simultaneity]." She then infers from this "co-reporting" nature of the two utterances that "an event's having an A-determination—

its being past, present, or future—is identical with that event's bearing a temporal relation [a B-relation] to some moment of time." Gale's argument is relevantly similar.

But their argument is invalid, for its premises, that the tensed and tenseless utterances differ in sense but refer to the same relata and ascribe the same relation, do not entail that the tensed utterance does not ascribe an irreducible A-property (where an irreducible A-property is one that is not identical with a B-relation). The locus of the invalidity is Beer's and Gale's admission that the two utterances *differ in sense*. Neither Beer nor Gale explains the exact nature of this difference in sense, and the premise that the two utterances both refer to t_7 and say of t_7 that it is simultaneous with itself is compatible with the supposition that their difference in sense consists in the fact that the tensed utterance alone conveys the information that t_7 has an irreducible A-property of presentness. In order to establish the conclusion that the A-property is reducible to the B-relation, Beer and Gale need to produce some argument that the difference in sense between the tensed and tenseless utterances cannot be explained by the fact that the tensed utterance alone ascribes an irreducible A-property.

It might be thought that this deficiency in their theory could easily be remedied. It might be said, for example, that the utterance of "It is now t_7" refers to t_7 *via* an indexical sense, whereas the utterance of "It is t_7 at t_7" does not. But this does not solve the problem but merely names it, for the question at hand concerns the nature of the indexical sense expressed by the utterance of "It is now t_7." The tenser argues that this indexical sense is an irreducible A-property of presentness and that, since this supposition is consistent with the co-reporting character of the two utterances, Beer's and Gale's inference that A-properties are reducible to B-relations is unjustified.

But perhaps this is not entirely fair to Beer and Gale. There is a passage in Beer's essay that may be taken as a clue as to the sense of the indexical "now." She writes that "'t_7's being now' is synonymous with 't_7's being simultaneous with this time'" (p. 91). This remark lends itself to the interpretation that the sense of "this time" is the indexical sense of "now." Gale seems to hold the same view: "'now' . . . means the same as 'this time,'" something also claimed earlier by Paul Fitzgerald and Oaklander.[5] But this specification of the sense of "now" is unacceptable, for "this time" is manifestly not synonymous with "now." Suppose we are both looking at a calendar and that today is 27 August. You ask, "When does the meeting start?" I point to 28 August on the calendar and say, "The meeting starts at this time." Clearly, this does not mean "The meeting starts now." Likewise, in giving a lecture on some historical period, such as World War I, I may ask the class, "At which time did the Battle of Verdun take place?" I may then point to different dates written on the blackboard: 1914, 1915, 1916, and finally let my pointer rest on 1916 and say "It took place at this time." Obviously this does not mean "It took place now." The expression "this time" serves to pick out the time that the speaker is demonstrating (for example, pointing at); to use David Kaplan's terminology, "this time" is a demonstrative, not an indexical.[6] But "now" is an indexical, not a demonstrative, for it is not accompanied by some act of demonstration. The indexical "now" conveys some information not conveyed by the demonstrative "this time," that the event or time is present. As Beer herself admits, "the proposition expressed by now using 't_7 is now' informs us that t_7 is present or now," but I would add that the proposition expressed by saying "t_7 is this time" does not convey this information, since I may be referring to some past or future time.

Perhaps there is another way to salvage Beer's and Gale's theory. They assume a sense/reference distinction for indexicals, which is reminiscent of Frege's theory; thus it might be thought that an appeal to Frege's theory of indexicals might help to justify the co-reporting theory. But, as is well known, Frege's theory of indexicals has been subjected to devastating criticisms by David Kaplan, John Perry, Hector-Neri Castañeda, Howard Wettstein, and many others,[7] and an "appeal to Frege" brings along with it the obligation to answer these criticisms. I will not presuppose any of these criticisms of Frege, here, however, but will confine myself to showing merely that an appeal to Frege cannot be used to justify the co-reporting theory, since the two theories are logically incompatible.

Frege said very little about the senses of indexicals, and his most explicit claims are made in his 1918–19 essay on "The Thought." In the case of indexical utterances,

> the mere wording, which can be grasped by writing or the gramophone does not suffice for the expression of the thought. The present tense is used in two ways: first, in order to give a date, second in order to eliminate any temporal restriction where timelessness or eternity is part of the thought. Think, for instance, of the laws of mathematics. Which of the two cases occurs is not expressed but must be guessed. If a time indication is needed by the present tense one must know when the sentence was uttered to apprehend the thought correctly. Therefore the time of utterance is part of the expression of the thought. If someone wants to say the same today as he expressed yesterday using the word "today," he must replace this word with "yesterday." Although the thought is the same, its

verbal expression must be different so that the sense, which would otherwise be affected by the differing times of utterance, is readjusted.[8]

According to Frege, the time of utterance of the temporal indexical, such as the time t_7, in conjunction with the utterance of the indexical itself, expresses the sense. Thus, the utterance of "now," in conjunction with the fact that this utterance occurs at t_7, expresses the sense, *at* t_7. But this is the same sense that is expressed by "at t_7" in the nonindexical sentence "It is t_7 at t_7." This shows why an appeal to Frege cannot save the co-reporting theory. The difference between the indexical and the nonindexical expression, for Frege, is a difference not in the sense expressed but in the means of expressing it. It is expressed solely by words (namely, "at t_7") in the nonindexical sentence and is expressed by words (namely, "now") in conjunction with the time of utterance in the indexical sentence. The Fregean theory is inconsistent with the co-reporting theory, for the latter states that *there is a difference in sense* between the indexical and nonindexical utterances, and the Fregean theory says that there is the same sense and merely a difference in the manner of expressing it. In short, Frege is a proponent of *the old tenseless theory of time,* and Beer and Gale are proponents of a version of the new tenseless theory of time.

But the tradition of the tenseless theory of time suggests that there are other alternatives for the senses of "now" than such senses as *at* t_7. For example, there is the theory that "now" has a token-reflexive sense, that "now" conveys the information that the time or event reported is simultaneous with the sentence-token to which "now" belongs. But the supposition that "now" has a token-reflexive sense is untenable. Let us call a certain utterance of "It is now t_7" at t_7 the utterance U_1, and suppose that "now," as used in this utterance, expresses the sense also expressible by "simultaneously with U_1," so that the utterance U_1 expresses the same proposition as

(1) It is t_7 simultaneously with U_1.

Suppose there is a second and simultaneous utterance U_2 of this sentence, which, according to the hypothesis under review, expresses the proposition

(2) It is t_7 simultaneously with U_2.

I take it to be an obvious semantic fact that the propositions expressed by two simultaneous utterances of "It is now t_7" are logically equivalent. If any semantic fact is obvious, this one is. However, the proposition expressed by (1) neither entails nor is entailed by the proposition expressed by (2), for U_1 could be uttered without U_2 being uttered, and vice versa. It follows that (1) and (2) cannot be the propositions expressed by these utterances, and therefore that the indexical sense of "now" cannot be a token-reflexive sense. (I would note parenthetically that these facts also show that John Searle's token-reflexive theory of "now," as advanced in his book *Intentionality,* cannot be true.[9]

A second possible candidate for the indexical sense of "now" is a sense that ascribes to t_7 simultaneity with some experience (sense-datum, mental act, and so on) of the utterer. This suggestion is similar to Eddy Zemach's, Lynne Rudder Baker's, Adolf Grünbaum's and one of Russell's theories.[10] Suppose that I am now experiencing sense-datum S_1 and that you are now experiencing sense-datum S_2. According to this suggestion, my utterance of "It is now t_7" at t_7 expresses the proposition expressed by

(3) t_7 is simultaneous with S_1,

and your utterance of the same sentence expresses the proposition

(4) t_7 is simultaneous with S_2.

But since (3) and (4) are not logically equivalent (it is not necessary for me to experience S_1 if you experience S_2, and vice versa), these also cannot be the propositions expressed by simultaneous utterances of "It is now t_7."

I cannot think of any other even remotely plausible candidates for a B-sense of "now," so I think it is reasonable to conclude that an utterance of "It is now t_7" differs in sense from the nonindexical sentence by virtue of the fact that it alone expresses the indexical sense comprising the irreducible A-property of presentness.[11]

Suppose, however, that I am wrong and that (1) and (2) or (3) and (4) or some other B-sentences express the same propositions as the tensed utterances: There would still be a problem, for it is part and parcel of the new tenseless theory of time that the propositions expressed by tensed utterances are not identical with propositions expressed by tenseless utterances. If the tenseless sentence (1) expresses the same proposition as the tensed utterance, then the old tenseless theory of time is true. Beer and Gale reject the old tenseless theory of time, which they characterize as a theory of "linguistic reduction." Beer writes:

> That the A-series can be reduced to the B-series—hereafter called "the B-reduction thesis"—has usually been based on a linguistic reduction in which a proposition describing an event as past, present, or future is claimed to be identical with a proposition describing an event as earlier than, simultaneous with, or later than another event or a moment of time. . . . In what follows, the thesis that every A-proposition is identical with a B-proposition will, for various reasons, be found to be wanting. I shall

therefore put forth a new version of the B-reduction thesis which does not make any claim for the linguistic reducibility of A-propositions. (pp. 87–88)

She then goes on to present arguments that the propositions expressed by utterances of A-sentences are not identical with those expressed by utterances of B-sentences. If her arguments are sound, then utterances of (1)–(4) or any other B-sentences do not express the propositions expressed by utterances of "It is now t_7," for the latter utterances express A-propositions and thus express propositions different from the B-propositions expressed by (1)–(4) or any other tenseless sentences.

Suppose one tries to defend the co-reporting theory by saying that the indexical B-sense of "now" cannot be expressed otherwise than by "now," and that this sense is not any of the senses expressed by the B-expressions (for example, "simultaneously with U_1") and is not the irreducible A-property of presentness. But this would not be a defense or justification of the co-reporting thesis but a mere assertion that it is true. What is needed is some argument that the sense of "now" is not an irreducible A-property, and a simple assertion to this effect will convince nobody but the faithful. Unless some justification for this assertion is advanced, Beer and Gale cannot maintain that they have *demonstrated* that A-properties are reducible to B-relations.

Perhaps, however, these remarks do not do full justice to Beer and Gale, for there is an argument to be found in their writings to the effect that "the moving present" cannot be the referent of "now." They state this argument as follows: "That 'now', like 'this time' and 'the present,' is used to refer to a time shows the absurdity of the common account of temporal becoming, used by both friends and foes of becoming, that it involves the shift or advance of the present or now to ever later times or events. For how can a moment of time cease to be identical with itself or even have the possibility of being a different time!"[12] This argument is transparently unsound, even if we accept the premise that "now" refers to the moment of its utterance. Suppose "now" as used at t_7 refers to t_7. Does this entail that an "advancing present" cannot be a part of the semantic content of "now"? Not at all, for it is perfectly consistent to hold both that

(5) the use of "now" at t_7 refers to t_7

and

(6) the use of "now" at t_7 ascribes to its referent, t_7, the irreducible monadic A-property of presentness.

The referent of this use of "now," t_7, does not advance to later and later times, but the monadic property ascribed to this referent, presentness, does advance to later times, t_8, t_9, . . . t_n, as they successively become present. First t_7 exemplifies this monadic property, then t_8 exemplifies it, and so on. Gale and Beer have mistakenly supposed that the disjunction "either uses of 'now' refer to times, such as t_7, or their semantic content includes the irreducible A-property of presentness" is an exclusive disjunction, whereas in fact it is an inclusive one. That is, they have failed to see that uses of "now" can both refer to times and ascribe to these times an irreducible property of presentness.

A major reason why the above-mentioned unsound argument seemed sound to Gale and Beer is that they did not distinguish between the referring and predicative aspects of A-expressions. This distinction is perhaps best clarified and substantiated in terms of Gilbert Plumer's insightful discussion of the two different uses of "present" in a recent article.[13] Plumer shows that Gale himself was tacitly committed to a distinction of this sort in his early tenser period. Plumer quotes the following passage from Gale's *The Language of Time:* "[T]he present (now), unlike here, shifts inexorably, independently of what we do. *Every event later than the present will become present* and every event earlier than the present did become present."[14] Plumer notes that in the italicized clause the second token of "present" cannot refer back to the time referred to by the first token of "present." If it did refer back to this time, we would get a contradiction of the sort to which Gale appeals in his above-mentioned argument that "now" or "the present" cannot refer to a moving present. If the first token of "present" in the italicized clause refers to some moment, say t_7, and the second token refers to this same moment, we get the contradiction that later events (or moments) will become t_7. But if we suppose that the second token of "present" refers to some different moment, say t_8, we get the equally absurd result that every event (or moment) later than t_7 will become t_8. To avoid these absurdities, we must suppose that "present" in its second occurrence in the italicized clause is being used in a fundamentally different way from its first occurrence. Plumer goes on to explain this second usage of "present," but his explanation is not entirely satisfactory. He explains the difference between the two uses in terms of a difference between two sentences that ascribe different sorts of temporal reference to "now":

ID Each temporally successive tokening of the word "now" refers to a different time.

IS Each temporally successive tokening of the word "now" refers to the same time.

ID is supposed to capture the use of "present" in its first occurrence in the italicized clause from Gale's book, and IS is supposed to capture the use of "present" in its second occurrence. I have no qualms about Plumer's interpretation of ID; he correctly notes that in ID "the different times indexed are particular instants," such as t_7 and t_8.[15] Plumer also correctly notes that "something else" is meant by "the same time" in IS; that is, something other than "the same instant" or "the same moment." But he rather mysteriously says that what is meant by "the same time" in IS is "the NOW."[16] But what is "the NOW"? He says that it is not a particular or a property of particulars but "a condition of the possibility of there being particulars in general."[17] But this statement is both unjustified and ineffable. First of all, the fact that something is a condition of the possibility of there being particulars does not entail that it is not a first-order property, for, arguably, *self-identity, oneness, being a particular, being extended or unextended,* are first-order properties and conditions of the possibility of there being particulars in general. Secondly, if the NOW is not a particular or a property of particulars, what is it? A property of universals? A relation? A set? A noumenon? Plumer gives us no hint. It seems to me that the sort of ineffable claim that Plumer is making about the NOW well motivates the dubious tone of such remarks as Beer's reference (Essay 7) to the tenser theory of "a mysterious, transcendent present or now." I suggest that what Plumer is trying to get at with his notion of the NOW is best understood as *the first-order property of presentness.* The statement IS that is supposed to capture the second use of "present" in the italicized clause from Gale's book is most perspicuously rephrased as

IS′ Each temporally successive tokening of the word "now" predicates the same temporal property, the property of presentness,

so that "the same time" in IS is parsed as "the same temporal property," and "refers to" in IS is replaced by "predicates." In other words, the second occurrence of "present" in the clause from Gale's book is a *predicative use* of "present," a use in which it serves solely to express the property of presentness and ascribe this property to something. Only the first occurrence of "present" is *a referential use,* in which "present" refers to the moment of its utterance. The italicized clause "*Every event later than the present will become present,*" if uttered at t_7, means "Every event later than t_7 will acquire the property of presentness." However, if we suppose that "present" or "the present" in its first occurrence in the italicized clause is synonymous with "now," then we should interpret this use of "the present" as being in a relevant sense *both* a referential and a predicative use; that is, as both referring to the time of its utterance, t_7, and predicating of this time the property of presentness, so that the italicized clause means "Every event later than t_7, which has the property of presentness, will acquire the property of presentness." Once these two different aspects of the use of "now" or "present" are distinguished, the difficulty alluded to by Gale and Beer disappears, for we can then coherently say that "now" refers to a time but that what shifts to ever later times is not what "now" refers to but what "now" predicates; namely, the property of presentness. Of course, some philosophers, such as Christensen, Levison, and Gale himself, believe that the thesis that *presentness is a property* entails McTaggart's paradox;[18] but I think that this belief is unjustified and that the conception of presentness as a property entails no paradoxical consequences (See Essays 14 and 15).[19]

Notes

1. Beer first expounded the co-reporting theory in her doctoral dissertation "Temporal Indexicals and the B-Theory of Time" (Pittsburgh, 1981). See also Gale and Beer, "Identity Theory of the A- and B-Series."
2. The phrase "the old tenseless theory of time" is used by Oaklander and myself to refer to the theory that tensed sentence-tokens are translated by tenseless sentence-tokens and therefore do not ascribe A-properties. See Essays 2 and 5.
3. Smart, "Time and Becoming"; Mellor, *Real Time.*
4. In his reply (Essay 5) Oaklander charges me with misunderstanding the import of the new tenseless theory of time. In a work in progress on language, time, and monism I argue that Oaklander has in fact misunderstood its import and has confused the new tenseless theory of time of Smart and Mellor with the *nonsemantic* tenseless theory of time espoused by Adolf Grünbaum and by Oaklander in *Temporal Relations,* chap. 4.
5. Gale and Beer, "Identity Theory of the A- and B-Series"; Paul Fitzgerald, "Nowness and the Understanding of Time," *Boston Studies in the Philosophy of Time* 20 (1974): 274; Oaklander, *Temporal Relations,* p. 127.
6. David Kaplan, "Demonstratives," draft no. 2 (mimeograph, 1977).
7. See ibid.; John Perry, "Frege on Demonstratives," *Philosophical Review* 86 (1977): 474–497; Hector-Neri Castañeda, *Thinking, Language, Experience* (Minneapolis, 1989); Wettstein, "Indexical Reference."
8. G. Frege, "The Thought," in E. Klempke, ed., *Essays on Frege* (Urbana, Ill., 1968), p. 516.
9. John Searle, *Intentionality* (Cambridge, 1983), pp. 223 ff. Searle does not use sentences of the forms (1) and (2) to translate the A-sentence-tokens, but sentences that include his technical phrase "at *cotemporal"; however, the facts about the nonequivalence of (1) and (2) also apply to Searle's sentences. For further criticisms of Searle's theory and of Smart's and Reichenbach's token-reflexive theories see Smith, "Impossibility of Token-Reflexive Analyses."
10. Eddy Zemach, "De Se and Descartes," *Noûs* 19 (1985): 181–204; Lynne Rudder Baker, "Temporal Becoming: The Argument from Physics," *Philosophical Forum* 6 (1974–75): 218–236; Grünbaum, *Modern Science and Zeno's Paradoxes;* Bertrand Russell, "On the Experience of Time," *Monist* 25 (1915): 212–233.
11. Some further reasons for this conclusion are presented in Smith, "Mind-Inde-

pendence of Temporal Becoming"; idem, "Sentences about Time"; and Essays 2, 12, and 34.

12. Gale and Beer, "Identity Theory of the A- and B-Series." Beer's statement of the argument in Essay 7 reads: "That 'now' is a referring term that denotes a time shows, moreover, the absurdity of one way of characterizing the passage of time, according to which the present shifts to ever later times. For this notion of temporal passage would require that this very time—the present or now—become a different time at ever later times. This way of conceiving the passage of time is clearly absurd; for how is it possible that this time should cease to be identical with itself?"
13. Gilbert Plumer, "Expressions of Passage," *Philosophical Quarterly* 37 (1987): 341–354.
14. Gale, *Language of Time,* p. 214; my emphasis.
15. Plumer, "Expressions of Passage," p. 351. At least, I agree that *different moments* are indexed; I do not necessarily endorse Plumer's thesis that these moments are instants rather than intervals. See his interesting article on this subject, "The Myth of the Specious Present," *Mind* 94 (1985): 19–35.
16. Plumer, "Expressions of Passage," p. 351.
17. Ibid., p. 352.
18. Christensen, "McTaggart's Paradox and the Nature of Time," pp. 289–299; Arnold B. Levison, "Events and Time's Flow," *Mind* 46 (1987): 341–353; Richard Gale, "Becoming" (mimeograph).
19. I argue for this view in Essay 15. Oaklander criticizes my argument in Essay 16, and I respond to Oaklander's criticism in Essay 17. I am grateful to Neil Cooper and the anonymous referee for *Philosophical Quarterly,* in which this was first published, for helpful comments on an earlier version of the essay.

ESSAY 9

The Date Analysis of Tensed Sentences

CLIFFORD WILLIAMS

The A-theory of time holds that pastness, presentness, and futurity are mind-independent properties of events over and above the relations of earlier than, simultaneous with, and later than. The B-theory of time asserts that time consists of nothing more than these relations. Recent discussion of these two theories has centered on the question of whether tensed sentences—those purportedly used to ascribe A-properties to events—are equivalent in some way to tenseless sentences—those used to ascribe only B-relations to events. Advocates of the A-theory often argue that the semantic nonequivalence of tensed sentences and tenseless date-sentences shows that the A-theory is true. If "tensed sentences and tokens semantically differ from date-sentences and tokens," writes Quentin Smith, then "tensed sentences/tokens do not ascribe to events the very same properties or relations that are described by date-sentences, viz., B-relations to the birth of Christ. They ascribe determinations of a different order."[1] These determinations, he says, are pastness, presentness, and futurity.

Proponents of the A-theory do not, however, always distinguish the idea of tensed and tenseless sentences being semantically nonequivalent from the idea of these sentences being used to describe different states of affairs. The distinction is crucial, for the A-theory requires not only that tensed and tenseless sentences be semantically nonequivalent but also that they be used to describe different states of affairs. In this essay, I shall argue that the semantic nonequivalence of these sentences does not show that they are used to describe different states of affairs and, therefore, does not support the A-theory.

A-theorists have used several arguments to demonstrate the semantic nonequivalence of tensed sentences and tenseless date-sentences. In the truth-condition argument, the A-theorist argues that the tensed sentence

(1) Linda is reading a poem now

is not semantically equivalent to

(2) Linda read (tenseless) a poem at 7:30 P.M. on 3 July 1990,

because (1) and (2) have different truth conditions. In order to know that (1) is true, we must know that Linda is reading at the time (1) is uttered; but in order to know that (2) is true, we do not need to know that Linda is reading at the time (2) is uttered. Since having different truth conditions is sufficient for having different meanings, (1) and (2) must have different meanings. Consequently, concludes the A-theorist, the A-determination expressed in (1) is not reducible to the B-relation expressed in (2).

Let us agree with the A-theorist that when two sentences have different truth conditions, they have different meanings. It does not follow that the sentences are used to describe different states of affairs. The meaning of a sentence may permit, or require, that the sentence be used to describe a different state of affairs each different time it is uttered. This would be the case if, as with (1), part of the sentence's meaning is that it be used to describe a state of affairs existing simultaneously with its utterance. In such a case, the meaning of (1) would remain the same at different times, but the state of affairs described through the use of (1) would be different at different times. And the meaning of (1) would be different from the meaning of (2), but the state of affairs described through the use of (1) and (2) might be the same. So having different truth conditions, though demonstrating difference of meaning, does not demonstrate difference of state of affairs.

The best way to see how these things can be so is to note that the meanings of sentences and words may contain a pragmatic component—a component telling us not the content of the propositions that we use the sentences to express or the content of the ideas associated with the words, but the conditions for their correct use. The pragmatic component of the meaning of some sentences may call for them to be used on different occasions to describe different states of affairs. Similarly, the pragmatic component of the meaning of some words may require them to be used at different times to refer to a different time ("now") or at different places to refer to a different place ("here") or by different persons to refer to a different person ("I"). Thus, to know that (1) and (2) are used to describe different states of affairs, we would have to examine the nonpragmatic component of their meanings. Simply knowing that their meanings are different is not enough to know that the states of affairs described are different.

One way to pinpoint the distinction between meaning and reference here is to see that when the date analysis of tensed sentences says that we use "now" to refer to the time at which "now" is uttered, the B-theorist does not mean either that the simultaneity of a time with an utterance is part of what we use "now" to refer to or that the utterance of "now" itself is part of what we use "now" to refer to. The only part of the phrase, "the time at which 'now' is uttered," that refers to what "now" is used to refer to is "the time." Neither the simultaneity of the time with "now's" utterance nor the clause, "at which 'now' is uttered," is referred to by "now." This latter clause, according to the B-theorist, expresses part of the conditions for the correct use of "now." It tells us which time we use "now" to refer to.

Although the distinction between meaning and reference has become prominent in the philosophy of language, it has not always been prominent in the writings of either A-theorists or B-theorists. A-theorists who use the truth-condition argument to refute the B-theorists' date analysis of tensed sentences sometimes seem unaware that the distinction is relevant to the argument. B-theorists themselves do not always make the distinction clearly. Nelson Goodman writes:

> The "now" in question [the 937th word uttered by George Washington in 1776] is translated by any "The period referred to by the 937th word uttered by George Washington in 1776" . . . Or we may seek a translation that contains no name of the indicator itself, but rather another name for what the indicator names. Thus a certain "here" is translated by any "Philadelphia," and a certain "ran" is translated by any "runs [tenseless] on January 7, 1948, at noon E.S.T."[2]

Goodman does not make it clear here whether his proposed translation is an identity of meaning or an identity of reference. His use of "translation" makes one think he meant the former, but his use of "name" makes one think he meant the latter. So perhaps part of the reason why A-theorists are not always aware that the distinction is relevant to the issue is the B-theorists' ambiguity.

Whatever ambiguity B-theorists' language contains, the date analysis of tensed sentences ought to be put in terms of states of affairs, and the date analysis of "now" ought to be put in terms of reference. For even though meaning equivalence is refuted by the A-theorists' truth-condition argument, the state of affairs or reference equivalence might still obtain and, if it did, would undermine the A-theory. The A-theory requires that the states of affairs described through the use of tensed sentences be different from the states of affairs described through the use of tenseless date-sentences. If they were the same, A-determinations would be reducible to B-relations. And if "now" were used to refer just to a time, nowness would be nothing more than the time; it would not be a fleeting property of the time.

Beer recognizes these points in putting forward her co-reporting thesis.

"The co-reporting thesis," she writes, "does not claim that every A-proposition is identical with a B-proposition but rather that every A-proposition reports one and the same event as does some B-proposition" (p. 90). Beer's distinction between the assertion that A-propositions and B-propositions are identical and the assertion that they report the same event forestalls A-theorists' use of the denial of the first assertion to undermine the B-theory. It is the second assertion that A-theorists must deny in order to undermine the B-theory.

The importance of the distinction between meaning and states of affairs becomes more evident when we see how it applies to "here," "this," and "I." Imagine that an A-theorist of space, one who holds that hereness and thereness are mind-independent properties of objects over and above the locations of the objects, were to argue for the falsity of the place analysis of here-sentences as follows. The here-sentence

(3) Linda is reading poems here

is not semantically equivalent to

(4) Linda is reading poems in her home's living room,

because (3) and (4) have different truth conditions. This semantic nonequivalence shows, say A-theorists of space, that here-sentences such as (3) express mind-independent properties that are not reducible to the locations referred to in place-sentences such as (4). Our first reaction to this odd argument would be that although having different truth conditions shows meaning nonequivalence, it does not show state-of-affairs nonequivalence, and it is this latter that must be shown in order to refute B-theorists of space, who hold that hereness and thereness are not mind-independent properties of objects over and above the locations of the objects. B-theorists of space believe that hereness is reducible to locations in space, because we use here-sentences to describe the same states of affairs as place-sentences, and "here" to refer to places.

We would react as I have indicated to this hypothetical argument because we are all B-theorists of space. And we are all B-theorists for "thisness" and "I-ness." We use "this" and "I" to refer to objects and persons, not to peculiar, mind-independent properties of those objects and persons. This linguistic fact shows, we believe, that thisness and I-ness are not mind-independent properties of objects and persons but are just the objects and persons. Note that it is a fact about the objects we use "this" and "I" to refer to and about the states of affairs we use this-sentences and I-sentences to describe that founds our belief in the B-theory of thisness and I-ness. We would agree with the A-theory of thisness that this-sentences have different meanings from object-sentences but would not infer, as A-theorists of thisness do, that thisness is a mind-independent feature of objects over and above the objects themselves.

These points about "here," "this," and "I" do not entail that we use "now" to refer to B-times. Nor does the fact that we use "now" to refer to something existing simultaneously with its utterance entail that the something referred to is a B-time; it may be an A-property. The meaning of "now" may be, as A-theorists claim, that we use it to refer to the A-property of nowness that characterizes events occurring simultaneously with the utterance of "now." (A-theorists here distinguish between meaning and reference.) What these points about "here" show is that A-theorists cannot infer that "now" has this A-theory meaning from the sole fact that its meaning differs from that of tenseless date-expressions. For the referents of certain uses of "now" may be just the same as those of certain date-expressions, even though the meanings differ. I conclude that the truth-condition argument for the semantic nonequivalence of tensed sentences and date-sentences does not support their state-of-affairs nonequivalence, and therefore does not support the A-theory.

This conclusion also holds for two other arguments that A-theorists have used. They sometimes argue that the date analysis of tensed sentences is mistaken because

(5) It is now 3:00 P.M.

is not a tautology, whereas its alleged equivalent,

(6) It is [tenseless] 3:00 P.M. at 3:00 P.M.,

is a tautology. Since a tautology cannot be equivalent to a nontautology, (5) and (6) are not equivalent. So, A-theorists conclude, (5) expresses a feature of times, their presentness, that (6) does not express.

The second argument states that the date analysis is mistaken because tensed sentences convey more information than date-sentences. If all we knew was that Linda reads (tenseless) poetry at 7:30 P.M. on 3 July 1990, we would not know when she is reading, because we would not know when 7:30 P.M. is present. Tensed sentences are not, A-theorists infer, equivalent to tenseless date-sentences (see Essay 32).

The way in which the distinction between meaning and reference applies to these arguments becomes clear when we look at the "here," "this," and "I" analogues to these two arguments. Suppose A-theorists of space were to argue that the place analysis of "here" is mistaken, because

(7) Here's Chicago

(a remark a traveler makes to a traveling companion) is not a tautology, whereas

(8) Chicago is at Chicago

is a tautology, or because (7) conveys more information than (8). Because we are convinced B-theorists of space, these arguments would not move us to believe that hereness is a mind-independent feature of Chicago. We would, rather, look for a way both to retain the thesis that "here" refers to places and to account for the differences between (7) and (8). That way is readily available in the fact that the meaning of "here" requires it to be used to refer to the place at which it is used, whereas the meaning of "Chicago" does not require that it be used in the city to which it refers. (7) conveys more information than (8) because the utterer and the listener know this fact. The utterer intends to convey the extra information, and the listener has it conveyed to her. The conveyance by the utterer to the listener takes place not by means of what "here" is used to refer to but by means of the common knowledge of the conditions for the correct use of "here."

Similar remarks are true for "this." We would not give up our belief that "this" is used to refer to objects and be moved to believe that thisness is a mind-independent feature of objects, just because

(9) This is Chicago

is nontautologous and conveys more information than

(10) Chicago is Chicago.

This conveyance of more information by (9) comes not from its content but from the convention that "this" is used to refer to an object near its utterer's location.

What A-theorists of time must do, therefore, is show that the nontautologous character of (5) and the extra information conveyed by it are due not just to the users' and hearers' assumptions about the difference between the ways (5) and (6) are used but to the difference between the states of affairs described through the use of (5) and (6). The nontautologous character of (5) and the extra information conveyed by it can, to be sure, be accounted for by the A-theorists' claim that (5) is used to ascribe the mind-independent property of nowness to a time, but they are not sufficient to demonstrate this claim.[3]

Notes

1. Smith, "Sentences about Time," p. 50.
2. Goodman, *Structure of Appearance,* p. 369.
3. I am indebted to Quentin Smith and to an anonymous referee for the *Australasian Journal of Philosophy,* where this essay was first published, for their comments on an earlier version.

ESSAY 10

Williams's Defense of the New Tenseless Theory of Time

QUENTIN SMITH

The debate among Mellor, Oaklander, and myself in the first six essays concerned whether the truth conditions of tensed sentences (types or tokens) were different from those of tenseless sentences and involved tensed facts. In Essay 9, Clifford Williams entered into the debate by presenting an argument that if tensed sentence-types have different truth conditions from tenseless sentence-types, that does not entail that the truth conditions of the tensed sentence-types involve tensed states of affairs (tensed facts). Williams also further developed the argument of Beer (Essay 7) that now-sentences express different propositions from tenseless date-sentences, but that it is nevertheless the case that uses of "now" refer to the time of their use rather than ascribe a property of presentness. I shall show in this paper that Williams's arguments are insufficient to defend the new tenseless theory of time.

Williams begins by arguing that

(1) The tensed sentence-type "Linda is reading a poem now" has different truth conditions from the tenseless sentence-type "Linda read [tenseless] a poem at 7:30 P.M. on 3 July 1990,"

does not entail that

(2) The two sentence-types are always used to describe different states of affairs.

The tensed sentence-type (let us call it *S*) may have the truth conditions that in part are specified by "*S* is true if and only if it describes a state of affairs existing simultaneously with its utterance," whereas the tenseless sentence-type will not include these conditions. (Williams suggests, regarding the tensed sentence-type, that "part of the sentence's meaning is that it be used to describe a state of affairs existing simultaneous with its utterance' (p. 105). The tenseless sentence-type will instead have the truth conditions specifying that it is true if and only if Linda reads a poem at 7:30 P.M. on 3 July 1990. Williams argues that this difference in truth conditions is consistent with the sentence-type *S* being used at 7:30 P.M. on 3 July 1990 to describe the very same state of affairs that is described by the tenseless sentence-type.

The argument is unsound, however, if the tensed sentence-type has tensed truth conditions, that is, truth conditions that involve tensed facts (which are states of affairs involving something's possession of an A-property such as presentness). It would follow from this that *S* as used on 7:30 P.M. on 3 July 1990 does not describe only the state of affairs described by the tenseless sentence-type.

I argued in essays 2, 4, 6, and 8 that the truth conditions of tensed sentences cannot be merely tenseless token-reflexive truth conditions or date-involving truth conditions but must also include tensed truth conditions, that is, conditions that involve tensed facts or tensed states of affairs, such as the presentness of Linda's reading. If these arguments are sound, then the addition of them to premise 1 would enable premise 2 to be derived. Williams admits that "the meaning of 'now' may be, as A-theorists claim, that we use it to refer to the A-property of nowness that characterizes events occurring simultaneously with the utterance of 'now'" (p. 108). But he emphasizes that "A-theorists cannot infer that 'now' has this A-theory meaning from the sole fact that its meaning differs from that of tenseless date-expressions" (p. 108). I would agree with Williams that this inference is invalid; but I would add that there are good arguments that the meaning of "now" does ascribe an A-property of presentness (see Essays 2, 4, and 8) and that the addition of these arguments to the premise about difference in meaning would render the inference valid.

Since Williams does not address these additional arguments but contents himself with making the minimal point that the inference from mere meaning difference is invalid, he does not offer an adequate defense of the new tenseless theory of time.

However, Williams does offer an independent argument against the thesis that "now" ascribes a property of presentness by pointing to some alleged analogies among "now," "I," and "here." However, even if these analogies obtain (exactly as specified by Williams), this is not sufficient to refute the tensed theory of time, since this theory may rest its case on the use of tenses ("is," "was," "will be") rather than indexicals ("now," "yesterday"), and there is no linguistic analogue in spatial or personal language for the tenses. There are spatial indexicals ("here," "there") but no spatial tenses. (The difference between tenses and temporal indexicals is discussed in a Essay 12.)

But I believe that Williams's alleged analogies between spatial, personal, and temporal indexicals do not support his claims. He presents the alternative that indexicals either merely refer to items or merely ascribe mind-independent properties. Since it is obvious that uses of "I" and "here" do not ascribe mind-independent properties of *me-ness or hereness,* it follows that, instead, they merely refer to persons and places. Williams suggests that analogous reasoning should make us incline towards the view that uses of "now" refer to times rather than ascribe a mind-independent property of presentness.

However, this argument is based on a number of false alternatives. First, Williams does not consider, and offers no argument against, the view that uses of "here" and "I" ascribe *mind-dependent* properties of *being here* (*here-ness*) and *being me* (*me-ness*). Castañeda's and Patrick Grim's examples of uses of personal indexicals have shown that *de se* uses are different from *de re* uses and that "I" is not used *de re,* that is, simply to refer to the speaker.[1] I believe that "I" ascribes a primitive property of *me-ness* (just as "now" ascribes a primitive property of presentness), but a detailed defense of this theory is not required for my limited purpose of showing that Williams cannot simply take it for granted that "I" either ascribes a mind-independent property of me-ness or merely refers to the speaker, as if there were no other possible theories of the meaning of this word. Williams writes: "We use 'this' and 'I' to refer to objects and persons, not to peculiar, mind-independent properties of those objects and persons" (p. 107), but this ignores the possibility that "this" and "I" refer to objects and persons and ascribe to them mind-dependent properties. (Williams oddly talks of these expressions "referring" to properties; but this is a mistake, since these expressions would instead *ascribe* or *attribute* the properties to something.)

It is also arguable that when I say in Chicago "Here's Chicago," my use of "here" refers to Chicago and ascribes to it the mind-dependent property of *being here* (or *here-ness*). The analogy may be continued for temporal indexicals. When I say at 7:30 P.M. "Linda is reading now," my use of "now" may refer to 7:30 P.M. and ascribe to this time the property of presentness.

One argument that Williams offers concerns "It is now 3:00 P.M." and "It is 3:00 P.M. at 3:00 P.M." He suggests that we need not try to explain the informativeness of the first sentence by saying it ascribes a property of presentness, since we would not attempt to do the same for "Here's Chicago" as contrasted with "Chicago is at Chicago." But my remarks above suggest that Williams cannot take it for granted that the informativeness of the here-sentence cannot be explained in terms of a property ascribed by this indexical.

He says nothing to rule out the possibility that the indexical sentence describes the state of affairs *Chicago exemplifies being here* (taken as a mind-dependent property) and that the nonindexical sentence describes *Chicago exemplifies being identical with Chicago.*

On the account I have suggested, the difference among the spatial, personal, and temporal states of affairs is that *hereness* and *me-ness* are mind-dependent properties whereas presentness is mind-independent. This is reflected in the structure of our language and beliefs. Many people (arguably all people except philosophers who are detensers) believe that presentness is mind-independent, but nobody believes that *hereness* or *me-ness* is mind-independent. "Being here" means roughly "being where I am" (or "where I am pointing"), but "being present" does not mean (even roughly) "being simultaneous with my utterance," as I have argued in earlier essays. "Here's Chicago" means, roughly, "Chicago is the place where I am now located (or where I am now pointing)," but I have previously argued that sentences such as "It is now 1980" do not mean (even roughly) "1980 is simultaneous with this utterance." Accordingly, Williams needs more argumentation before his appeals to alleged analogies among "now," "here," and "I" can serve to defend the new tenseless theory of time against the criticisms that have been advanced.

Note

1. Castañeda, *Thinking, Language, and Experience;* Patrick Grim, "Against Omniscience," *Noûs* 19 (1985): 151–180.

ESSAY 11

Demonstratives

DAVID KAPLAN

I believe my theory of demonstratives to be uncontrovertible and largely uncontroversial. This is not a tribute to the power of my theory but a concession to its obviousness. In the past, no one seems to have followed these obvious facts out to their obvious conclusions. I do that. What is original with me is some terminology to help fix ideas when things get complicated. It has been fascinating to see how interesting the obvious consequences of obvious principles can be.[1]

Demonstratives, Indexicals, and Pure Indexicals

I tend to describe my theory as "a theory of demonstratives," but that is poor usage. It stems from the fact that I began my investigations by asking what is said when a speaker points at someone and says, "He is suspicious."[2] The word "he," so used, is a demonstrative, and the accompanying pointing is the requisite associated demonstration. I hypothesized a certain semantic theory for such demonstratives, and then I invented a new demonstrative, "dthat," and stipulated that its semantics be in accord with my theory. I was so delighted with this methodological sleight of hand for my demonstrative "dthat" that when I generalized the theory to apply to words like "I," "now," "here," and so on—words which do *not* require an associated demonstration—I continued to call my theory a "theory of demonstratives," and I referred to these words as "demonstratives."

That terminological practice conflicts with what I preach, and I will try to correct it. (But I tend to backslide.) The group of words for which I propose a semantic theory includes the pronouns "I," "my," "you," "he," "his," "she," "it"; the demonstrative pronouns "that," "this"; the adverbs "here," "now," "tomorrow," "yesterday"; and the adjectives "actual," "present," and others. These words have uses other than those in which I am interested (or, perhaps, depending on how you individuate words, we should say that they have homonyms in which I am not interested). For example, the pronouns "he" and "his" are used not as demonstratives but as bound variables in "For what is a man profited, if he shall gain the whole world, and lose his own soul?"

What is common to the words or usages in which I am interested is that the referent is dependent on the context of use and that the meaning of the word provides a rule which determines the referent in terms of certain aspects of the context. The term I now favor for these words is "indexical." Other authors have used other terms; Russell used "egocentric particular," and Reichenbach used "token reflexive." I prefer "indexical" (which, I believe, is due to Pierce) because it seems less theory-laden than the others, and because I regard Russell's and Reichenbach's theories as defective.

Some of the indexicals require, in order to determine their referents, an associated demonstration: typically, though not invariably, a (visual) presentation of a local object discriminated by a pointing.[3] These indexicals are the true demonstratives, and "that" is their paradigm. The demonstra*tive* (an expression) refers to that which the demon*stration* demonstrates. I call that which is demonstrated the "demonstratum."

A demonstrative without an associated demonstration is incomplete. The linguistic rules which govern the use of the true demonstratives "that," "he," and so forth are not sufficient to determine their referent in all contexts of use. Something else—an associated demonstration—must be provided. The linguistic rules assume that such a demonstration accompanies each (demonstrative) use of a demonstrative. An incomplete demonstrative is not *vacuous* like an improper definite description. A demonstrative *can* be vacuous in various cases: for example, when its associated demonstration has no demonstratum (a hallucination) or the wrong kind of demonstratum (pointing to a flower and saying "he" in the belief that one is pointing to a man disguised as a flower[4]) or too many demonstrata (pointing to two intertwined vines and saying "that vine"). But it is clear that one can distinguish a demonstrative with a vacuous demonstration—no referent—from a demonstrative with no associated demonstration—one that is incomplete.

All this is by way of contrasting true demonstratives with pure indexicals. For the latter, *no associated demonstration is required, and any demonstration supplied is either for emphasis or is irrelevant.*[5] Among the pure indexicals are "I," "now," "here" (in one sense), "tomorrow", and others. The linguistic rules which govern *their* use fully determine the referent for each context.[6] No supplementary actions or intentions are needed. The speaker refers to himself when he uses "I," and no pointing to another or believing that he is another or intending to refer to another can defeat his reference.[7]

Michael Bennett has noted that some indexicals have both a pure *and* a demonstrative use. "Here" is a pure indexical in

I am in here

and a demonstrative in

In two weeks, I will be here (pointing at a city on a map).

Two Obvious Principles

So much for preliminaries. My theory is based on two obvious principles. The first has been noted in every discussion of the subject.

Principle 1 The referent of a pure indexical depends on the context, and the referent of a demonstrative depends on the associated demonstration.

If you and I both say "I," we refer to different persons. The demonstratives "that" and "he" can be correctly used to refer to any one of a wide variety of objects simply by adjusting the accompanying demonstration.

The second obvious principle has been formulated explicitly less often.

Principle 2 Indexicals, pure and demonstrative alike, are directly referential.

Remarks on Rigid Designators

In an earlier draft I adopted the terminology of Kripke, called indexicals "rigid designators," and tried to explain that my usage differed from his. I am now shying away from that terminology. But because it is so well known, I will make some comments on the notion or notions involved.

The term "rigid designator" was coined by Saul Kripke to characterize those expressions which designate the same thing in every possible world in which that thing exists and which designate nothing elsewhere. He uses it in connection with his controversial, though, I believe, correct claim that proper names, as well as many common nouns, are rigid designators. There is an unfortunate confusion in the idea that a proper name would designate nothing if the bearer of the name were not to exist.[8] Kripke himself adopts positions which seem inconsistent with this feature of rigid designators. In arguing that the object designated by a rigid designator need not exist in every possible world, he seems to assert that, under certain circumstances, what is expressed by "Hitler does not exist" would have been true not because "Hitler" would have designated nothing (in *that* case we might have given the sentence *no* truth-value) but because what "Hitler" would have designated—namely, Hitler—would not have existed.[9] Furthermore, it is a striking and important feature of the possible-world semantics for quantified intensional logics, which Kripke did so much to create and popularize, that variables, those paradigms of rigid designation, designate the same individual in *all* possible worlds, whether the individual "exists" or not.[10]

Whatever Kripke's intentions (did he, as I suspect, misdescribe his own concept?) and whatever associations or even meaning the phrase "rigid designator" may have, I intend to use "*directly referential*" for an expression whose referent, once determined, is taken as fixed for all possible circumstances, that is, is taken as *being* the propositional component.

For me, the intuitive idea is not that of an expression which *turns out* to designate the same object in all possible circumstances, but an expression whose semantical *rules* provide *directly* that the referent in all possible circumstances is fixed to be the actual referent. In typical cases the semantical rules will do this only implicitly, by providing a way of determining the *actual* referent and no way of determining any other propositional component.[11]

We should beware of a certain confusion in interpreting the phrase "designates the same object in all circumstances." We do not mean that the expression *could not have been used* to designate a different object. We mean rather that, given a *use* of the expression, we may ask of *what has been said* whether *it* would have been true or false in various counterfactual circumstances and in such counterfactual circumstances, which are the individuals relevant to determining truth-value. Thus we must distinguish possible occasions of *use*—which I call *contexts*—from possible circumstances of *evaluation* of what was said on a given occasion of use. Possible circumstances of evaluation I call *circumstances* or, sometimes, just *counterfactual situations*. A directly referential term *may* designate different objects when used in different *contexts*. But when evaluating what was said in a given context, only a single object will be relevant to the evaluation in all circumstances. This sharp distinction between *contexts of use* and *circumstances of evaluation* must be kept in mind if we are to avoid a seeming conflict between principles 1 and 2.[12] To look at the matter from another point of view, once we recognize the obviousness of both principles (I have not yet argued for principle 2), the distinction between contexts of use and circumstances of evaluation is forced upon us.

If I may wax metaphysical in order to fix an image, let us think of the vehicles of evaluation—the what-is-said in a given context—as propositions.

Don't think of propositions as sets of possible worlds, but rather as structured entities looking something like the sentences which express them. For each occurrence of a singular term in a sentence, there will be a corresponding constituent in the proposition expressed. The constituent of the proposition determines, for each circumstance of evaluation, the object relevant to evaluating the proposition in that circumstance. In general, the constituent of the proposition will be some sort of complex, constructed from various attributes by logical composition. But in the case of a singular term which is directly referential, the constituent of the proposition is just the object itself. Thus it is that it does not just *turn out* that the constituent determines the same object in every circumstance; the constituent (corresponding to a rigid designator) just *is* the object. *There is no determining to do at all.* On this picture—and this is *really* a picture and not a theory—the definite description

$$\text{The } n[(\text{Snow is slight} \wedge n^2 = 9) \vee (\sim\text{Snow is slight} \wedge 2^2 = n + 1)]^{13} \qquad (1)$$

would yield a constituent which is complex, although it would determine the same object in all circumstances. Thus (1), though a rigid designator, is not directly referential from this (metaphysical) point of view. Note, however, that every proposition which contains the complex expressed by (1) is *equivalent* to some singular proposition which contains just the number three itself as constituent.[14]

The semantic feature that *I* wish to highlight in calling an expression *directly referential* is not the *fact* that it designates the same object in every circumstance but the *way* in which it designates an object in any circumstance. Such an expression is a *device of direct reference*. This does not imply that it has no conventionally fixed semantic rules which determine its referent in each context of use; quite the opposite. There are semantic rules which determine the referent in each context of use—but that is all. *The rules do not provide a complex which together with a circumstance of evaluation yields an object. They just provide an object.*

If we keep in mind our sharp distinction between contexts of use and circumstances of evaluation, we will not be tempted to confuse a rule which assigns an object to each *context* with a "complex" which assigns an object to each *circumstance*. For example, each context has an *agent* (loosely, a speaker). Thus an appropriate designation rule for a directly referential term would be:

(2) In each possible context of use the given term refers to the agent of the context.

But this rule could not be used to assign a relevant object to each circumstance of evaluation. Circumstances of evaluation do not, in general, have agents. Suppose I say,

(3) I do not exist.

Under what circumstances would *what I said* be true? It would be true in circumstances in which I did not exist. Among such circumstances are those in which no one, and thus, no speakers, no agents exist. To search for a circumstance of evaluation for a speaker in order to (mis)apply rule 2 would be to go off on an irrelevant chase.

Three paragraphs ago I sketched a metaphysical picture of the structure of a proposition. The picture is taken from the semantic parts of Russell's *Principles of Mathematics*.[15] Two years later, in "On Denoting,"[16] even Russell rejected that picture. But I still like it. It is not part of my theory, but it conveys well my conception of a directly referential expression and of the semantics of direct reference. (The picture needs *some* modification in order to avoid the difficulties which Russell noted later—though he attributed them to Frege's theory rather than to his own earlier theory.)[17]

If we adopt a possible worlds semantics, all directly referential terms will be regarded as rigid designators in the *modified* sense of an expression which designates the same thing in *all* possible worlds (irrespective of whether the thing exists in the possible world or not).[18] However, as already noted, I do not regard all rigid designators—not even all strongly rigid designators (those that designate something that exists in all possible worlds) or all rigid designators in the modified sense—as directly referential. I believe that proper names, like variables, are directly referential. They are not, in general, strongly rigid designators; nor are they rigid designators in the original sense.[19] What is characteristic of directly referential terms is that the designatum (referent) determines the propositional component, rather than the propositional component, along with a circumstance, determining the designatum. It is for this reason that a directly referential term that designates a contingently existing object will still be a rigid designator in the modified sense. The propositional component need not choose its designatum from those offered by a passing circumstance; it has already secured its designatum before the encounter with the circumstance.

When we think in terms of possible world semantics, this fundamental distinction becomes subliminal. This is because the style of the semantic rules obscures the distinction and makes it appear that directly referential terms differ from ordinary definite descriptions only in that the propositional component in the former case must be a *constant* function of circumstances. In actual fact, the referent, in a circumstance, of a directly referential term is simply *independent* of the circumstance and is no more a function (constant or otherwise) of circumstance than my action is a function of your desires when I decide to do it whether you like it or not. The distinction that is obscured by the style of possible-world semantics is dramatized by the structured propositions picture. That is part of the reason why I like it.

Some directly referential terms, like proper names, may have no semantically relevant descriptive meaning, or at least none that is specific, that distinguishes one such term from another. Others, like the indexicals, may have a limited kind of specific descriptive meaning relevant to the features of a context of use. Still others, like "dthat" terms (see below), may be associated with full-blown Fregean senses used to fix the referent. But in any case, the descriptive meaning of a directly referential term is no part of the propositional content.

Argument for Principle 2: Pure Indexicals

As stated earlier, I believe this principle is uncontroversial. But I had best distinguish it from similar principles that are false. I am *not* claiming, as has been claimed for proper names, that indexicals lack anything that might be called "descriptive meaning." Indexicals in general have a rather easily statable descriptive meaning. But it is clear that this meaning is relevant only to determining a referent in a context of use and *not* to determining a relevant individual in a circumstance of evaluation. Let us return to the example in connection with sentence 3 and the indexical "I." The bizarre result of taking the descriptive meaning of the indexical to be the propositional constituent is that what I said in uttering (3) would be true in a circumstance of evaluation if and only if the speaker (assuming there is one) of the circumstance does not exist in the circumstance. Nonsense! If *that* were the correct analysis, what I said could not be true. From which it follows that

It is impossible that I do not exist.

Here is another example showing that the descriptive meaning of an indexical may be entirely *inapplicable* in the circumstance of evaluation. When I say,

I wish I were not speaking now,

the circumstances desired do not involve contexts of *use* and *agents* who are not speaking. The *actual* context of use is used to determine the relevant individual—*me*—and time—*now*—and then we query the various circumstances of evaluation with respect to *that* individual and *that* time.

Here is another example, not of the inapplicability of the descriptive meaning to circumstances but of its irrelevance. Suppose I say at t_0, "It will soon be the case that all that is now beautiful is faded." Consider what was said in the subsentence,

All that is now beautiful is faded.

I wish to evaluate that content at some time in the near future, t_1. What is the relevant time associated with the indexical "now"? Is it the future time t_1? No, it is t_0, of course, the time of the context of use.

See how rigidly the indexical clings to the referent determined in the context of use:

(4) It is possible that in Pakistan, in five years, only those who are actually here now are envied.

The point of (4) is that the circumstance, place, and time referred to by the indexicals "actually," "here," and "now" are the circumstance, place, and time of the *context,* not a circumstance, place, and time determined by the modal, locational, and temporal operators within whose scope the indexicals lie.

It may be objected that this shows only that indexicals always take *primary* scope (in the sense of Russell's scope of a definite description). This objection attempts to relegate all direct reference to implicit use of the paradigm of the semantics of direct reference, the variable. Thus (4) is transformed into:

The actual circumstances, here and now, are such that it is possible that in Pakistan in five years only those who, in the first, are located at the second, during the third, are envied.

Although this may not be the most felicitous form of expression, its meaning and, in particular, its symbolization should be clear to those familiar with quantified intensional logics. The pronouns "the first," "the second," and "the third" are to be represented by distinct variables bound to existential quantifiers at the beginning and identified with "the actual circumstance," "here," and "now," respectively.

(5) $(\exists w)\ (\exists P)\ (\exists t)\ [w =$ the actual circumstance $\wedge\ p =$ *here* $\wedge\ t =$ *now* $\wedge\ \Diamond$ In Pakistan In five years $\forall x$ (x is envied $\rightarrow x$ is located at p during t in w)].

But such transformations, when thought of as representing the claim that indexicals take primary scope, do not provide an *alternative* to principle 2, since we may still ask of an utterance of (5) in a context *c*, when evaluating it with respect to an arbitrary circumstance, to what do the indexicals "actual," "here," and "now" refer. The answer, as always, is: the relevant features of the context *c*. (In fact, although (4) is equivalent to (5), neither indexicals nor quantification across intensional operators are dispensable in favor of the other.)

Perhaps enough has been said to establish the following:

T1 The descriptive meaning of a pure indexical determines the referent of the indexical with respect to a context of use but is either inapplicable or irrelevant to determining a referent with respect to a circumstance of evaluation.

I hope that your intuition will agree with mine that it is for this reason that:

T2 When what was said in using a pure indexical in a context *c* is to be evaluated with respect to an arbitrary circumstance, the relevant object is always the referent of the indexical with respect to the context *c*.

This is just a slightly elaborated version of principle 2.

Before turning to true demonstratives, we will adopt some terminology.

Terminological Remarks

Principles 1 and 2, taken together, imply that sentences containing pure indexicals have two kinds of meaning.

Content and Circumstance

What is said in using a given indexical in different contexts may be different. Thus if I say today:

I was insulted yesterday,

and you utter the same words tomorrow, what is said is different. If what we say differs in truth-value, that is enough to show that we say different things. But even if the truth-values were the same, it is clear that there are possible circumstances in which what I said would be true, but what you said would be false. Thus we say different things.

Let us call this first kind of meaning—what is said—*content*. The content of a sentence in a given context is what has traditionally been called a proposition. Strawson, in noting that the sentence

The present king of France is bald

could be used on different occasions to make different statements, used "statement" in a way similar to our use of *content of a sentence*. If we wish to express the same content in different contexts, we may have to change indexicals. Frege, here using "thought" for content of a sentence, expresses the point well: "If someone wants to say the same today as he expressed yesterday using the word 'today,' he must replace this word with 'yesterday.' Although the thought is the same, its verbal expression must be different so that the sense, which would otherwise be affected by the differing times of utterance, is readjusted.[20]

I take *content* as a notion applying not only to sentences taken in a context but to any meaningful part of speech taken in a context. Thus we can speak of the content of a definite description, an indexical, a predicate, and so forth. It is *contents* that are evaluated in circumstances of evaluation. If the content is a proposition (that is, the content of a sentence taken in some context), the result of the evaluation will be a truth-value. The result of evaluating the content of a singular term in a circumstance will be an object (what I earlier called "the relevant object"). In general, the result of evaluating the content of a well-formed expression α in a circumstance will be an appropriate extension for α (that is, for a sentence, a truth-value; for a term, an individual; for an n-place predicate, a set of n-tuples of individuals, and so on). This suggests that we can represent a content by a function from circumstances of evaluation to an appropriate extension. Carnap called such functions *intensions*.

The representation is a handy one, and I will often speak of contents in terms of it; but we should note that contents which are distinct but equivalent (that is, share a value in all circumstances) are represented by the same intension. Among other things, this results in the loss of my distinction between terms which are devices of direct reference and descriptions which *turn out* to be rigid designators. (Recall the metaphysical paragraph earlier.) I wanted the content of an indexical to be just the referent itself, but the intension of such a content will be a constant function. Use of representing intensions does not mean that I am abandoning that idea—just ignoring it temporarily.

A *fixed content* is one represented by a constant function. All directly referential expressions (as well as all rigid designators) have a fixed content—what I elsewhere call a *stable* content.

Let us settle on *circumstances* for possible circumstances of evaluation. By this I mean both actual and counterfactual situations with respect to which it is appropriate to ask for the extensions of a given well-formed expression. A circumstance will usually include a possible state or history of the world, a time, and perhaps other features as well. The amount of information we require from a circumstance is linked to the degree of specificity of contents, and thus to the kinds of operators in the language.

Operators of the familiar kind treated in intensional logic (modal, temporal, and the rest) operate on contents. (Since we represent contents by intensions, it is not surprising that intensional operators operate on contents.) Thus an appropriate extension for an intensional operator is a function from intensions to extensions.[21] A modal operator when applied to an intension will look at the behavior of the intension with respect to the possible state of the world feature of the circumstances of evaluation. A temporal operator, similarly, will be concerned with the time of the circumstance. If we built the time of evaluation into the contents (thus removing time from the circumstances, leaving only, say, a possible world history and making contents *specific* as to time), it would make no sense to have temporal operators. To put the point another way, if *what is said* is thought of as incorporating reference to a specific time or state of the world or whatever, it is otiose to ask whether what is said would have been true at another time, in another state of the world, or whatever. Temporal operators applied to eternal sentences (those whose contents incorporate a specific time of evaluation) are redundant. Any intensional operators applied to *perfect* sentences (those whose contents incorporate specific values for all features of circumstances) are redundant.[22]

What sorts of intensional operators to admit seems to me largely a matter of language engineering. It is a question of which features of what we intuitively think of as possible circumstances can be sufficiently well defined and isolated. If we wish to isolate location and regard it as a feature of possible circumstances, we can introduce locational operators: "Two miles north it is the case that," and so on. Such operators can be iterated and can be mixed with modal and temporal operators. However, to make such operators interesting, we must have contents which are locationally neutral. That is, it must be appropriate to ask if *what is said* would be true in Pakistan. (For example, "It is raining" seems to be locationally, as well as temporally and modally, neutral.)

This function notion of the content of a sentence in a context may not, because of the neutrality of content with respect to time and place, say, exactly correspond to the classical conception of a proposition. But the classical conception can be introduced by adding the demonstratives "now" and "here" to the sentence and taking the content of the result. I will continue to refer to the content of a sentence as a proposition, ignoring the classical use.

Before leaving the subject of circumstances of evaluation, I should, perhaps, note that the mere attempt to show that an expression is directly referential requires that it be meaningful to ask of an individual in one circumstance whether and with what properties it exists in another circumstance. If such questions cannot be raised because they are regarded as metaphysically meaningless, the question of whether a particular expression is directly referential (or even, a rigid designator) cannot be raised. I have elsewhere referred to the view that such questions are meaningful as *haecceitism,* and I have described other metaphysical manifestations of this view.[23] I advocate this position, although I am uncomfortable with some of its seeming consequences (for example, that the world might be in a state qualitatively exactly as it is but with a permutation of individuals).

It is hard to see how one could think about the semantics of indexicals and modality without adopting such a view.

Character

The second kind of meaning, most prominent in the case of indexicals, is that which determines the content in varying contexts. The rule

"I" refers to the speaker or writer

is a meaning rule of the second kind. The phrase "the speaker or writer" is not supposed to be a complete description; nor is it supposed to refer to the speaker or writer of the *word* "I." (There are many such.) It refers to the speaker or writer of the relevant *occurrence* of the word "I," that is, the agent of the context.

Unfortunately, as usually stated, these meaning rules are incomplete, in that they do not explicitly specify that the indexical is directly referential and thus do not completely determine the content in each context. I will return to this later.

Let us call the second kind of meaning *character.* The character of an expression is set by linguistic conventions and, in turn, determines the content of the expression in every context.[24] Because character is what is set by linguistic conventions, it is natural to think of it as *meaning* in the sense of what is known by the competent language-user.

Just as it was convenient to represent contents by functions from possible circumstances to extensions (Carnap's intentions), so it is convenient to represent characters by functions from possible contexts to contents. (As before, we have the drawback that equivalent characters are identified.[25]) This gives us the following picture:

Character: Contexts ⇒ Contents
Content: Circumstances ⇒ Extensions

or, in more familiar language,

Meaning + Context ⇒ Intension

Intension + Possible World ⇒ Extension.

Indexicals have a *context-sensitive* character. It is characteristic of an indexical that its content varies with context. Nonindexicals have a *fixed* character. The same content is invoked in all contexts. This content will typically be sensitive to circumstances; that is, nonindexicals are typically not rigid designators but vary in extension from circumstance to circumstance. Eternal sentences are generally good examples of expressions with a fixed character.

All persons alive in 1977 will have died by 2077

expresses the same proposition no matter when said, by whom, or under what circumstances. The truth-value of that proposition may, of course, vary with possible circumstances, but the character is fixed. Sentences with fixed character are very useful to those wishing to leave historical records.

Now that we have two kinds of meaning in addition to extension, Frege's principle of intensional interchange[26] becomes two principles:

F1 The character of the whole is a function of the character of the parts. That is, if two compound well-formed expressions differ only with respect to components which have the same character, then the character of the compounds is the same.

F2 The content of the whole is a function of the content of the parts. That is, if two compound well-formed expressions, each set in (possibly different) contexts differ only with respect to components which *when taken in their respective contexts* have the same content, then the content of the two compounds *each taken in its own context* is the same.

It is the second principle that accounts for the often noted fact that speakers in different contexts can say the same thing by switching indexicals. (And indeed, they often *must* switch indexicals to do so.) Frege illustrated this point with respect to "today" and "yesterday" in "The Thought." (But note that his treatment of "I" suggests that he does not believe that utterances of "I" and "you" could be similarly related!)

Earlier, in my metaphysical phase, I suggested that we should think of the content of an indexical as being just the referent itself, and I resented the fact that the representation of contents as intensions forced us to regard such contents as constant functions. A similar remark applies here. If we are not overly concerned with standardized representations (which certainly have their value for model-theoretic investigations), we might be inclined to say that the character of an indexical-free word or phrase just *is* its (constant) content.

The Meaning of Indexicals

In order to correctly and more explicitly state the semantic rule which the Dictionary attempts to capture by the entry

I: the person who is speaking or writing,

we would have to develop our semantic theory—the semantics of direct reference—and then state that

D1 "I" is an indexical, different utterances of which may have different contents.
D2 "I" is, in each of its utterances, directly referential.
D3 In each of its utterances, "I" refers to the person who utters it.

Errors have been found in the Fregean analysis of demonstratives and in Reichenbach's analysis of indexicals, all of which stem from failure to realize that these words are directly referential. When we say that a word is directly referential, are we saying that its meaning *is* its reference (its only meaning is its reference, its meaning is nothing more than its reference)? Certainly not.[27] Insofar as meaning is given by the rules of a language and is what is known by competent speakers, I would be more inclined to say in the case of directly referential words and phrases that their reference is *no* part of their meaning. The meaning of the word "I" does not change when different persons use it. The meaning of "I" is given by the rules D1, D2, and D3 above.

Meanings tell us how the content of a word or phrase is determined by the context of use. Thus, the meaning of a word or phrase is what I have called its *character*. (Words and phrases with no indexical element express the same content in every context; they have a fixed character.) To supply a synonym for a word or phrase is to find another with the same *character;* finding another with the same *content* in a particular context certainly won't do. The content of "I" used by me may be identical with the content of "you" used by you. This doesn't make "I" and "you" synonyms. Frege noticed that if one wishes to say again what one said yesterday using "today," today one must use "yesterday." (Incidentally the relevant passage, quoted above, propounds what I take to be a direct reference theory of the indexicals "today" and "yesterday.") But "today" and "yesterday" are not synonyms. For two words or phrases to be synonyms, they must have the same content in every context. In general, for indexicals, it is not possible to find synonyms. This is because indexicals are directly referential, and the compound phrases which can be used to give their reference ("the person who is speaking," "the individual being demonstrated") are not.

Summary of Findings (So Far): Pure Indexicals

Let me now try to summarize my findings regarding the semantics of demonstratives and other indexicals. First, let us consider the nondemonstrative indexicals such as "I," "here" (in its nondemonstrative sense), "now," "today," "yesterday," and the rest. In the case of these words, the linguistic conventions which constitute *meaning* consist of rules specifying the referent of a given *occurrence* of the word (we might say, a given token, or even utterance, of the word, if we are willing to be somewhat less abstract) in terms of various features of the context of the occurrence. Although these rules fix the referent and, in a very special sense, might be said to define the indexical, the way in which the rules are given does not provide a synonym for the indexical. The rules tell us for any possible occurrence of the indexical what the referent would be, but they do *not* constitute the content of such an occurrence. Indexicals are directly referential. The rules tell us what it is that is referred to. Thus, they *determine* the content (the propositional constituent) for a particular occurrence of an indexical. But they are not a *part* of the content (they constitute no part of the propositional constituent). In order to keep clear on a topic where ambiguities constantly threaten, I have introduced two technical terms, *content and character,* for the two kinds of meaning (in addition to extension) I associate with indexicals. Distinct occurrences of an indexical (in distinct contexts) may not only have distinct referents; they may have distinct meanings in the sense of *content.* If I say "I am tired today" today and Montgomery Furth says "I am tired today" tomorrow, our utterances have different contents, in that the factors which are relevant to determining the truth-value of what Furth said in both actual and counterfactual circumstances are quite different from the factors which are relevant to determining the truth-value of what I said. Our two utterances are as different in content as are the sentences "David Kaplan is tired on 26 March 1977" and "Montgomery Furth is tired on 27 March 1977." But there is another sense of meaning in which, absent lexical or syntactic ambiguities, two occurrences of the *same* word or phrase *must* mean the same. (Otherwise how could we learn and communicate with language?) This sense of meaning—which I call *character*—is what determines the content of an occurrence of a word or phrase in a given context. For indexicals, the rules of language constitute the meaning in the sense of *character.* As normally expressed, in dictionaries and the like, these rules are incomplete in that, by omitting to mention that indexicals are directly referential, they fail to specify the full content of an occurrence of an indexical.

Three important features to keep in mind about these two kinds of meaning are:

1. Character applies only to words and phrases as types, content to occurrences of words and phrases in contexts.
2. Occurrences of two phrases can agree in content although the phrases differ in character, and two phrases can agree in character but differ in content in distinct contexts.
3. The relationship of character to content is something like that traditionally regarded as the relationship of sense to denotation; character is a way of presenting content.

Notes

1. Not everything I assert is part of my theory. At places I make judgments about the correct use of certain words, and I propose detailed analyses of certain notions. I recognize that these matters may be controversial. I do not regard them as part of the basic, obvious theory.
2. See Kaplan, "Dthat," p. 320.
3. However, a demonstration may also be opportune and require no special action on the speaker's part, as when someone shouts "Stop that man" while only one man is rushing toward the door. My notion of a demonstration is a theoretical concept. I do not, in the present essay, undertake a detailed "operational" analysis of this notion, although there are scattered remarks relevant to the issue. I do consider elsewhere (David Kaplan, "Demonstratives," in Joseph Almog et al., eds., *Themes from Kaplan* (New York,1989).
4. I am aware (1) that in some languages the so-called masculine-gender pronoun may be appropriate for flowers, but it is not so in English; (2) that a background story can be provided that will make pointing at the flower a contextually appropriate, though deviant, way of referring to a man; e.g., if we are talking of great hybridizers; and (3) that it is possible to treat the example as a *referential use* of the demonstrative "he" on the model of Donnellan's referential use of a definite description (see "Reference and Definite Descriptions"). Under the referential use treatment we would assign as referent for "he" whatever the speaker *intended* to demonstrate. I intended the example to exemplify a failed demonstration, thus a case in which the speaker, falsely believing the flower to be some man or other in disguise, but having no particular man in mind, and certainly not intending to refer to anything other than that man, says, pointing at the flower, "He has been following me around all day."
5. I have in mind such cases at pointing at oneself while saying "I" (emphasis) or pointing at someone else while saying "I" (irrelevance or madness or what?).
6. There are certain uses of pure indexicals that might be called "messages recorded for later broadcast," which exhibit a special uncertainty as to the referent of "here" and "now." If the message "I am not here now" is recorded on a telephone

answering device, it is to be assumed that the time referred to by "now" is the time of playback rather than the time of recording. Donnellan has suggested that if there were typically a significant lag between our production of speech and its audition (e.g., if sound traveled very slowly), our language might contain two forms of "now": one for the time of production, another for the time of audition. The indexicals "here" and "now" also suffer from vagueness regarding the size of the spatial and temporal neighborhoods to which they refer. These facts do not seem to me to slur the difference between demonstratives and pure indexicals.

7. Of course, it is certain intentions on the part of the speaker that make a particular vocable the first-person singular pronoun rather than a nickname for Irving. My semantic theory is a theory of word meaning, not speaker's meaning. It is based on linguistic rules known, explicitly or implicitly, by all competent users of the language.
8. I have discussed this and related issues in "Bob and Carol and Ted and Alice," in J. Hintikka et al., eds., *Approaches to Natural Language* (Dordrecht, 1973), esp. appendix 10.
9. Saul Kripke, *Naming and Necessity* (Oxford, 1972), p. 78.
10. The matter is even more complicated. There are two "definitions" of "rigid designator" in *Naming and Necessity* (pp. 48–9). The first conforms to what seems to me to have been the intended concept—same designation in *all* possible worlds—the second, scarcely a page later, conforms to the more widely held view that a rigid designator need not designate the object, or any object, at worlds in which the object does not exist. According to this conception, a designator cannot, at a given world, designate something which does not exist in that world. The introduction of the notion of a *strongly* rigid designator, a rigid designator whose designatum exists in all possible worlds, suggests that the latter idea was uppermost in Kripke's mind. (The second definition is given, unequivocally, in "Identity and Necessity," in M. K. Munitz, ed., *Identity and Individuation* (New York, 1971), p. 146.) In spite of the textual evidence, systematic considerations, including the fact that variables cannot be accounted for otherwise, leave me with the conviction that the former notion was intended.
11. Here, and in the preceding paragraph, in attempting to convey my notion of a directly referential singular term, I slide back and forth between two metaphysical pictures: that of possible worlds and that of structured propositions. It seems to me that a truly semantic idea should presuppose neither picture and be expressible in terms of either. Kripke's discussion of rigid designators is, I believe, distorted by an excessive dependence on the possible worlds picture and the associated semantic style. For more on the relationship between the two pictures, see David Kaplan, "How to Russell a Frege-Church," *Journal of Philosophy* 72 (1975): 716–29, esp. 724–5.
12. I think it likely that it was just the failure to notice this distinction that led to a failure to recognize principle 2. Some of the history and consequences of the conflation of context and circumstance are discussed in Kaplan, "Demonstratives," pp. 507–514.
13. I would have used "Snow is white," but I wanted a contingent clause, and so many people (possibly including me) nowadays seem to have views which allow that "Snow is white" may be necessary.

14. I am ignoring propositions expressed by sentences containing epistemic operators or others for which equivalence is not a sufficient condition for interchange of operand.
15. Russell, *Principles of Mathematics.*
16. Bertrand Russell, "On Denoting," *Mind* 14 (1905): 479–493.
17. Here is a difficulty in Russell's 1903 picture that has some historical interest. Consider the proposition expressed by the sentence "The center of mass of the Solar System is a point." Call the proposition *P*. *P* has in its subject place a certain complex, expressed by the definite description. Call the complex "Plexy." We can describe Plexy as "the complex expressed by 'the center of mass of the solar system'." Can we produce a directly referential term which designates Plexy? Leaving aside for the moment the controversial question of whether "Plexy" is such a term, let us imagine, as Russell believed, that we can directly refer to Plexy by affixing a kind of *meaning marks* (on the analogy of quotation marks) to the description itself. Now consider the sentence "“"The center of mass of the solar system"” is a point." Because the subject of this sentence is directly referential and refers to Plexy, the proposition the sentence expresses will have as its subject constituent Plexy itself. A moment's reflection will reveal that this proposition is simply *P* again. But this is absurd, since the two sentences speak about radically different objects.

    I believe the foregoing argument lies behind some of the largely incomprehensible arguments mounted by Russell against Frege in "On Denoting," though there are certainly other difficulties in that argument. It is not surprising that Russell there confused Frege's theory with his own of *Principle of Mathematics*. The first footnote of "On Denoting" asserts that the two theories are "very nearly the same.")

    The solution to the difficulty is simple. Regard the "object" places of a singular proposition as marked by some operation which cannot mark a complex. (There will always be some such operation.) For example, suppose that no complex is (represented by) a set containing a single member. Then we need only add {. . .} to mark the places in a singular proposition which correspond to directly referential terms. We no longer need to worry about confusing a complex with a propositional constituent corresponding to a directly referring term, because no complex will have the form {*x*}. In particular, Plexy ≠ {Plexy}. This technique can also be used to resolve another confusion in Russell. He argued that a sentence containing a nondenoting directly referential term (he would have called it a nondenoting "logically proper name") would be meaningless, presumably because the purported singular proposition would be incomplete. But the braces themselves can fill out the singular proposition, and if they contain nothing, no more anomalies need result than what the development of Free Logic has already inured us to.
18. This is the *first sense* of n. 10.
19. This is the *second sense* of n. 10.
20. Frege, "The Thought," p. 296. If Frege had only supplemented these comments with the observation that indexicals are devices of direct reference, the whole theory of indexicals would have been his. But his theory of meaning blinded him to this obvious point. Frege, I believe, mixed together the two kinds of meaning in what he called *Sinn*. A *thought* is, for him, the *Sinn* of a sentence,

or perhaps we should say a *complete* sentence. *Sinn* is to contain both "the manner and context of presentation [of the denotation]," according to "Über Sinn und Bedeutung," *Zeitschrift für Philosophie und philosophische Kritik* 100 (1892) (trans. as "On Sense and Nominatum," in I. Copi and S. Gould, eds., *Contemporary Readings in Logical Theory* (New York, 1967); mistrans. as "On Sense and Meaning," in Martinich, ed., *The Philosophy of Language* (New York, 1985), pp. 190–202). *Sinn* is first introduced to represent the cognitive significance of a sign, and thus to solve Frege's problem: how can [$\alpha = \beta$] if true differ in cognitive significance from [$\alpha = \alpha$]. However, it also is taken to represent the truth conditions or *content* (in our sense). Frege felt the pull of the two notions, which he reflects in some tortured passages about "I" in "The Thought" (quoted in Kaplan, "Demonstratives," p. 533). If one says "Today is beautiful" on Tuesday and "Yesterday was beautiful" on Wednesday, one expresses the same thought according to the passage quoted. Yet one can clearly lose track of the days and not realize that one is expressing the same thought. It seems, then, that thoughts are not appropriate bearers of cognitive significance. For further discussion of this topic see ibid., pp. 529–540. A detailed examination of Frege on demonstratives is contained in Perry, "Frege on Demonstratives."

21. As we shall see, indexical operators such as "It is now the case that," "It is actually the case that," and "dthat" (the last takes a term rather than a sentence as argument) are also intensional operators. They differ from the familiar operators in only two ways: first, their extension (the function from intensions to extensions) depends on context, and, second, they are directly referential (thus they have a fixed content). I argue elsewhere (ibid., pp. 510–512) that all operators that can be given an English reading are "at most" intensional. Note that when discussing issues in terms of the formal representations of the model-theoretic semantics, I tend to speak in terms of intensions and intensional operators rather than contents and content operators.

22. The notion of redundancy involved would be made precise. When I speak of building the time of evaluation into contents or making contents specific as to time or taking what is said to incorporate reference to a specific time, what I have in mind is this. Given a sentence *S*, "I am writing," in the present context *c*, which of the following should we take as the content: (1) the proposition that David Kaplan is writing at 10 A.M. on 26 March 77 or (2) the "proposition" that David Kaplan is writing? Proposition 1 is specific as to time; "proposition" 2 (the scare-quotes reflect my feeling that this is not the traditional notion of a proposition) is neutral with respect to time. If we take the content of *S* in *c* to be (2), we can ask whether it would be true at times other than the time of *c*. Thus we think of the temporally neutral "proposition" as changing its truth-value over time. Note that it is not just the noneternal sentence *S* that changes its truth-value over time, but the "proposition" itself. Since the sentence *S* contains an indexical "I," it will express different "propositions" in different contexts. But since *S* contains no *temporal* indexical, the time of the context will not influence the "proposition" expressed. An alternative (and more traditional) view is to say that the verb tense in *S* involves an implicit temporal indexical, so that *S* is understood as synonymous with S', "I am writing now." If we take this point of view, we will take the content of *S* in *c* to be (1). In this case *what*

is said is eternal; it does not change its truth-value over time, although S will express different propositions at different times.

There are both technical and philosophical issues involved in choosing between (1) and (2). Philosophically, we may ask why the temporal indexical should be taken to be implicit (making the proposition eternal) when no modal indexical is taken to be implicit. After all, we *could* understand S as synonymous with S'', "I am actually writing now." The content of S'' in c is not only eternal; it is perfect. Its truth changes through neither time nor possibility. Is there some good philosophical reason for preferring contents which are neutral with respect to possibility but draw fixed values from the context for all other features of a possible circumstance regardless of whether the sentence contains an explicit indexical? (It may be that the traditional view was abetted by one of the delightful anomalies of the logic of indexicals, namely, that S, S', and S'' are all logically equivalent! See Kaplan, "Demonstratives," p. 547.) Technically, we should note that intensional operators must, if they are not to be vacuous, operate on contents which are neutral with respect to the feature of circumstance the operator is interested in. Thus, e.g., if we take the content of S to be (1), the application of a temporal operator to such a content would have no effect; the operator would be vacuous. Furthermore, if we do not wish the iteration of such operators to be vacuous, the content of the compound sentence containing the operator must again be neutral with respect to the relevant feature of circumstance. This is not to say that no such operator can have the effect of *fixing* the relevant feature and thus, in effect, rendering subsequent operations vacuous; indexical operators do just this. It is just that this must not be the general situation. A content must be the *kind* of entity that is subject to modification in the feature relevant to the operator. (The textual material to which this note is appended is too cryptic and should be rewritten.)

23. "How to Russell a Frege-Church." The pronunciation is Heḱ-ee-i-tis-m. The epithet was suggested by Robert Adams. It is no accident that it is derived from a demonstrative.
24. This does not imply that if you know the character and are in first one and then another context, you can *decide* whether the contents are the same. I may twice use "here" on separate occasions and not recognize that the place is the same or twice hear "I" and not know if the content is the same. What I do know is this: if it was the same person speaking, then the content was the same. (For more on this epistemological stuff see Kaplan, "Demonstratives," pp. 529–540.
25. I am, at this stage, deliberately ignoring Kripke's theory of proper names in order to see whether the revisions in Fregean semantic theory, which seem plainly required to accommodate indexicals (this is the "obviousness" of my theory), can throw any light on it. Here we assume that aside from indexicals, Frege's theory is correct: roughly, that words and phrases have a kind of descriptive meaning or sense which at one and the same time constitutes their cognitive significance and their conditions of applicability.

   Kripke says repeatedly in *Naming and Necessity* that he is providing only a picture of how proper names refer and that he does not have an exact theory. His picture yields some startling results. In the case of indexicals we do have a rather precise theory, which avoids the difficulty of specifying a chain of communication and which yields many analogous results. In facing the vastly more difficult problems associated with a theory of reference for proper names, the theory of indexicals may prove useful, if only to show—as I believe—that proper names are not indexicals and have no meaning in the sense in which indexicals have meaning (viz., a "cognitive content" which fixes the references in all contexts). (The issues that arise, involving token-reflexives, homonymous words with distinct character, and homonymous token-reflexives with the same character are best saved for later.)

26. See Rudolf Carnap, *Meaning and Necessity* (Chicago, 1917), §28.
27. We see here a drawback to the terminology "direct reference." It suggests, falsely, that the reference is not mediated by a meaning, which it is. The meaning (character) is directly associated, by convention, with the word. The meaning determines the referent; and the referent determines the content. It is this to which I allude in the parenthetical remark following the picture in Kaplan, "Demonstratives," p. 486. Note, however, that the kind of descriptive meaning involved in giving the character of indexicals like "I," "now," etc., is, because of the focus on context rather than circumstance, unlike that traditionally thought of as the Fregean sense. It is the idea that the referent determines the content—that, *contra* Frege, there *is* a road back—that I wish to capture. This is the importance of principle 2.

ESSAY 12

Temporal Indexicals

QUENTIN SMITH

The prevalent theory of indexicals today is the direct reference theory, which is based on the idea that indexicals do not refer to items indirectly via a sense, but possess an unmediated reference relation to items. Temporal indexicals ("now," "yesterday") are supposed to refer directly to times. If "now" is uttered at noon, 1 June 1986, it refers directly to noon, 1 June 1986.

I believe that the direct reference theory (hereafter "DR theory") is based on some genuine insights but that it goes too far in its rejection of the indirect reference theory. A number of intractable problems confronting the DR theory can be resolved if we allow that temporal indexicals refer directly *and* express senses that characterize the direct referents. This resolution preserves the insights of both the classical, or Fregean, indirect reference theories and the new DR theories, without adopting the "extreme" or "purist" position of either one (see Sections 1 and 2).

A second problem with the DR theory is that it assumes that one of its central theses—namely, that uses of indexicals are rigid designators—does not entail specific or controversial metaphysical theses about the nature of the designata. But I shall argue that the most widely accepted metaphysical theory about the nature of time, the relational theory, according to which times are sets of events, is incompatible with the thesis of rigid designation. If the rigid designation thesis is true, so must be the absolute theory of time, according to which times are event-independent moments (see Section 3). I will conclude by contrasting the theory of rigidity that I develop with another theory of temporal indexicals that also combines sense with rigid designation (see Section 4).

Although my discussion primarily concerns temporal indexicals, I shall in the appropriate places (at the end of Sections 1 and 3) draw some analogous conclusions about spatial indexicals.

The principal foil in my critical discussions is the most widely accepted DR theory, the theory developed by David Kaplan, John Perry, Howard Wettstein, and Joseph Almog—although I do not mean to imply that they agree with each other on every issue (especially not since Wettstein's recent paper[1]). My arguments also apply, *mutatis mutandis,* to other DR theories, such as Boer's and Lycan's,[2] but I shall not pause to show this here.

1. The Adverbial and Pronominal Uses of Temporal Indexicals

DR theorists in the Kaplan-Perry tradition hold that the proposition expressed by some sentence-token containing an indexical includes the referent of the indexical as a constituent. The indexical contributes to the proposition only by supplying the referent of the indexical; the indexical does not also supply a sense. "Now" in the noon utterance of "The meeting starts now" does not contribute a sense that refers to noon, but noon itself. The only sense that belongs to the proposition is contributed by the rest of the sentence-token, "The meeting starts." In Perry's words: "To have a thought [proposition] we need an object and an incomplete sense. The demonstrative in context gives us the one, the rest of the sentence the other."[3]

But this thesis breaks down once we recognize the difference between the pronominal and adverbial uses of temporal and spatial indexicals. John Pollock was the first to draw attention to this distinction in regard to temporal indexicals.[4] He gives as examples

(1) He is coming now.
(2) I did not believe it until now.

"Now" is used as an adverb in (1) and as a pronoun in (2). Pollock points out that "now" in (1) is paraphrased by "at the present time," but "now" in (2) is paraphrased instead by "the present time." Armed with this distinction, let us compare noon, 1 June 1986 utterances of

(3) The meeting starts now.
(4) Now is when the meeting starts.

The noon utterance of (4), which involves a pronominal use of "now," expresses (according to the theory being examined) the proposition

(5) ⟨Noon, 1 June 1986, the sense of "is when the meeting starts"⟩.

This is an ordered pair consisting of noon itself and the sense of "is when the meeting starts." It is also expressed by the sentence

(6) Noon, 1 June 1986 is when the meeting starts,

where "noon, 1 June 1986" is a referentially used definite description that refers directly to noon, 1 June 1986. (I am here relying upon William Lycan's theory of the referential use of date-descriptions.[5]) This description must be used referentially rather than attributively, because if it were used attributively, it would introduce a sense into the proposition rather than noon itself. Used attributively, "noon, 1 June 1986" expresses the sense *whatever time is* 1,985 *years,* 5 *months, and* 12 *hours later than Christ's birth.* Used referentially, it does not express this sense but refers directly to the time that is in fact 1,985 years, 5 months, and 12 hours later than Christ's birth.

Given that this date-description is used referentially in (6), we can say that (6) expresses the same proposition as the noon, 1 June 1986 utterance of "Now is when the meeting starts." But this is not to say that (6) and sentence 4 have the same rule of use—the same "role" (in Perry's terminology) or "character" (in Kaplan's terminology). (4) is an indexical sentence, whereas (6) is an eternal sentence. In Kaplan's terminology, (6) has the same content as the noon utterance of (4)—it expresses the same proposition—but it has a different character from (4). We may say that (6) is a *content-preserving translation* of the noon utterance of (4) but not a *character-preserving translation* of this utterance.

Nothing appears wrong on the face of it with the idea that (6) and the noon utterance of (4) express the same proposition. But now consider the content-preserving translation of the noon utterance of (3). If we adopt the hypothesis of Perry and other DR theorists that indexicals contribute no sense, but only their referents to propositions, then the noon utterance of (3) will express the proposition

(7) ⟨Noon, 1 June 1986, the sense of "the meeting starts"⟩.

Let us replace "now" with a referentially used date-description that directly refers to the same date as the noon utterance of "now." If we replace "now" with "noon, 1 June 1986," as referentially used, then we have the following, content-preserving translation of the noon utterance of the "The meeting starts now":

(8) The meeting starts noon, 1 June 1986.

But this is syntactically incomplete. What has gone wrong? Clearly what is missing is a word or phrase with the sense of "at" or "simultaneously with" that relates the start of the meeting to noon, 1 June 1986.[6] But according to the DR theory, "now" does not contribute any sense to the proposition and *eo ipso* does not contribute the sense of "at." But this supposition produces the unpalatable result that the noon utterance of "The meeting starts now" is translated in respect of its content by the incomplete sentence (8). "Now," *when used as an adverb rather than a pronoun, must express the sense of "at" or "simultaneously with"* in addition to referring directly to the date.

It is instructive to see how the DR theorists deal with this problem. Perry implicitly reacts to it by regarding the sense of "at" or "simultaneously with" as part of the sense of "starts." This enables him to write, in reference to (3), that "only at twelve noon can someone think the thought [proposition] consisting of noon and the sense of 'The meeting starts at ( )' by entertaining the sense of 'The meeting starts now.'"[7] But this maneuver fails, since "starts" and "starts simultaneously with" do not have the same sense. If they did, they would be intersubstitutable *salva veritate* in extensional contexts. But compare "Noon is when the meeting starts" with "Noon is when the meeting starts simultaneously with." Truth-value cannot be preserved, since the second expression is ill-formed and consequently cannot possess any truth-value.

A defender of Perry might object that "starts" has different senses, depending upon whether the temporal indexical with which it is associated is used pronominally or adverbially. If "now" is used pronominally, as in "Now is when the meeting starts," then "starts" does not express the sense of "simultaneously with." But if "now" is used adverbially, as in "The meeting starts now," "starts" does express this sense.

This response seems implausible and *ad hoc.* "Starts" is used in the same way and as a verb in each sentence, and there is no reason to think that it changes its sense depending upon whether "now" is used pronominally or adverbially. Since the change is in the use of "now," from a pronominal to an adverbial use, it is more reasonable to suppose that the change of semantic content pertains to it. Indeed, this is precisely what we would expect in change from a pronominal to an adverbial use. A pronoun is simply a referring device, but an adverb is used to express something that *qualifies* what is expressed by a verb. Now a time by itself, such as noon, 1 June 1986, does not qualify anything. Thus it is implausible to suppose that "now" in its adverbial use does nothing more than refer directly to a time. But note that the relational property, ( ) *simultaneously with noon, 1 June 1986,* does qualify something, namely, that of which it is the property. If the adverb "now" is used to express this property, this fits in with our expectation that this adverb is used to express something that qualifies what is expressed by the verb "starts." The adverb is used to express a property of the meeting's start, namely, the property of *being simultaneous with noon, 1 June 1986*.

The assertion that "now" in its adverbial use expresses this relational property is consistent with the claim that this use of "now" directly refers to a time as well as conveys the sense of "simultaneously with." This relational property consists of a time, noon, 1 June 1986, and a sense, the dyadic property expressed by "simultaneously with." The adverb's expression of this relational property consists in part of its direct reference to the time and in part of its conveyance of the sense of "simultaneously with." But of course the adverb does not express a mere heap or aggregate of the time and the sense; rather, it expresses them as linked to each other as relation (the sense) to relational term (the time). By this means, the adverb is able to express something, a relational property, that qualifies what is expressed by the sentence's verb.

This analysis of the difference between the adverbial and pronominal uses of "now" gains in plausibility if it can be shown that an analogous distinction holds in regard to "here," which also admits of this twofold use.[8] Compare

(9) Here is the place to settle.
(10) There are over 5,000 nuclear warheads here.

If (9) is tokened in Charleston, South Carolina, it expresses the same proposition that is expressed by

(11) Charleston, South Carolina, is the place to settle,

where "Charleston, South Carolina" directly refers to the city and is a content-preserving translation of the token of "here." But "Charleston, South Carolina" is not a content-preserving translation of the Charleston-token of "here" in (10). If it were, (10) would be translated by

(12) There are over 5,000 nuclear warheads Charleston, South Carolina.

Clearly, the Charleston-token of (10) demands the translation

(13) There are over 5,000 nuclear warheads *in* Charleston, South Carolina.

The adverb "here" directly refers to Charleston, South Carolina, and expresses the sense of "in" or "located at." Surely it is counterintuitive to suppose that the sense of "located at" is a part of the sense of "There are" or "warheads!" Indeed, the fact that "here" is used adverbially rather than pronominally entails that it is used not as a mere referring device to pick out a place but that it is also used to express a relational property, ( ) *located at Charleston, South Carolina,* that qualifies what is expressed by the verb of the sentence.

But such senses as of "at" and "in" are not the only senses that need to be included in the content of the relevant indexical utterances. It will be shown in the next section that temporal indexicals, whether used adverbially or pronominally, express additional senses.

2. The Entailment of Sentences without Temporal Indexicals by Sentences That Include Temporal Indexicals and the Problem of Cognitive Significance

A second problem affecting temporal indexicals that requires additional senses to be introduced into their contents concerns the entailment relations between sentences containing a temporal indexical and sentences not containing one. A case in point is the entailment of

(1) The meeting is starting

by

(2) The meeting starts now.

In order to see why this poses a problem for the DR theory, it is necessary to explain the differences between these two sentences. (1) contains a present-tensed copula but no temporal indexical. Temporal indexicals are words used adverbially or pronominally to refer to dates (times), words such as "now," "today," and "yesterday." Their essential character is that they refer to a different date at each different date they are used; "today" as uttered on Monday refers to Monday and as uttered on Tuesday to Tuesday. Tensed copulae ("was," "is," "will be") lack this characteristic. The present tense of the copula "is" is neutral with respect to dates; it does not refer to the date of its utterance or any other date. Unlike a temporal indexical, the present tense of the copula has a constant semantic content, the same content on each occasion of use. This is reflected in the fact that sentences such as (1) express the same proposition on each occasion of use, whereas indexical-containing sentences such as (2) express a different proposition at each different date they are uttered (see my remark in note 3 about my elliptical use of "the meeting"). This difference between indexical-free present-tensed sentences and sentences with temporal indexicals has been noted by Hans Kamp, A. N. Prior, and Kaplan.[9] Kaplan gives the example of "I am writing" as a sentence that is constant with respect to its temporally pertinent semantic content and the example of "I am writing now" as a sentence that is variable in this respect.

What is the constant semantic content of the present tense of the copula? Kaplan is silent on this issue, but I have argued in other places[10] that the only theory consistent with all the data about the present tense of the copula is that which maintains that this tensed aspect serves to ascribe the monadic property of presentness to the relevant items denoted by the rest of the sentence. The present tense of the "is" in "The meeting is starting" serves to ascribe presentness to the start of the meeting. The idea that presentness is a monadic property carries with it some difficulties, such as those articulated by E. Sosa and P. Yourgrau,[11] but I believe that these difficulties can be resolved by the theory that presentness is an unusual sort of monadic property, an infinitely reflexive property, as explain in Essay 15. Assuming, then, that the notion that presentness is a monadic property is acceptable, we can represent the constant semantic content of (1) by

(3) ⟨⟨The sense of "the meeting," the property of starting⟩ the property of presentness⟩.

(3) shows that the property of starting is ascribed to the referent of the sense of "the meeting" and that the property of presentness is ascribed to the exemplification of the property of starting by the referent of the sense of "the meeting." In other words, it proposes that the meeting has the property of starting and that its *having* of this property itself *has presentness*. Whenever (1) is uttered, it expresses this semantic content.

This theory of the present tense of the copula provides an explanation for the semantic difference between sentences such as (1) and (2) that was noted but left unexplained by Kamp, Prior, and Kaplan. They point out that the semantic constancy of sentences such as (1) and the inconstancy of sentences such as (2) are clearly evinced when these sentences are embedded in temporal contexts. Suppose the meeting starts at noon, 1 June. In that case, a noon, 1 June utterance of

(4) It will be true tomorrow that the meeting starts now

is true, but a noon, 1 June utterance of

(5) It will be true tomorrow that the meeting is starting

is false. This is because the noon, 1 June utterance of (4) refers to the date of its utterance, noon, 1 June, and indicates that it will be true tomorrow—on 2 June—that the meeting starts at noon, 1 June. Certainly if it is true at noon, 1 June that the meeting starts at noon, 1 June, this will be true on 2 June as well. The noon, 1 June utterance of (5) is false because it does not refer to the date of its utterance and therefore does not indicate that the meeting starts at noon, 1 June. "The meeting is starting" merely conveys that the start of the meeting has presentness, without indicating the date at which it has presentness. Thus, what will be true tomorrow, according to the noon, 1 June utterance of (5), is not that the start of the meeting is present on a particular date (namely, noon, 1 June) but merely that it is present. But in fact this will not be true tomorrow, since on 2 June the meeting does not have presentness but pastness. This difference between "The meeting starts now" and "The meeting is starting" does not prevent the former from entailing the latter in extensional contexts, where neither is prefixed by an operator such as "It will be true tomorrow that." The extensional sentence (1) is entailed by the extensional sentence (2). What this means, precisely put, is that for any time *t*, if (2) is uttered at *t* and is true, then (as a matter of logical necessity) (1) is also utterable at *t* with truth. Manifestly, it is self-contradictory to respond to "The meeting starts now" by saying "I agree. But I don't agree that the meeting is starting." In terms of propositions, we may say that the proposition expressible at *t* by (2) entails the proposition expressible by (1).

I now come to the crux of my argument. This entailment relation is inexplicable if the DR theory is correct, even in the modified version developed in Section 1. Suppose the proposition expressed by the noon, 1 June utterance of (2) is also expressible by

(6) The meeting starts at noon, 1 June.

The proposition expressed by (6) does not entail the proposition expressed by "The meeting is starting." The proposition expressed by (6) is true when 2 June or 31 May is present no less than when noon, 1 June is present, for it does not ascribe a property of presentness, pastness, or futurity to the start of the meeting, but merely a relation to noon, 1 June of being simultaneous with it, a relation that holds at every time if it holds at any time. But the proposition expressed by "The meeting is starting" is true only when noon, 1 June is present, for only then does the start of the meeting possess presentness. On 2 June the start of the meeting has pastness, and on 31 May it has futurity, and consequently this proposition is false at these times.

These results suggest that the DR theory needs to be revised so as to explain the entailment relation between (2) and (1). The revision consists in introducing another sense into the content of temporal indexicals. In Section 1, I introduced the sense of "at" or "simultaneously with" into the content of "now" as adverbially used. Here I shall introduce the sense of "is present" or "has presentness" into the adverbial and pronominal uses of "now." The noon, 1 June token of "The meeting starts now" refers to noon, 1 June, *but it also ascribes presentness to this date*. This token has the same content as the sentence

(7) The meeting starts at noon, 1 June, which is present.

What is said (the proposition expressed) by (7) *does* entail what is said by "The meeting is starting." What is said by (7) is true only if the start of the meeting has presentness rather than pastness or futurity—that is, only if the meeting *is* starting rather than was or will be starting—and thus is true only if the proposition expressed by (1) is simultaneously true.

The hypothesis that the noon, 1 June token of "The meeting starts now" has the same content as (7) requires that intensional sentences such as (4) be given a different analysis than the one given to them by the DR theorists. According to the DR theory, the noon, 1 June utterance of (4) has the same content as

(8) It will be true tomorrow that the meeting starts at noon, 1 June.

(8) can be analyzed as saying that the proposition expressed by "The meeting starts at noon, 1 June" has the value of true tomorrow. But if the noon utterance of "The meeting starts now" expresses the proposition expressed by (7) (let us call this proposition *P*), the noon utterance of (4) will *entail* that the proposition expressed by "The meeting starts at noon, 1 June" will be true tomorrow, but it will *state* something else. What the noon utterance of (4) states is something about *P*. But the noon utterance of (4) cannot be analyzed as stating that *P* will be true tomorrow. For clearly *P* will be false tomorrow—on 2 June, noon, 1 June will *not* be present. The noon utterance of (4) must instead be analyzed as stating that *P*'s present truth is entailed by some of tomorrow's truths. That is, the noon utterance of (4) asserts that there are some propositions which will be true tomorrow and whose being true tomorrow entails that *P* is true at present, at noon, 1 June. Such a proposition is that expressible by "The meeting starts at noon, 1 June, which was present over twelve hours ago." If this proposition is true on 2 June, its being true then entails that *P* is true at noon, 1 June.

This analysis of the noon utterance of (4) is more complex than the DR analysis of this utterance, but it alone has the virtue of being consistent with the fact that (2) entails (1). Moreover, this analysis mirrors perfectly the analysis of the noon utterance of (4) in terms of *sentential* (rather than propositional) truth-values. If we look at the noon utterance of (4) from the viewpoint that the operator "It will be true tomorrow that" operates not only on the proposition expressed by the utterance of "The meeting starts now" but also on the sentence of which this utterance is an utterance, we shall arrive at a parallel analysis to the one I have given. The noon utterance of (4) cannot be viewed as saying that the sentence "The meeting starts now" will be true tomorrow (that is, will be utterable with truth tomorrow), for it will not be. Instead, it must be viewed as saying that the present truth of this sentence is entailed by the being true tomorrow of some other sentence, such as the sentence "The meeting starts at noon, 1 June, which was present over twelve hours ago."

I believe that the introduction of the sense of "is present" into the semantic content of "now" is buttressed by considerations of the *cognitive significance* of "now," what the utterer of this word believes of its referent when she utters it. According to Kaplan and Perry, its cognitive significance is its character, or role, *that "now" is used to refer directly to the time of its use*. But there are two reasons why this rule of use cannot account for the (complete) cognitive significance of "now." The first reason was noted by Castañeda,[12] that the rule of use is a general formula for determining the referent of any use of "now" and that what is believed by the utterer is something singular about the time to which reference is made by her use of "now." The utterer believes of the time that it uniquely possesses a certain property, one that is attributed to the time by her use of "now." The second reason is that each utterer of "now" believes of the time denoted that it has presentness rather than pastness or futurity, and the presentness of this time is not stated or entailed by the rule of use. This relates to the first reason, in that the belief in the presentness of the time can be identified with the utterer's singular belief about this time. The utterer believes of this time that it is the only time that then possesses the property attributed to it by her use of "now," namely, the property of presentness; every other time then possesses some degree of pastness or futurity. In this manner, the introduction of the sense of "is present" into "now" enables these two problems regarding the cognitive significance of "now" to be solved.

3. *The Use of Temporal Indexicals in Modal Contexts*

The idea that uses of temporal indexicals are rigid designators is part and parcel of the DR theory,[13] but proponents of this theory seem to think that the semantics of direct reference is metaphysically uncommitted, or at least does not entail specific and highly controversial metaphysical positions about the nature of time. But this metaphysical naiveté of the DR theorists quickly leads them into trouble, for their semantic theory brings with it an (unrecognized) commitment to the absolutist theory of time. This is troublesome to the DR theory, for any argument against the absolutist theory of time is, by implication, an argument against the rigid designation thesis, which saddles the DR theorist with the imposing task of having to settle the metaphysical debate about the relational or absolute nature of time before he draws any conclusion about the rigidity of uses of temporal indexicals.

Let me proceed immediately to my argument that

(1) Uses of temporal indexicals rigidly designate times

entails

(2) The absolutist theory of time is true.

Three theories of time have been advocated: (a) the relational theory that times are sets of simultaneous events, (b) the absolutist theory that times are moments which are independent of the events that occupy them, and (c) the propositional theory (of Prior) that times are conjunctions of simultaneously true propositions. Only the second of these theories is compatible with the rigid designation thesis, as becomes apparent from a consideration of the use of temporal indexicals in modal contexts. Let us first examine the implications that counterfactuals have for the widely accepted relational theory of time.[14] Consider an utterance of the sentence

(3) David is writing now, but he might not have been

that occurs at noon, 1 June 1987, when David is in fact writing. If the use of "now" directly and rigidly refers to this time, then (given the modifications made in Sections 1 and 2) this utterance will have the same semantic content as

(4) David is writing at noon, 1 June 1987, which is present, but he might not have been,

where "noon, 1 June 1987" is a directly referring expression. Suppose that this expression rigidly refers to the set of all and only those events that are actually 1,986 years, 5 months, and 12 hours later than Christ's birth. Call this set *S*. (4) then entails

(5) *S* actually includes the event of David's writing, but there is some possible world in which *S* does not include this event.

Now sets have their members essentially. In each possible world in which *S* exists, it includes all and only the events it includes in the actual world. If *S* actually includes David's writing, it includes this event in every world in which *S* exists. This shows that (5) is self-contradictory, for it asserts that *S* both actually includes David's writing and does not include it in some merely possible world. Since (4) is true, it cannot entail (5), and consequently its use of "noon, 1 June 1987" cannot rigidly refer to a set of events.

The Priorian theory of times as propositions is also incompatible with the semantic thesis of rigid designation. According to Prior, the appropriate way to define an instant *a* is to "equate the instant *a* with a conjunction of all those propositions which would ordinarily be said to be true at that instant."[15] The instant *a* at which (3) began to be uttered is a proposition one of whose conjuncts is *that David is writing*. Let us suppose that the time rigidly designated by the token of "now" that belongs to the utterance of (3) is the instant *a* (it might be more plausible to say that this token designates some interval, but we can ignore this complication for present purposes). If *a* is the rigid designatum of the token of "now," then the utterance of (3) entails

(6) David is writing when *a* is true, but there is some possible world in which David is not writing when *a* is true.

It is evident that (6) is no less self-contradictory than (5), for if *a* is true, it is true *that David is writing,* since this latter proposition is one of the conjuncts of *a*.

Only the absolutist theory of time, that times are moments which are independent of, and accidentally related to, the events that occupy them, is compatible with the thesis that uses of "now" designate rigidly. The moment *m* that is actually occupied by David's writing and actually is 1,986 years, 5 months, and 12 hours later than Christ's birth is not occupied by this event and does not possess this temporal relation to Christ's birth in some merely possible worlds. Indeed, in some of these worlds (those which contain "empty times"), *m* is not occupied by any events at all and is not related to Christ's birth. If the "now" in the utterance of (3) refers rigidly to *m,* the utterance is true, since *m* is actually occupied by David's writing but is not occupied by this event in some merely possible world. Manifestly, the problems present in (5) and (6) are not present in

(7) David's writing actually occupies *m,* but there is some possible world in which *m* is not occupied by David's writing.

The thesis that uses of "now" rigidly designate times, then, entails the absolutist theory of time. If this metaphysical theory of time is false, then the rigid designator thesis, and hence the DR theory itself, is false. For if times are sets of events (or Priorian propositions), then uses of "now" must refer to them nonrigidly, and therefore indirectly, by means of descriptive senses that are satisfied by different sets of events in different worlds. The use of "now" in (3) must express a descriptive sense that is actually satisfied by a set that includes David's writing but is satisfied in a merely possible world by a different set that does not include his writing. An example of such a descriptive sense is the sense expressible by an attributive use of a definite description of a calendrical date; for instance, the utterance of (3) could be read as expressing the same proposition as

(8) David is writing at whatever set of events is present and has the relational property of being 1,986 years, 5 months, and 12 hours later than Christ's birth, but he might not have been.

The set that actually has this temporal relation to Christ's birth includes David's writing, but in some possible world the set that has this relation to Christ's birth does not include the event of his writing. In world *W,* for example, the set of events that is 1,986 years, 5 months, and 12 hours later than Christ's birth and is present includes instead the event of David's reading. But this example is not meant to imply that senses expressible by attributive uses of definite descriptions of calendrical dates *must* be identified with the descriptive senses of uses of "now" if the relational theory of time is true. There are serious problems with this view of the sense of "now," as I have argued elsewhere.[16] My point is that descriptive senses of *some* sort that are satisfied by different sets of events in different worlds must be identified with the mechanisms by which "now" refers to times if the relational theory of time is true.

The above argument regarding the connection between the rigidity of uses of "now" and the absolute theory of time is paralleled by an argument regarding the rigidity of uses of "here" and the absolute theory of space. The relational theory of space identifies places with sets of physical objects spatially arranged in a certain way. The absolute theory identifies places with spaces that are independent of, and accidentally related to, the physical objects that occupy them (if any objects do occupy them). Consider a counterfactual utterance involving "here." Suppose John, standing in the midst of a pile of boxes on the sidewalk, exclaims

(9) The box of books was left here, but it might not have been.

Let us assume that his token of "here" rigidly designates a place and that the relational theory of space is true. In that case, his token of "here" rigidly designates the set of objects as spatially arranged in a certain way that constitutes the place where John is located. This set, call it *T,* includes John himself, the pavement on the sidewalk, the pile of boxes on the pavement, and the air molecules in between the boxes, such that these items stand in certain spatial relations—a distance and direction—to each other. One member of this set is the box of books which, say, is on top of a box of kitchen utensils and underneath a box of clothes. If John's use of "here" rigidly designates *T,* then his utterance of (9) is logically equivalent to

(10) The box of books is a member of T, but there is some possible world in which *T* does not include this box.

This, manifestly, is self-contradictory, since the set *T* cannot include the box of books in one world and not include it in another world. *T*'s inclusion of this box is essential to it. John's use of "here," then, cannot rigidly refer to *T*.

The rigidity of his use of "here" can be preserved if we assume that the absolute theory of space is true. Suppose John's token of "here" rigidly designates the space *U* that is occupied by himself, the relevant parts of the sidewalk, the pile of boxes, and the intervening air molecules. This space contingently contains the box of books and in some other worlds does not contain it. On this interpretation, the utterance of (9) is logically equivalent to

(11) The box of books occupies a part of *U,* but there is some possible world in which *U* is not occupied by the box of books.

This is self-consistent and thus permits rigidity to be attributed to John's use of "here."

These arguments suggest that the semantic theory of "now" and "here" is not independent of, but is partly based upon, the metaphysical theory of times and places and thus that the correct semantic theory can be determined only *after* the correct metaphysical theory has been determined. What are the prospects for a decision in these matters? I have argued elsewhere[17] that there is reason to think that the absolute theory of time is sound; consequently I think that the rigidity of "now" can be plausibly maintained.[18] The semantic content of an adverbial use of "now" is then captured by

(12) at *m,* which is present,

and the content of a pronominal use is captured by

(10) *m,* which is present,

where *m* directly refers to the moment of time that is occupied by the use of "now." If Earman, Nerlich, and others are correct in arguing that the absolute theory of space is preferable to the relational theory of space,[19] then the rigidity of uses of "here" can be plausibly maintained as well.

4. Conclusion: Theories that Combine Sense with Rigid Designation

The theory of temporal indexicals that I am advocating asserts that uses of these indexicals directly and rigidly refer to moments of time and express senses that (a) characterize these moments as present or as past or future to some degree and (in the case of adverbial uses) that (b) relate the moments to the events designated by the rest of the sentence via the relations of simultaneity, earlier than, or later than. It is worthwhile comparing this theory with another theory that also combines sense with rigid designation. The most developed alternative theory is the theory of *modally stable senses,* whose leading advocate is A. Plantinga.[20] Plantinga's theory was developed for proper names, but it can easily be applied to temporal indexicals. Plantinga guaranteed the rigidity of proper names by introducing the sense of "actually" or "in the actual world" into the descriptive sense expressed by names. For example, the proper name "George Washington" expresses the sense *whoever is actually the first president of the United States,* which refers to the same person in every world to which it refers, namely, to the person who *in the actual world* is the first president of the United States.

This theory can be applied to uses of "now" in a way that retains the three theses articulated in this essay, the theses that (1) adverbial uses express the sense of "simultaneously with," (2) all uses ascribe the monadic property of presentness, and (3) all uses rigidly designate moments of time. The result of this application is that all adverbial uses of "now" express the modally stable sense that is expressible by "at whatever moment is actually present," and all pronominal uses express the sense of "whatever moment is actually present." This theory retains the rigidity of uses of "now," for these senses refer, in respect of every possible world to which they refer, to the moment that is present *in the actual world.* If the moment *m* is actually present, then *m* is the referent of the modally stable senses in every world to which these senses refer. The main difference between this theory of "now" and the theory developed in the preceding sections is that this theory guarantees rigidity via a modally stable sense, whereas the theory developed in earlier sections guarantees rigidity via direct reference.

In order to evaluate the relative merits of these two theories that combine (in different ways) sense with rigid designation, it is necessary to say a few words about the behavior of the modally unstable description "whatever moment is present" in temporal contexts. Consider

(1) It will be true at noon tomorrow that the meeting starts at whatever moment is present.

(1) behaves similarly to sentence 5 of Section 2: "It will be true tomorrow that the meeting is starting." Sentence 1 says, in effect, that whatever moment is present *at noon tomorrow* is the moment at which the meeting starts. The temporal operator "It will be true tomorrow that" determines the description "whatever moment is present" to refer to the time to which the operator

refers. Thus (1) is true only if the meeting starts at this time, at noon, tomorrow. But we know from Section 2 that different truth conditions belong to

(2) It will be true at noon tomorrow that the meeting starts now.

This is true only if the meeting starts *now*, at the moment (2) is uttered. The temporal operator does not affect the reference of "now."[21]

It is a short step from here to the realization that the modally stable sense theory of "now" cannot be correct. The insertion of "actually" between "is" and "present" in (1) ensures that "whatever moment is actually present" refers, in respect to every merely possible world to which it refers, to the same moment to which it refers in the actual world. But it does not determine this description to refer to the moment that is present at the time of utterance of the sentence. "Whatever moment is actually present" is still determined by the temporal operator to refer to the moment that is present at noon tomorrow. Thus, the modally stable version of (1) has different truth conditions from the "now"-sentence (2). This entails that "whatever moment is actually present" does not have the same semantic content as "now." If sense is to be combined with rigidity in a way that captures the behavior of "now," the path of modally stable senses is not the path that should be taken. The viable path is the one taken in Sections 1–3, which combines the rigidity of *direct reference* with senses that *characterize* (but do not *refer* to) the direct referent. It remains to be seen if this combined *direct reference/characterizing sense* theory also applies to all other indexicals (for example, "I" and "this"), proper names, and natural-kind terms.[22]

Notes

1. Howard Wettstein, "Has Semantics Rested on a Mistake?" *Journal of Philosophy* 83 (1986): 185–209.
2. S. E. Boer and W. G. Lycan, "Who Me?" *Philosophical Review* 89 (1980): 427–466.
3. Perry, "Frege on Demonstratives," p. 493. "The meeting" is an improper definite description, and there is reason to think that such descriptions do not (ordinarily) introduce senses and propositions but are used referentially to introduce their referents into propositions. Cf. Wettstein, "Demonstrative Reference and Definite Descriptions," *Philosophical Studies* 40 (1981): 241–257. To render this view consistent with my adoption of Perry's assumption that "the meeting" introduces a sense into the proposition, I shall henceforth use "the meeting" as elliptical for the proper definite description "the one and the only meeting of the 1986 admissions committee of the Stanford University philosophy department."
4. John Pollock, *Language and Thought* (Princeton, N.J., 1982), pp. 120–121.

5. William Lycan, "Eternal Sentences Again," *Philosophical Studies* 26 (1974): 411–418.
6. This is a little more obvious if we shorten (8) to "The meeting starts noon." It is even more obvious in other translations; e.g., the translations of "She is running now" by "She is running noon."
7. Perry, "Frege on Demonstratives," p. 494.
8. Pollock denies that "here" admits of a pronominal use. Cf. Pollock, *Language and Thought,* pp. 124–127. Used adverbially, as in "He is here," "here" is paraphrasable by "in this place." But, Pollock contends, "here" is not used pronominally to mean "this place." However, I believe it does have a pronominal use. "Here is where we shall settle" is paraphrased by "This place is where we shall settle."
9. See Hans Kamp, "Formal Properties of 'Now'," *Theoria* 37 (1971): 227–273; A. N. Prior, "Now," *Noûs* 2 (1986): 101–119; Kaplan, "Demonstratives," pp. 23 and 105–106.
10. See Quentin Smith, "Mind-Independence of Temporal Becoming"; idem, Impossibility of Token-Reflexive Analyses"; idem, "Sentences about Time"; Essay 2; and Essay 34.
11. Ernest Sosa, "The Status of Becoming: What is Happening Now?" *Journal of Philosophy* 77 (1979): 26–30; Palle Yourgrau, "Frege, Perry, and Demonstratives," *Canadian Journal of Philosophy* 12 (1982): 744.
12. Hector-Neri Castañeda, "Direct Reference, Realism, and Guise Theory" (mimeograph, 1984), pp. 20–28. Also see Yourgrau, "Frege, Perry, and Demonstratives," p. 728.
13. By the "DR theory" I mean the semantic theory that includes the essentialist semantics (what Almog calls "metaphysical semantics": Joseph Almog, "Naming without Necessity," *Journal of Philosophy* 83 (1986): 215 ff.) on which the notion of rigid designation is based. Thus, I am not disagreeing with Nathan Salmon's arguments (*Reference and Essence* (Princeton, N.J., 1981)) that the DR theory minus the essentialist semantics does not entail nontrivial essentialism.
14. It is commonly accepted that the relational theory of time identifies a time with a set of events. Hugh Lacey ("The Causal Theory of Time: A Critique of Grünbaum's Version," *Philosophy of Science* 35 (1968): 332) correctly observes that, "according to relational theories time is analyzed simply as the set of point-events ordered by the relations . . . 'is simultaneous with' . . . 'is earlier than' . . . 'is later than'." W. H. Newton-Smith calls the relational theories "reductionist" and notes that "we can form the collection of all events simultaneous with any particular event used in identifying the moment. This collection, the reductionist claims, just is the moment" (W. H. Newton-Smith, *The Structure of Time* (London, 1980), p. 6). Examples of relational theories include Hans Reichenbach, *The Philosophy of Space and Time* (New York, 1950); Bertrand Russell, *Our Knowledge of the External World* (London, 1914); Adolf Grünbaum, *Philosophical Problems of Space and Time* (Dordrecht, 1973); M. Bunge, *Foundations of Physics* (New York, 1967); and I. Hinckfuss, *The Existence of Space and Time* (Oxford, 1975).
15. A. N. Prior, *Papers on Time and Tense* (Oxford, 1968), pp. 122–123.
16. Smith, "Sentences about Time," and Essays 2 and 34.

17. Smith, *Felt Meanings,* chaps. 4 and 6, and idem, "Kant and the Beginning of the World," *New Scholasticism* 59 (1985): 339–346.
18. In Essay 2 I state that uses of "now" are not rigid designators. This statement is made on the basis of the assumption provisionally adopted there that the widely accepted theory that times are sets of events is true.
19. John Earman, "Who's Afraid of Absolute Space?" *Australasian Journal of Philosophy* 50 (1970): 287–319; Graham Nerlich, *The Shape of Space* (Cambridge, 1976).
20. Alvin Plantinga, "The Boethian Compromise," *American Philosophical Quarterly* 15 (1978): 129–138. It should be noted that Plantinga's theory of possible worlds is unable to deal with tensed modal sentences and needs to be replaced by a different theory (cf. Quentin Smith, "Tensed States of Affairs and Possible Worlds," *Grazer Philosophische Studien* 31 (1988): 225–235).
21. Both Pavel Tichy ("The Transiency of Truth," *Theoria* 46 (1980): 165–182) and Schlesinger (*Aspects of Time,* pp. 134–135) claim that "now" means "whatever time is present" (or something logically equivalent to this description). The difference in truth conditions between (1) and (2) shows that they are mistaken. It also refutes the claim of E. J. Lowe ("The Indexical Fallacy in McTaggart's Proof of the Unreality of Time," *Mind* 96 (1987): 65) that "'E is present' means, of course, 'E is happening *now*'."
22. Contrary to the received view, the indexical "now" is used in many different ways; only one of these, the standardly discussed "present time use" of "now," has been the subject of this essay. The other uses are discussed in Quentin Smith, "The Multiple Uses of Indexicals," *Synthese* 78, no. 2 (1989): 167–191. I am grateful to a referee for *Erkenntnis,* in which this essay was first published, for comments that led to improvement.

PART II

McTaggart's Paradox and the Passage of Time

Introduction: McTaggart's Paradox and the Tensed Theory of Time

L. NATHAN OAKLANDER

The development of the new theory of time has brought with it a resurgence of interest in McTaggart's famous argument for the unreality of time first propounded in 1908. It is not difficult to see why. Since the question of whether tensed statements can be translated into tenseless statements without loss of meaning is no longer thought by detensers to be crucial to the debate about the metaphysical nature of time, the discussion has begun to focus on the more explicitly metaphysical problem of change—more specifically, on the problem of how we are to understand time if we are to make sense of the possibility of change. McTaggart argued that change requires temporal becoming—events changing with respect to the nonrelational properties of pastness, presentness, and futurity—and that since temporal becoming is inherently contradictory, neither time nor change is real. Part of the fascination surrounding McTaggart's argument is the degree of disagreement as to its validity. Broad, for example, claimed that McTaggart's argument for the unreality of time was a "philosophical howler."[1] Mellor, by contrast, maintained that McTaggart's proof of contradiction in the tensed account of change "seems beyond all reasonable doubt" (p. 165). A brief summary of McTaggart's argument will help set the stage for the debate that follows.

McTaggart begins his argument for the unreality of time by distinguishing between two ways in which we ordinarily conceive and talk about time. On the one hand, we think of time and events in time as *moving,* or *passing,* from the far future to the near future, from the near future to the present, and then from the present receding into the more and more distant past. Events in a series of terms which are either past, present, or future are said to be located in an A-series. On the other hand, we speak and think of moments or events as being earlier than, later than, and simultaneous with other events, and we believe that these relations are unchanging, permanent, and fixed. Thus, for example, if the event of Caesar's crossing the Rubicon occurs before the event of Caesar's death, then it is always the case that he crosses the Rubicon before his death, and it can never be the case that Caesar dies before he crosses the Rubicon. More generally, if one event is *ever* earlier (later) than another event, then it is *always* earlier (later) than the other event; events do not change their positions in the B-series which runs from earlier to later.

McTaggart then argues that although both the A-series and the B-series are essential to our ordinary thinking about, and experience of, time, the A-series and temporal becoming are more fundamental, since temporal or B-relations are dependent upon temporal becoming (or A-properties). He reasons that time involves change, and therefore that for the B-series alone to constitute time (as the detenser maintains), it too must involve change. But, he continues, there is nothing in the B-series that can change. Since sentences which describe temporal relations between events are always true, it follows, according to McTaggart, that events in the B-series do not change by coming into existence and going out of existence; nor do they change their relations to each other. Consequently, if the B-series is to be a time series, then its terms (events) must exemplify the temporal characteristics of pastness, presentness, and futurity and change with respect to them as time passes. But this is to say that time and change require an A-series and temporal becoming.

This leads to the main argument whereby McTaggart attempts to prove that time is unreal. He argues as follows:

1. If the application of a concept to reality implies a contradiction, then that concept cannot be true of reality.
2. Time involves (stands or falls with) the A-series and temporal becoming; that is, if the A-series involves a contradiction, then time involves a contradiction.
3. The application of the A-series and temporal becoming to reality involves a contradiction.
4. Therefore, neither the A-series nor time can be true of reality; thus time is unreal.

For present purposes the crucial premise is (3). In support of it, McTaggart argues simply that if events move through time from the future to the present to the past, then every event in time must *be* past, present, and future. However, past, present, and future are incompatible properties; if an event is present, then it is not past or future, and if it is past, it is not present and future, and if it is future, it is not present or past. Thus, the existence of the moving NOW entails a contradiction—that every event both is and is not past, present, and future—and so time is unreal.

McTaggart was aware of course that the contradiction appears to have an obvious resolution if we specify the various *times* at which events have these incompatible temporal properties. Thus, instead of saying that, for example, event E is past, present, and future, we should say that E is past at time t_3, present at time t_2, and future at time t_1. But McTaggart claims that to introduce time in this way (or by saying that E is present *before* E is past) involves a vicious circle. It assumes time, either in the form of a B-series of moments (t_1 is earlier than t_2) or in the form of second-order events (E's being future is earlier than E's being present), in order to explain the possibility of an A-series and temporal passage. But, given his earlier reasoning, in order for there to be a B-series at all, we must assume the existence of an A-series.

Furthermore, the introduction of time in the form of moments or temporal relations is self-defeating, since it does away with the fact of change that the A-series and temporal becoming sought to capture. This becomes clear when we recognize that on the tensed theory of time, "Event E is in the future" expresses a proposition that changes its truth-value with the passage of time, whereas "Event E is future at t_1" has an unchanging truth-value, meaning no more and no less than "Event E is later than t_1."

Of course, we could reintroduce time and change into reality by subjecting the times, or moments, at which events are past, present, and future to a change in their transitory properties. That is, we could say that E is present at t_2 and that t_2 is past, present, and future; that E is past at t_3 and that t_3 is past, present, and future; and so on. Indeed, attributing different A-characteristics to moments is necessary since, if t_1, t_2, and t_3 are genuinely *temporal* entities, then they must be terms in a changing A-series. But unfortunately, with that move, the contradiction in temporal predication rears its ugly head once again, this time with respect to moments, not events. It is obvious, according to McTaggart, that the appeal to time to explain how moments can have incompatible temporal properties is just another step in an infinite chain that fails to remove the paradox with which we began. Thus, he concludes, whether we stop at a contradiction or at the denial of genuine (A-series) change, time is unreal.

Critics of McTaggart, such as Broad, Prior, Schlesinger, Smith, and others have been quick to point out that since there is no contradiction to be avoided in the first place, in that no event ever *is* (nonsuccessively) past, present, and future (or present and not present), there is no need to get embroiled in an infinite regress in order to avoid such. As Broad puts it:

> I cannot myself see that there is any contradiction to be avoided. When it is said that pastness, presentness, and futurity are incompatible predicates, this is true only in the sense that no one term could have two of them *simultaneously* or *timelessly*. Now no term ever appears to have any of them simultaneously. What appears to be the case is that certain terms have them *successively*. Thus, there is nothing in the temporal appearance to suggest that there is a contradiction to be avoided.[2]

Of course, whether that response is adequate depends on how one unpacks the notion of *succession*. According to tensers, if we take tense seriously, then *E* is past, present, and future is never true. What is true is that

> *E is* past, *was* present, and *was* (still earlier) future; *E is* future, *will be* present, and *will* (still later) *be* past; or *E is* present, *was* future, and *will be* past.

The dominant issue in recent discussions of McTaggart's paradox is whether the appeal to tense can be given an interpretation capable of resolving the difficulties with temporal passage. According to defenders of the new theory of time, none of the solutions to McTaggart's paradox and related problems that have been proffered are successful, so we must give up the idea that events form a real A-series and change with respect to their temporal location in it. For detensers, "past," "present," and "future" are indexical expressions whose referent cannot be separated from the time of utterance or from the utterance itself. Thus, for Mellor, an event is past if and only if it is earlier than the judgment about it, present if and only if it is simultaneous with the judgment about it, and future if and only if it is later than the judgment about it. Tensers cannot accept this solution to McTaggart's paradox, however, since it leaves out something essential to time—namely, temporal passage.

The essays in this part of the book describe various recent attempts to circumvent McTaggart's paradox and other related problems and assess the varying degrees to which they succeed. They fall into two groups. The first group (Essays 13–18) deal explicitly with the logical cogency of McTaggart's negative argument, and the second group (Essays 19–25) concern problems with temporal becoming first raised by J. J. C. Smart and D. C. Williams.

The first essay in Part II, by Mellor, constitutes a defense of McTaggart's argument for the unreality of tense. Mellor argues that there is no way of specifying in tensed terms *when* an event or moment has incompatible A-properties (or *when* a tensed sentence or proposition has incompatible truth-values) that avoids a vicious infinite regress. Smith then takes issue with Mellor's argument (Essay 14) and offers his own account of the "logic of A-properties" (Essay 15). This admittedly entails an infinite regress of temporal attributions (presentness inheres in event *E*, and it also inheres in the inherence of presentness in *E*, and so on), but a regress that is not vicious, for at no stage is there a contradiction to be avoided. The subsequent debate between Smith and Oaklander (Essays 16–18) concerns the adequacy (or inadequacy) of Smith's analysis of the tenses.

J. J. C. Smart and D. C. Williams have pointed out difficulties with the notion of temporal passage that closely parallel McTaggart's argument.[3] Smart argues that if we think of time as a river, or some kind of particular or property that moves, then it must make sense to ask how fast it is moving. Yet the question, How fast did time flow yesterday?, seems to be senseless. If we try to make sense of the flow of time by appealing to a $time_2$ through which $time_1$ moves, then, as Williams puts it, "The history of the new moving present, in $time_2$, then composes a new and higher time dimension again, which cries to be vitalized by a newer level of passage, and so on forever."[4] Thus, we are stuck with either an unintelligible notion of temporal motion, of the moving NOW, or an infinite regress or temporal series. The adequacy of various tenser responses to these objections is debated in the essays by Schlesinger, Oaklander, and Zeilicovici.

In the selection from his book *Aspects of Time* included here as Essay 19, George Schlesinger argues that the hypothesis that time has two dimensions can make sense of the notion of the rate of temporal flow, or becoming, and resolve McTaggart's paradox (which he takes to be the problem of understanding how an instantaneous event can persist through a change of incompatible A-properties). Schlesinger's view is explained and modified in the subsequent debate (Essays 20–22) with Oaklander, who argues that neither the two-dimensional hypothesis nor its revision can avoid McTaggart's difficulties or make temporal becoming intelligible.

David Zeilicovici (Essay 23) offers a novel view of temporal becoming, which attempts to account for the transient aspect of time, but without the idea of the NOW as a property, or particular, that moves along the series of events. Oaklander then critiques this view (Essay 24).

Schlesinger then offers yet another account of time, one that attempts to answer both Smart's question about the rate of temporal flow and McTaggart's question about incompatible temporal properties. Rather than relativizing past, present, and future to times, Schlesinger proposes relativizing them to different worlds. Whether Schlesinger's version of the tensed theory is any more successful than its predecessors is left open.

In any event, the essays in Part II should give the reader an accurate picture of the main difficulties that need to be addressed in any attempt to defend the tenser position and supply a major impetus for the new theory of time.[5]

Notes

1. Broad, *Examination of McTaggart's Philosophy*, vol. 2, pt. 1, p. 316. See J. E. McTaggart, "The Unreality of Time," *Mind* 18 (1908): 457–474, rpt. in S. V. Keeling, ed., *Philosophical Studies* (London, 1934), pp. 110–131; and idem, *The Nature of Existence*, vol. 2, ed. C. D. Broad (New York, 1927), pp. 9–31.
2. Broad, *Examination of McTaggart's Philosophy*, p. 313.
3. J. J. C. Smart, "The River of Time," *Mind* 58 (1949): 483–494; Williams, "Myth of Passage."
4. Williams, "Myth of Passage," pp. 453–464.
5. A recent debate over the cogency of McTaggart's argument not discussed in this book began with E. J. Lowe, "The Indexical Fallacy in McTaggart's Proof of the Unreality of Time," *Mind* 96 (1987): 62–70, and continued with R. Le Poidevin and D. H. Mellor, "Time, Change, and the 'Indexical Fallacy,'" *Mind* 96 (1987): 534–538; E. J. Lowe, "Reply to Le Poidevin and Mellor," *Mind* 96 (1987): 539–542; idem, "McTaggart's Paradox Revisited," *Mind* 101 (1992): 323–326; Le Poidevin, "Lowe on McTaggart," *Mind* 102 (1993): 163–170; "Comment on Le Poidevin," *Mind* 102 (1993): 171–173.

ESSAY 13

The Unreality of Tense

D. H. MELLOR

In chapters 4 and 5 (reproduced here as Essay 1) of *Real Time,* I discredited three faulty inferences drawn from the tenseless view of time. Here my aim is to discredit the rival view, which I do by discrediting the tensed idea of change. Change is clearly of time's essence, and many have thought it to be the downfall of tenseless time—that only a tensed view of time can account for it. In fact, the opposite is true. The A-series is disproved by a contradiction inherent in the idea that tenses change.

Change, obviously (if vaguely), is having a property at one time and not at another. More specifically, it refers to something having incompatible properties at different dates, such as being at different temperatures or in different places. Cooling is a change of temperature, being first hot and then cold; movement is a change of place, being first somewhere and then somewhere else. Similarly, there are changing sizes, shapes, colors, and other properties of things. In each case something has one of several mutually incompatible properties at one B-series time and another one later.

This tenseless idea of change is basically right, as well as being obvious. But there are objections to it, of which two especially have long preserved a tensed alternative. The first is that it does not really distinguish change through time from change across space. Properties can after all vary from place to place as well as from time to time. A poker, for example, may be at once hot at one end and cold at the other: why is that not change as much as the whole poker being hot one day and cold the next? We could, of course, define change to be variation in time as opposed to variation in space—but only given some other way of distinguishing time from space. If time is marked off only as the dimension of change, we should be arguing in an indecently small circle. But it is not obvious how else to distinguish it from spatial dimensions.

Consider again a clock's second hand passing successively the figures "1" and "2." The latter event is both later than the former and to the right of it. We see this as change (namely, movement), which is how we distinguish the temporal and spatial relations of the two events. Specifically, it is how we tell a thin hand moving across the clock face from a fat one spread statically across it in two spatial dimensions. In short, we perceive the temporal relation in this case by perceiving change. Similarly for changes of place, temperature, color, and everything else. To see that one event is later than another is to see something change. How else could time be perceived or understood, except as the dimension of change? But change cannot then be defined as variation in time. And the objection to tenseless time is that it has no other way of defining change, and so, in particular, no way of distinguishing change from spatial variation.

This objection is reinforced by the other one: namely, that the tenseless view reduces change to changeless facts. If a poker is hot one day and cold the next, then those always were and always will be its temperatures on those two days. B-series facts of this kind do not change with time; that is the mark of the B- as opposed to the A-series.[1] And as for time, so for space. There is no spatial change in a poker being at once hot at one end and cold at the other. The hot and cold ends of a poker are not a case of change, because they coexist: the spatially tenseless world contains them both, but located in different parts of tenseless space. Likewise, the hot poker and the cold coexist in a temporally tenseless world. It contains them both, only located in different parts of B-series time. And if, as everyone agrees, coexistence prevents change in the spatial case, how can it be compatible with change in time?

The A-series provides a ready answer to all these questions. Change is the changing tense of things and events moving from future to past. It is peculiar to time because the A-series has no real spatial analogue. The token-reflexive facts of spatial tense are all there is to it; but on the tensed view they are not all there is to temporal tense. In the temporal case, over and above those tenseless facts, there is a present moment which is forever changing its date. The reality of the clock hand's movement consists ultimately in the events of its passing the figures "1" and "2" and these becoming successively present and then past; and similarly for all other changes.

On the tensed view, change is first of all the successive presence of earlier and later things and events. This defines the temporal relations *earlier than* and *later than:* one event is tenselessly earlier than another if and only if its ever changing tenses bring it to the present first. Since there are no real spatial tenses, this definition has no spatial analogue. The tenseless spatial relations of events and things are *sui generis,* and that is the difference between space and time. In time, but not in space, the tenseless B-series is supposed to be derived from the A-series—hence the idea of tenseless change that I started with. Change is still defined as variation through time; but by defining time first as the dimension of changing tense, the tensed view prevents this definition of change from begging the question against its spatial analogues.

This tensed view of change may be supported by further doctrines about what else turns on the difference between present and future. We have already noticed the view that existence turns on it; that is, that coming to be present is coming to exist. This provides a still more profound basis for distinguishing time from space, as the dimension in which things come successively into existence. However, it makes no odds what else turns on tense unless the idea of changing tenses actually does account for change; and, appearances to the contrary notwithstanding, the fact is that it doesn't.

To start with, one might accuse the tensed view itself of begging the question against spatial change, in denying the reality of spatial tenses. They, after all, vary across space, just as temporal tenses vary through time. If change is different events becoming temporally present from time to time, why is it not different things becoming spatially present—here—from place to place? To such accusations, upholders of tense reply that we have a direct intuition of temporal presence which is lacking in the spatial case. We see things laid out tenselessly in space, whereas we do not see things laid out tenselessly in time. But we have already disposed of this supposed intuition of temporal tense.[2] We have seen how observing tenseless temporal relations is independent of observing tenses and how the apparent temporal presence of our experiences can be tenselessly accounted for.

Although the tensed view of change could easily be convicted of distinguishing time from space no better than the tenseless view, I will not press the charge. To prefer debatable and relatively trifling charges is pointless when a capital offense can be proved against the same party. The capital offense is self-contradiction, of which tensed views of both time and space are guilty. But, as the tensed view of space has been put away already, I need only prosecute the temporal case.

The proof of contradiction in the tensed account of change is not new. It was given by J. E. McTaggart in 1908[3] and has been much debated since. To me it seems beyond all reasonable doubt; but since it is still disputed, I fear I must prosecute it yet again. Two factors, however, encourage me to hope for more success than McTaggart had. One is that he prosecuted time itself. He thought that time needs change and change needs changing tense and so thought to convict time along with tense. But the obvious unimpeachability of time defeated him, and unfortunately drew suspicion on his whole case. Tense in fact has been wrongly acquitted to save innocent time. But we need not acquit the guilty in this case in order to save the innocent. What is wrong with McTaggart's prosecution of time is not his prosecution of tense but his contention that disposing of tense disposes of change. Change can be explained and distinguished from spatial variation without any appeal to tense.[4] And given that, the reality of changing tense can safely be denied without imperiling the reality of change and hence of time itself. Once that is realized, McTaggart's proof will, I hope, meet much less resistance.

The other factor encouraging me is the token-reflexive treatment of tense.[5] Although this factor is not new, it was not there in McTaggart's time, and it should make his proof more persuasive. On the one hand, it should make its validity more obvious and, on the other, its conclusions more palatable. Tense, it will turn out, is not being banished altogether; it is merely being replaced where it belongs—in our heads.

McTaggart's Proof

McTaggart's proof is very simple.[6] Many A-series positions are incompatible with each other. An event which is *yesterday*, for example, cannot also be *tomorrow*. Past, present, and future tenses are mutually incompatible properties of things and events. But because they are forever changing, everything has to have them all. Everything occupies every A-series position, from the remotest future through the present to the remotest past. But nothing can really have incompatible properties, so nothing in reality has tenses. The A-series is a myth.

The defense has an immediate and obvious riposte to this attack, and its rebuttal is unfortunately much less obvious; which is why McTaggart's proof has rarely carried the conviction it deserves. The riposte is that nothing has incompatible tenses at the same time. Nothing is present *when* it is past or future when it is present. Things and events only have these properties successively; first they are future, then present, then past. And nothing prevents things having incompatible properties at different times. On the contrary, that is how change is defined: the successive possession of incompatible properties. All McTaggart has shown is that changing tense fits that definition, as it should.

To rebut this riposte, McTaggart asks when, in tensed terms, things and events have their various tenses; and here it will help to use some symbols. Let P, N, and F be respectively the properties of being past, present (that is, now), and future, and let e be any event. Then e being past, present, and future I write respectively as Pe, Ne and Fe. Complex tenses I represent by repeated ascription of *P, N,* and *F:* thus *PFe* means *e was* future, *FPNe* means *e will have been* present, and so on. ~, &, and ⊢ as usual mean respectively "not," "and," and "entails."

Then McTaggart's basic argument is that, on the one hand, the three properties *P, N,* and *F* are mutually incompatible, so that for any event *e*

$$Pe \vdash {\sim}Ne;\ Ne \vdash {\sim}Fe;\ Fe \vdash {\sim}Pe; \text{ and so on.} \tag{1}$$

On the other hand, the inexorable change of tense means that every event has all three A-series positions; that is:

$$Pe \,\&\, Ne \,\&\, Fe. \tag{2}$$

But (1) and (2) cannot both be true, since if (1) is true, two of the statements in (2) must be false, so (2) as a whole must be false. But our concept of tense commits us to both (1) and (2); so it leads us inevitably into contradiction and thus cannot apply to reality. Reality, therefore, must be tenseless: there are no tensed facts.

To this the riposte is that *e* has no more than one of these incompatible properties at once, so there is no contradiction after all. Suppose, for example, that *e* is actually present; that is, *Ne*. Then *e* is neither past nor future; that is, both *Pe* and *Fe* are false, as (1) requires. The truth, rather, is that *E will be* past and *was* future; that is, not (2) but

$$FPe \,\&\, Ne \,\&\, PFe, \tag{3}$$

which is quite compatible with (1).

So it is. But, as McTaggart remarks, there are more complex tenses than those in (3), and not all combinations of them are so innocuous. Specifically, there are also *PP* and *PN, FF* and *FN,* and *NP, NN,* and *NF.* And just as every event has all A-series positions if it has any of them, so it also has all these other complex tenses. For example, whatever has any simple tense obviously also has it *now;* that is:

$$Pe \vdash NPe;\ Ne \vdash NNe;\ Fe \vdash NFe.$$

Obviously also, whatever is past *was* present and *was* future, and whatever is future *will be* present and *will be* past, so that

$$Pe \vdash PNe;\ Pe \vdash PFe;\ Fe \vdash FNe;\ Fe \vdash FPe.$$

Moreover, whatever is sufficiently past also *was* past; for example, what happened two days ago was already past yesterday. And sufficiently future events likewise also *will* be future, which gives us *PP* and *FF,* as well as *P* and *F.*

In place, then, of the original three simple tenses, we have the nine compound tenses *PP, PN, PF; NP, NN, NF; FP, FN, FF*. But McTaggart's argument applies just as well to them. Because of the way tense incessantly changes, every event that has any of these nine tenses has to have them all; but they are not all mutually compatible. For example, *FF* and *PP* are incompatible, since what will be future cannot also have been past. And *NP, NN,* and *NF* are even more clearly incompatible, because they are equivalent to the simple *P, N,* and *F*.

The riposte will again be made that events do not have these incompatible tenses all at once. But again, saying in tensed terms just when they do have them only generates another set of properties, including mutually incompatible ones like *PPP, NNN,* and *FFF,* all of which every event has to have. There is, in other words, an endless regress of ripostes and rebuttals, a regress that is vicious, because at no stage can all the supposed tensed facts be consistently stated.

The Defense of McTaggart

That, basically, is how McTaggart put his case. His critics have reacted by denying the viciousness of his regress. At every stage, they say, the appearance of contradiction is removed by distinguishing the different times at which events have different tenses. They ignore the fact that the tensed means they use to distinguish these times are also subject to the contradiction they are trying to remove. However, the debate is now too well worn to be settled by mere repetition of McTaggart's proof, sound though it is. To change the metaphor, too many people have managed to inoculate themselves against it. If it is to wipe out belief in real tenses, as it should, a more immediately virulent strain of it is needed, a strain that I believe is best nurtured on the token-reflexive facts that make tensed sentences true or false.

Before developing the new strain, however, it is worth neutralizing some antidotes that have been proposed to McTaggart's original proof. First, I should perhaps remark that although I have dealt only with the unqualified past, present, and future, the proof applies also to more precise A-series positions. *Yesterday* and *three days ago,* for example, are likewise incompatible properties of things and events of less than a day's duration, both of which they must all nevertheless possess. But there is no point in complicating the discussion by bringing all these other tenses into it explicitly. If the argument works for *P, N,* and *F,* it will work for all tenses; and if not, for none.

Second, I have followed McTaggart in ascribing these problematic A-series properties to events. Tense logicians mostly prefer to treat "*P,*" "*N,*" and "*F*" as "operators" (analogous to "It is not the case that" or "It may be the case that") prefixed to present-tense core sentences or propositions. This is tantamount to regarding *P, N,* and *F* as properties not of events, but of tensed facts. Where, for example, McTaggart and I start with a thunderstorm, tense logic starts with the sentence or proposition saying that a thunderstorm is happening now. Where we say the thunderstorm is two days past, they say the fact of its happening now is two days past; that is, the present-tense sentence or proposition saying that it is happening now was true two days ago. In the symbolism above, this amounts to replacing "*e*" throughout by "*Ne*." But it makes no odds to the argument, as readers may verify for themselves; facts are no better at being at once both and not both past and present, present and future, and so on than events are.

Nor does it help to distinguish the "object language," in which events are said to be past, present, or future, from a "metalanguage" in which object language sentences are said to be true or false. At least, it helps only if the metalanguage sentences are the tenseless ones used to give the token-reflexive truth conditions laid out in chapter 2 of *Real Time*. When the metalanguage sentences are themselves tensed, the problem simply reappears in a new guise. Truth and falsity are now the incompatible properties of the object language sentence-types (since tensed truth conditions are supposed not to be token-reflexive; if they were, as we shall see later, they would not be tensed). But to say that a particular sentence-type is true—that is, that any token of it would be true—is to say that it is not false, and vice versa. Yet any true contingent tensed sentence-type must also be (sometime) false—otherwise it would not be tensed. This now is McTaggart's basic contradiction, and the riposte to it is the same: no tensed sentence-type is both true and false at the same time. Metalanguage sentences then say when these object language sentences are true and when false. But if the metalanguage sentences are themselves tensed, they too will be both true and false. The contradiction simply recurs in the metalanguage. Removing it from there by using a tensed meta-metalanguage to say when metalanguage sentences are true and when false only leads to McTaggart's regress. Iterating tensed metalanguages no more refutes McTaggart than iterating tensed properties of events or facts does, or than iterating tensed operators on propositions or sentences within a single language.

The plain fact is that nothing can have mutually incompatible properties, whether they be tenses or truth-values: neither events, things, facts, propositions, sentences, nor anything else. I prefer, therefore, to stick to events and things, as being the natural inhabitants of A-series positions. I will eschew translations of the problem into other and more fashionable idioms, since these only pander to the erroneous conviction that McTaggart is thus easily answered.

Third, however, I must deal with the complaint that in symbolizing McTaggart's argument I have myself begged the question against tenses. Specifically, in using *Pe* to say that *e* is past, I have left out the verb "is." By so doing, I have tacitly treated the "is" in "*e* is past" as a tenseless copula, which is why *e*'s being past, present, and future appears to be contradictory. For in fact the verb "is" in "*e* is past" is tensed; that is, it really means that *e* is *now* past. And given that, the contradiction vanishes, since if *e* is now past, it is not also now present or now future. Of course, it *was* future, and it *was* present, but that is quite compatible with *e* being now past. In short, the supposed contradiction has been artificially generated by suppressing the essentially tensed verbs used in ascribing to *e* the properties *P, N,* and *F.*

This complaint misses the point of tense completely. The A-series is supposed to be a feature of the world, not of verbs. As we have seen,[7] adverbs and phrases like "yesterday," "this week," and "next year" make verbal tenses redundant, and so do "in the past," "now," and "in the future"; that is, "*P*," "*N*," and "*F*." That is their function: to take over the roles respectively of the past, present, and future verbal tenses to which, by definition, they are equivalent. Given these expressions, verbs might as well be tenseless, that is, take the same form regardless of the A-series position of the events they refer to. Suppose, for example, that "happens" is made such a tenseless form. Then "It happened" means "It happens in the past." If the past-tense form of the verb in "It happened in the past" were not redundant, it would have to mean *PP* rather than *P,* which it clearly doesn't. It simply means "It happened." Just as "in the past" is superfluous given the past verbal tense, so the verbal tense is superfluous given "in the past."

As for the past, so for the present. Adding "now" to "It happens" or "It is happening" makes the present-tense connotations of the verb superfluous. Given the "now," "happens" is as tenseless in "It happens now" as in "It happens in the past" and "It happens in the future." And as for "happens," so for "is." The temporal meanings of "*e* is past," "*e* is present," and "*e* is future" are supplied entirely by the words "past," "present," and "future." The "is" *is* a mere tenseless copula, present only because English grammar gives sentences verbs even when, as here, they contribute nothing to the content. Nothing is left out, therefore, by abbreviating these sentences to "*Pe*," "*Ne*," and "*Fe*." So the abbreviation can generate no contradiction that was not already there.

Anyone who still says that the "is" in "*e* is past" is present tense, so that "*e* is past" means "*e* is now past," will have to say what tense "is" then has in "*e* is now past." It is clearly either tenseless or present tense—and if tenseless, McTaggart's contradiction reappears at once, because "*e* is now past" is not always true. It is true only when "*e* is past" is true (which is why the two sentences mean the same). So all we have to do to regenerate McTaggart's proof, as readers may easily verify, is replace "*P*," "*N*," and "*F*" throughout by "*NP*," "*NN*," and "*NF*."

But if the "is" in "*e* is now past" is tensed, as in "*e* is past," the same vicious regress appears in the form of the verb itself. For "*e* is past," meaning "*e* is now past," must now also mean "*e* is now now past," in which, again, the "is" must either be tenseless or present tense. If tenseless, we again get McTaggart's argument, starting this time with "*NNP*," "*NNN*," and "*NNF*"; and if present tense, the regress continues with "*e* is now now now past," "*e* is now now now now past," and so on, *ad infinitum*. And for no sentence-type in this endless sequence can we consistently give tensed truth conditions. It is no use saying, at any stage in the sequence, that the last sentence-type in it is true *now*, because whether that is so depends on when *now* is. Saying that merely generates the next type in the sequence, concerning which the same question arises. To stop and give a definite answer at any stage only produces a contradiction, because if the sentence is true (at some present time), it is also false (at some other). The only way to avoid contradiction is never to stop at all, which is tantamount to admitting that the original sentence-type has *no* tensed truth conditions, that is, cannot be made true or false by any tensed fact such as that *e* is past, *e* is now past, *e* is now now past, and so on. In short, supposing that there are such facts is either self-contradictory or useless for saying what makes tensed sentences true or false.

McTaggart and Token-Reflexives

So much by way of reinforcing McTaggart's own proof. But in case it still does not convince, I will now put it explicitly in terms of token-reflexive truth conditions. First, recall that tensed facts are supposed to provide *non*-token-reflexive truth conditions for tensed sentences and judgments. Just as all tokens of "Snow is white," whenever and wherever they occur, are made true by the single fact that snow is white, so all true tokens of "*e* is past," whenever they occur, are supposed to be made true by the single fact that *e* is past or, if for some reason that won't work, by the fact that *e* is now past, or that *e* is now now past, and so forth. All tokens of the same tensed sentence-type are supposed to have the same tensed truth conditions, however much their tenseless truth conditions may vary from token to token.

Now Essay 1 showed that tensed sentences have no tenseless translations, because tokens of the translation would all have the same tenseless truth conditions, whereas those of tensed sentence-tokens vary from time to time. That is, no one tenseless fact can make true all true tokens of a given tensed type regardless of when they occur. However, *e* being past or being now past, and so on, should be able to do this because, unlike tenseless facts, they are not facts at all dates. They should be able to make tokens of "*e* is past" true just when it is a fact that *e* is past, is now past, or whatever.

Now we know when it is a fact (if it is) that *e* is past, is now past, and the rest: namely, at all and only dates later than *e* itself. Tokens of "*e* is past," "*e* is now past," and so on, are all alike true if and only if they occur later than *e*. In tenseless terms, all these sentences are merely different ways of saying "*e* is past." But in tensed terms, all this means is that true tokens of "*e* is past" could be made true, consistently with their agreed tenseless truth conditions, by any one of these facts, that *e* is past, *e* is now past, and so on. And in tensed terms these are, arguably at least, different facts. For although *e* is admittedly not just—that is, not always—past, it is supposed to be *now* past; and if not always now past, at least now now past; and so on. Adding "now" to any of these claims is supposed to help to make it—now—true. So the additions presumably somehow affect the claims' tensed truth conditions, which is to say that *e*'s being past, being now past, and so forth are somehow different facts.

Yet in B-series terms none of them will do the job. None gives "*e* is past" non-token-reflexive truth conditions, because even when they *are* facts, its tokens' truth-values still depend on their own dates. Thus at any date *t* later than *e,* the facts are that *e* is past, now past, now now past, and so on. But none of these facts makes tokens of "*e* is past" which occurred before *e* true at *t*. They are false, then, just as they always were and always will be. The whole point of the type/token distinction is that tensed tokens, as opposed to types, have definite and temporally unqualified truth-values.[8] True, long-lasting thing tokens can vary in truth-value during their lifetimes; for example, a token of "*e* is past" printed before *e* will start off false and end up true. But that does not change its truth-value at the earlier dates, any more than my slimming in 1984 will require me to rewrite, Newspeak fashion, the record of what I weighed in 1983. A saying or writing of "*e* is past" which occurs before *e* always was and always will be just plain false.

But can the tensed facts we are considering give "*e* is past" non-token-reflexive truth conditions when times are specified in A-series terms? Unfortunately for them, no. Even when it is *now* a fact that *e* is past, *e* is now past, and so forth, tokens of "*e* is past" are not all true now, regardless of their A-series position. Tokens that are now more past than *e* itself will be false now, as they always were and always will be. Even in A-series terms the truth conditions of "*e* is past" are token-reflexive; they depend on the A-series position of the token itself.

And as for "*e* is past," so for all tensed sentences. All tensed sentences have tenseless token-reflexive truth conditions. That is, some tokens of the same tensed type will differ in truth-value, depending on their date. But whatever has a date has at every moment a corresponding tense.[9] In particular, tensed tokens with diverse dates will always have correspondingly diverse A-series positions, with which their diverse truth-values will therefore also vary. Thus, if the tenseless truth conditions of tensed sentences are token-reflexive, so are their tensed truth conditions.

Giving any tensed sentence non-token-reflexive truth conditions, tensed or tenseless, always leads to contradiction. That, for tenseless truth conditions, is how I proved that tensed sentences have no tenseless translations; the translation would give all tokens of the type the same truth-value regardless of their date, thus inevitably contradicting the truth-value of some tokens of the tensed sentence, which differ from date to date. And the same goes for tensed truth conditions. Since the truth-value of tensed tokens is never independent of their A-series position, giving them now all the same truth-value will inevitably make some past or future tokens both true and false.

This, in token-reflexive terms, is McTaggart's contradiction. That it is so is most easily seen in the metalanguage version of his argument discussed above. Because the tensed truth conditions of "*e* is past" are token-reflexive, any attempt to state in a tensed metalanguage the one tensed fact that makes all its true tokens true is bound to fail. The alleged fact would, by definition, have to make all tokens of the type true, regardless of their A-series position, whereas in fact some are always true and others always false. Hence the contradiction. And it is, I hope, easier to see in this version of the argument that complex tenses are no better off; that is, that the regress of metalanguages which McTaggart's critics invoke is indeed vicious. For the above argument applies to tenses of any complexity. Provided they have tenseless token-reflexive truth conditions, their tensed truth conditions will also be token-reflexive.[10] And whatever doubts there may be about the token-reflexivity of some complex tenses, there can be none about those that McTaggart's critics resort to. Unless "*e* is now past," "*e* is now now past," and so on have the same token-reflexive truth conditions as "*e* is past," they cannot be substituted for it. And if they do have those truth conditions, McTaggart's argument disposes of the tensed facts that they allegedly state, just as it disposes of the alleged fact that *e* is past.

Finally, I suppose that defenders of tense might ask why tensed truth conditions cannot be token-reflexive if tenseless ones are. The answer is that they then cease to be tensed. Suppose, for example, that it is now *n* years after *e; e* is now past, but that fact alone does not suffice to make all tokens of "*e* is past" now true. However, consider a token that is only *m* years past, where *m* is less than *n*. The token is true, because it is less past than *e* itself. Those are the ostensibly tensed truth conditions of any token of the type: it is true if and only if, when it is present, *e* is past; that is, $n-m$ is positive. But if $n-m$ is ever positive, it is always positive, because the temporal distance between A-series positions never changes. The values of *n* and *m* continually increase, therefore, but always at the same rate, so the value of their difference does not. The fact is simply that the token is—tenselessly—$n-m$ years later than *e*. The variably tensed elements *n* and *m* in the supposedly tensed token-reflexive truth conditions cancel out, leaving the already familiar tenseless truth conditions: true if later than *e,* false otherwise.

Similarly for all other tensed sentence-types. Their tensed truth conditions are either self-contradictory or token-reflexive; and if token-reflexive, they reduce to tenseless truth conditions. As McTaggart saw, the truth conditions of tensed sentences are either tenseless or self-contradictory.

My version of McTaggart's proof started from the fact that all tensed sentences and judgments have tenseless token-reflexive truth conditions.[11] It was left open whether tensed sentences also state tensed facts; but we see now that this is not a real option after all. And while those immunized against McTaggart may need something like the above argument to convert them, there is a simpler argument which should serve to sway more open minds.

The sole function of tensed facts is to make tensed sentences and judgments true or false. But that job is already done by the tenseless facts that fix the truth-values of all tensed sentence- and judgment-tokens. Provided a token of "*e* is past" is later than *e,* it is true. Nothing else about *e* and it matters a jot; in particular, no tensed fact about them matters. It is immaterial, for a start, where *e* and the token are in the A-series; and if that is not material, no more *recherché* tensed fact can be. Similarly for tokens of all other tensed types. Their tenseless truth conditions leave tensed facts no scope for determining their truth-values. But these facts, by definition, determine their truth-values. So, in reality, there are no such facts.

Notes

1. Mellor, *Real Time,* chap. 1.
2. Ibid., chaps. 1 and 3.
3. McTaggart, "Unreality of Time."
4. Mellor, *Real Time,* chap. 7.

5. Ibid., chaps. 2–5.
6. McTaggart, "Unreality of Time"; idem, *Nature of Existence,* vol. 2, chap. 33.
7. Mellor, *Real Time,* chap. 1.
8. Ibid., chap. 2.
9. Ibid., chap. 1.
10. Ibid., chap. 2.
11. Ibid.

ESSAY 14

Mellor and McTaggart's Paradox

QUENTIN SMITH

Mellor's argument for the self-contradictory character of tensed facts is based on McTaggart's paradox, which involves the idea that P (pastness), N (nowness or presentness), and F (futurity) are incompatible properties and yet that every event e has them all. It is true that

$$Pe \ \& \ Ne \ \& \ Fe, \tag{1}$$

and yet it is also true that

$$Pe \rightarrow \sim Ne;\ Pe \rightarrow \sim Fe;\ Ne \rightarrow \sim Fe;\ Ne \rightarrow \sim Pe;\ Fe \rightarrow \sim Pe;\ Fe \rightarrow \sim Ne. \tag{2}$$

The attempt to reconcile (1) and (2) by specifying different past, present, or future times at which e possesses P, N, and F results in more complex temporal properties, and some of the latter are themselves incompatible. This specification results in the expansion of (1) to

$$PPe \ \& \ PNe \ \& \ PFe \ \& \ NPe \ \& \ NNe \ \& \ NFe \ \& \ FP3 \ \& \ FN3 \ \& \ FFe. \tag{3}$$

But (3) is no less self-contradictory than (1) since some of the conjuncts of (3), such as PNe (e was present) and FNe (e will be present), are incompatible. The resort to even more complex properties of e, such as $FPNe$ (e will have been present) and $PFNe$ (e will have been past), is of no avail, since many of these more complex properties are themselves incompatible. The regress of more and more complex properties is vicious, since at every level of complexity some of the properties are mutually incompatible.

I believe that the response of the tenser to this argument should be that there is no vicious regress since there is no contradiction to begin with. The tenser denies that his theory commits him to (1), since (1) implies that e *tenselessly* possesses P, N, and F. It is instead the case that

> Either *e* is past and was present and was (still earlier) future, or *e* will be past, is present, and was future, or *e* will be present and will be (still later) past and is future,

which embodies no contradiction and consequently does not originate a vicious regress.

Mellor is well aware of this possible response to his argument:

> I must deal with the complaint that in symbolizing McTaggart's argument I have myself begged the question against tenses. Specifically, in using *Pe* to say that *e* is past, I have left out the verb "is." By so doing, I have tacitly treated the "is" in "*e* is past" as a tenseless copula, which is why *e*'s being past, present, and future appears to be contradictory. For in fact the verb "is" in "*e* is past" is tensed; that is, it really means that *e* is *now* past. And given that, the contradiction vanishes, since if *e* is now past, it is not also now present or now future. Of course, it *was* future, and it *was* present, but that is quite compatible with *e* being now past. (p. 170)

He proceeds to argue that introducing a tensed copula cannot save the tensed theory of time from contradiction, since this introduction entails a vicious infinite regress of its own.

> Anyone who . . . says the "is" in "*e* is past" is present tense, so that "*e* is past" means "*e* is now past," will have to say what tense "is" then has in "*e* is now past" . . . [I]f the "is" in "*e* is now past" is tensed, as in "*e* is past," the same vicious regress appears in the form of the verb itself. For "*e* is past," meaning "*e* is now past," must now also mean "*e* is now now past," in which again the "is" must be either tenseless or present tense. If . . . present tense, the regress continues with "*e* is now now now past," "*e* is now now now now past," and so on, *ad infinitum*. . . . To stop and give a definite answer at any stage only produces a contradiction, because if the sentence is true (at some present time), it is also false (at some other). The only way to avoid contradiction is never to stop at all, which is tantamount to admitting that the original sentence-type has *no* tensed truth conditions, that is, cannot be made true or false by any tensed fact such as that *e* is past, *e* is now past, *e* is now now past, and so on. (pp. 170–171)

Several comments are in order. To begin with, even if the premises of this argument are true, the argument is invalid. For if "one never stops at all," that does not entail that the original sentence has no tensed truth conditions but merely that it has no finite tensed truth conditions. If one never stops, the entailment is that the sentence has an infinite tensed truth condition. The tenser, if he accepts Mellor's premises, may concede to Mellor that the sentence is not made true by any single tensed fact of the form *e* is now past or *e* is now now past, since tensed facts of this form are one and all finitely complex. But this concession is consistent with holding either (a) that the sentence is made true by a single tensed fact of the form *e* is past now now now . . . now . . . , *ad infinitum,* which is infinitely complex; or (b) that the sentence is made true by an infinite number of finitely complex tensed facts of the form *e* is now, . . . now past, where the number of nows in each such fact is less than aleph-zero. This infinite factual complexity avoids contradiction, since it involves a specification of when each temporal property is exemplified; *e* exemplifies pastness *now,* and *e*'s exemplification of pastness exemplifies nowness *now,* and the exemplification of nowness by *e*'s exemplification of pastness exemplifies nowness *now,* and so on without end. A contradiction is produced only if one stops somewhere in this regress. For example, if one stops after the exemplification of nowness by *e*'s exemplification of pastness, it will not be specified when *e*'s exemplification of pastness exemplifies nowness; thus, one will not be able to say that the exemplification occurs now rather than in the past or future, and the inference that it tenselessly obtains in the past, present, and future cannot be blocked. But if one never stops, it is specified when this exemplification occurs and when each other exemplification of a temporal property occurs, namely, now rather than in the past or future, and thereby one avoids predicating incompatible properties of any one of the exemplifications.

But this response concedes too much to Mellor. First, his assumption that if "*e* is past" is tensed, this means that "*e* is now past" is false, for the copula in "*e* is past" is not an indexical, whereas "now" is an indexical, and no nonindexical has the same meaning as an indexical. The regress can be stated in terms of nonindexical A-locutions, such as "The being past of *e* is present," "The being present of the being past of *e* is present," and so on. Second, his assumption that the original sentence "*e* is past" means the same as the sentences stating the further stages of the regress is also false, since it confuses meaning with entailment. "*E* is past" does not mean "The being present of the being past of *e* is present," but entails this. If sentences meant what they entailed, then "John is running" would mean "John, who is running, is neither the number one nor the number two nor the number three, and so on *ad infinitum,*" since "John is running" is true only if John is identical with himself and different from each number. Third, if we are to follow Mellor in holding that "the truth conditions of a sentence *S*" refers to *whatever* must obtain if *S* is true, then every sentence has infinitely complex truth conditions. For "John (is) walking at noon" is true only if 1 + 1 = 2 and 2 + 1 = 3 and 3 + 1 = 4 and so on, *ad infinitum,* and only if John is identical with himself, the property-tie tying John and identity is identical with itself, and so on, *ad infinitum*. But if we are to maintain this inflated notion of a truth condition, we must be careful to distinguish between the state of affairs that corresponds to a sentence (as used on some occasion) and the states of affairs the obtaining of which are the necessary conditions of the truth of the sentence. "John (is) walking at noon" does not correspond to the infinite number of states of affairs in the series 1 + 1 = 2, 2 + 1 = 3, . . . , but to the single, finitely complex state of affairs consisting of John, walking, and noon as ordered by property-ties. Likewise, "*e* is past" does not correspond to an infinite number of inherences of presentness in its own inherences, but to presentness's inherence in the pastness of *e*. "*E* is past" expresses a finitely complex proposition whose constituents bear correspondence relations to a finitely complex state of affairs, although this finite state of affairs cannot obtain unless an infinite number of other states of affairs also obtain.

The theory of infinite tensed facts outlined in this essay will be further developed in the following essay on the infinite regress of temporal attributions.

ESSAY 15

The Infinite Regress of Temporal Attributions

QUENTIN SMITH

The idea that presentness, pastness, and futurity are attributes of events is supposed to entail a vicious infinite regress of tenseless predications. This belief is held not only by McTaggart, its first advocate, but also by most twentieth-century philosophers of time, including both tensers and detensers. The widely adopted solution to this problem is to deny the premise that presentness, pastness, and futurity are attributes of events. Tensers argue that time does consist of presentness, pastness, and futurity, but that these are not *attributes* of events.[1] Detensers deny that presentness, pastness, and futurity are real aspects of time and maintain that time consists only of relations of simultaneity, earlier than, and later than.

The purpose of this essay is to challenge this dogma of twentieth-century philosophies of time and show that the idea that presentness, pastness, and futurity are properties does indeed entail an infinite regress, but that this regress is neither vicious nor constituted of tenseless predications (cf. Section 1). I then argue that the various attempts to show that presentness, pastness, and futurity are real but are neither attributes nor regressive are unsuccessful (cf. Section 2). Finally, I indicate that temporal attributes are not unique but share their infinitely regressive character with other reflexive properties (cf. Section 3).

McTaggart's Paradoxical Interpretation of the Infinite Regress of Temporal Attributions

"McTaggart's paradox" is not in truth the infinite regress of temporal attributions but his (and his critics') interpretation of this regress. That the regress in reality is logically unproblematic can be made apparent through an examination and critique of McTaggart's theory.

The following levels of temporal predication are implied by McTaggart's theory.[2] (In what follows, tenseless copulas are indicated by parentheses.)

First Level

The first level of temporal predications applies to events; every event (is) present, past, and future. Since presentness, pastness, and futurity are incompatible predicates, they cannot belong to the same event simultaneously, but only successively. In tensed language, this means that the event *is* present, *will be* past, and *has been* future, or that it *is* past, and *has been* future and present, or that it *is* future and *will be* present and past. The import of these sentences is made explicit in more complicated tenseless sentences; for example, the statement "The event *is* present, *will be* past, and *has been* future" means "The event (is) present at a moment of present time, (is) past at some moment of future time, and (is) future at some moment of past time." It is tenseless sentences of the latter sort that make explicit the nature of temporal predications on the first level.

Second Level

The second-level predications can be specified if it is pointed out that each of the moments of time introduced on the first level (is) present, past, and future. For example, in the statement "The event (is) present at a moment of present time," the moment of time referred to (is) not only a present moment but (is) also a past and future moment. Since it is contradictory to assert that the moment (is) present, past, and future simultaneously, it must be asserted instead that the moment (is) present at some higher-level present moment, (is) past at some higher-level future moment, and (is) future at some higher-level past moment.

Third Level

The third-level temporal predications are introduced through explicating the characteristics of the second-level moments. Each of the second-level moments (is) present, past, and future, not simultaneously but at different third-level moments. A second-level moment (is) present at some third-level present moment, (is) past at some third-level future moment, and (is) future at some third-level past moment.

Fourth Level

And the same applies to each of these third-level moments, and so on, infinitely.

McTaggart believed that this infinite regress is vicious, which he explained as follows: "Such an infinity is vicious. The attribution of the characteristics past, present, and future to the terms of any series leads to a contradiction, unless it is specified that they have them successively. This means, as we have seen, that they have them in relation to terms specified as past, present, and future. These again, to avoid a like contradiction, must in turn be specified as past, present, and future. And, since this continues infinitely, the first set of terms never escapes from contradiction at all."[3]

These remarks are paradoxical. McTaggart is indeed correct that the attribution of presentness, pastness, and futurity leads to a contradiction unless they are attributed successively. However, in each case it *is* specified that the terms to which they are attributed have them successively. For example, on the first level, it is specified that each event (is) present, past, and future, not simultaneously but at different moments of time; and on the second level it is specified that each moment has these attributes at different higher-level moments of time. It seems to follow from this that none of the attributions of presentness, pastness, and futurity leads to a contradiction. But McTaggart concludes that, since there are infinite levels of predication, "the first set of terms never escapes from contradiction at all." However could it *never escape* from contradiction if it *never was* contradictory? The first set of terms, the events, are contradictory only if it is *not* specified that these terms have presentness, pastness, and futurity successively. But it *is* specified that they have them successively!

McTaggart seems to believe that the first set of terms is contradictory, because he infers from

(1) The attribution of the characteristics past, present, and future to the terms of any series leads to a contradiction, unless it is specified that they have them successively

to

(2) The attribution of the characteristics past, present, and future to the terms of any series leads to a contradiction, which is subsequently resolved by specifying that they have them successively.

This is an invalid inference, for a statement of the form *A unless B* does not entail a statement of the form *A and B*.

If this inference were valid, the levels of predication would be increased, so as to include before each of the levels outlined above another level or sublevel consisting of contradictory predications. The hierarchy would be represented so:

First Level

(3) Each event (is) present, past, and future simultaneously.

Second Level

(4a) Each event (is) present at a present moment, past at a future moment, and future at a past moment.

(4b) However, each moment (is) present, past,and future simultaneously.

Third Level

(5a) Each moment (is) present at a higher-level present moment, past at a higher-level future moment, and future at a higher-level past moment.

(5b) However, each of these higher-level moments (is) present, past, and future simultaneously.

Fourth Level

And so on, *ad infinitum.*

The three simultaneous predications (on the first level and sublevels 4b and 5b are superfluous. Their presence in the hierarchy is unnecessary and unjustified, introduced only as a consequence of the tacit and fallacious inference from assertion 1 to assertion 2. Just because the temporal predicates *would* be predicated inconsistently if predicated simultaneously does not imply that on each level they *are* so predicated, such that the contradiction this involves must be resolved by a new successive predication of them. It is as if every time I predicate blackness and whiteness of something, like a zebra or a newspaper, I must first predicate blackness and whiteness *in the same respect* and then resolve this contradictory predication by a new consistent predication wherein blackness and whiteness are predicated *in different respects.*

Introducing these contradictory temporal predications leads to a misrepresentation of the relation that the genuine and necessary levels of temporal predication have to each level. Each higher level stands to the immediately lower level not as a *resolution* stands to a *contradiction* but as an *analysans* stands to an *analysandum.* If we remove the superfluous contradictory statements, then the first two levels are (4a) and (5a) (corresponding to the first two levels represented earlier). (5a) makes explicit what is implicit in (4a), namely, that the first-level moments are present, past, and future at higher-level present, past, and future moments. (5a) is an analysis of the meaning of the statement that the first-level moments are present, past, and future, a meaning that is not made explicit in (4a) itself.

These considerations support the contention that the infinite regress of genuine and necessary temporal predications is a regress of *analysandum* and *analysans,* not of contradictions and resolutions, and consequently lacks the viciousness that McTaggart attributed to it.

McTaggart makes a second unnecessary assumption: namely, that there is a *hierarchy of levels of moments.* He assumed that for a moment to be present, it must occupy a higher-level present moment such that the answer to the question "When does presentness inhere in the first-level moment?" is "At a second-level moment which (is) present." But a more parsimonious answer can be given to this question: namely, that "Presentness inheres in the first-level moment *at present,*" this answer meaning that the inherence of presentness in the first-level moment is itself present; presentness inheres not only in the moment but also in its own inherence in the moment. To the question "And when does presentness inhere in its own inherence in the moment?" a similar answer is given: "At present"; presentness inheres not only in its inherence in the moment, but also in its inherence in its inherence in the moment. This regress of the inherences of presentness in its inherences continues *ad infinitum.* Accordingly, the infinite regress of attributions of presentness is represented not as

(6) *E* (is) present at a moment which (is) present, and this moment (is) present at a second-level moment which (is) present, and the second-level moment (is) present at a third-level moment which (is) present, and so on, infinitely,

but as

(6a) *E* (is) present at a moment which (is) present, and the (being) present of this moment (is) present, and the (being) present of the (being) present of the moment (is) present, and so on, infinitely.

(6a) is more parsimonious than (6), for instead of postulating an infinite number of levels of moments and predications of presentness, it postulates only one level of moments and an infinite number of predications of presentness.

A third unnecessary assumption can now be eliminated from McTaggart's conception of the regress: namely, that *events occupy moments.* It is possible to do away not only with the levels of moments above the first level but also with the first level itself. Although it is arguable that the facts of immediate experience[4] and science[5] suggest that events in fact occupy moments, it is not logically necessary that they do. There is no contradiction in the idea that presentness inheres in an event and in its own inherence in the event, and so on, infinitely. Accordingly, the regress of attributions of presentness is most parsimoniously represented as

(6b) *E* (is) present, and the (being) present of *E* (is) present, and the (being) present of the (being) present of *E* (is) present, and so on, infinitely.

A fourth, and final, assumption of McTaggart's that can be rejected is that *the temporal predications are predicated by tenseless copulas*. This assumption is incompatible with the goal of providing an analysis of *tensed* predications. If (6b) is to be the complete *analysans* of the present-tensed sentence "*E* is present," it must convey all the information that the latter conveys. But it does not. "*E* is present" conveys the information that *E* is *now* present rather than *was* or *will be* present; but (6b), through predicating presentness tenselessly, does not convey this information. For example, the first conjunct of (6b), "*E* (is) present," does not indicate *when E* has the property of presentness; it does not indicate whether *E was* present, *is now* present, or *will be* present. Analogous considerations hold for each of the other conjuncts of (6b). Consequently, (6b) cannot be the complete *analysans* of "*E* is present."

This situation can be remedied by tensing the copulas in (6b), to read

(6c) *E* is present, and the being present of *E* is present, and the being present of the being present of *E* is present, and so on, infinitely.

(6c) clearly conveys all the information that "*E* is present" conveys, namely, that *E is now* present rather than was or will be present, and thereby can function as the complete *analysans* of "*E* is present."[6]

But it may be wondered whether the elimination of the tenseless copulas also eliminates the need for a regress. A complete analysis of the first conjunct of (6c) shows that this is not the case. In the sentence "*E* is present," *E* stands for an event, "present" stands for the property of presentness, and the present-tense copula stands for the present inherence of the property of presentness in the event. "*E* is present" means

(7) Presentness presently inheres in *E;*

or, more fully,

(7a) Presentness inheres in *E* and in its own inherence in *E*.

But (7a) does not make the sense of "*E* is present" fully explicit. If "*E* is present" is true, then the inherence of presentness in its own inherence in *E* cannot be past or future; it must be present, so that

(7b) Presentness inheres in *E* and in its own inherence in *E* and in its inherence in its inherence in *E*.

It can easily be seen that this inherence of presentness in its own inherences goes on *ad infinitum*.

It should be observed that "inheres" and "inherence" as they appear in (7), (7a), and (7b) must be tensed; for if they are not, these sentences will not convey the same information as "*E* is present." For instance, if the "inheres" in the first conjunct of (7b) were tenseless, this conjunct would not convey that presentness *now* inheres in *E,* and thus it, along with the remaining conjuncts of (7b), would fail to convey what "*E* is present" conveys. The "inheres" in the first conjunct of (7b) must be present-tensed, such that it designates what the second conjunct more explicitly designates, that presentness inheres in its own inherence in *E*.

These remarks about "inheres" and "inherence" show that the relation between "*E* is present" and (7), (7a), and (7b) and between "*E* is present" and its infinite analysis in (6c) is one of logical equivalence but not referential identity. That is, "*E* is present" and (6c), to take two of these sentences, entail each other, but they do not explicitly refer to the same items. Let us say that a sentence explicitly refers to an item if there is a word or phrase in the sentence that stands for it. An item is implicitly referred to by a sentence if there are no words or phrases in the sentence that stand for it but there is a different sentence with such a word or phrase that is logically equivalent to the sentence in question. Thus "The triangle (is) three-sided" explicitly refers to the triangle, the property of three-sidedness, and the inherence of this property in the triangle and implicitly refers to the property of being three-angled and the inherence of this property in the triangle; for "The triangle is three-sided" entails and is entailed by "The triangle is three-angled."[7] Two sentences are referentially identical if and only if they explicitly refer to the same items.

We can thus understand how "*E* is present" and (6c) are logically equivalent but not referentially identical. The former explicitly refers to an event *E,* the property of presentness, and the present inherence of this property in *E*. It implicitly refers to the infinite regress of inherences of presentness in its inherences that stems from the inherence of presentness in its inherence in *E,* for there are no words in this sentence standing for these infinite inherences of presentness; yet this sentence entails and is entailed by a sentence, (6c), in which there is a phrase ("and so on, infinitely") standing for these infinite inherences.

To clarify this notion, the difference between real and merely apparent referential distinctness must be noted. Two sentences merely appear to be referentially distinct if one of the sentences, S_1, has words or phrases in addition to those of the other sentence, S_2, such that these words or phrases do not succeed in explicitly referring to anything to which explicit reference is not made in S_2. The two sentences

(8) *E* is present,
(9) *E* is present, and *E* is present, and *E* is present, and so on, infinitely,

have a merely apparent referential distinctness. Each "*E* is present" in (9) explicitly refers to the same items, namely, the present inherence of presentness in *E*. There is not an infinite number of distinct items to which (9) explicitly refers, but the same few items, items that are referred to over and over again, an infinite number of times. Thus no item is explicitly referred to in (9) to which explicit reference is not made in (8).

Sentence 6c, on the other hand, is really referentially distinct from (8), for it explicitly refers to an infinity of items to which explicit reference is not made in (8). It is not the case that each explicit predication of presentness in (6c) is a predication of presentness of *E* or of the inherence of presentness in *E*. Rather, each explicit predication of presentness predicates presentness of a *different* item. The first conjunct of (6c) predicates presentness of the event *E*, and each of the remaining conjuncts predicates presentness of a *different* inherence of presentness; the second conjunct predicates presentness of the inherence_1 of presentness in *E;* the third conjunct predicates presentness of the inherence_2 of presentness in its inherence_1 in *E;* and so on.

Besides being logically equivalent to, and really referentially distinct from, (8), (6c) has a third relation to (8); it is its *complete explication*. It explicitly refers not only to what (8) explicitly refers to, but also to what (8) implicitly refers to. (By contrast, "The triangle is three-angled" is logically equivalent to, and referentially distinct from, "The triangle is three-sided," but it is not its *complete explication,* since it does not explicitly refer to every item to which the former sentence explicitly refers.)

What are the complete explications of "*E* is past" and "*E* is future"? They are not to be understood on the analogy of (6c): the explication of "*E* is past" is not

(10) *E* is past, and the being past of *E* is past, and the being past of the being past of *E* is past, and so on, infinitely.

This implies that the being past of *E* is itself past, which is false. Events are past not in the past but in the present; the events of a minute ago are not past a minute ago but past in the present minute. The answer to the question "When does pastness inhere in events?" is "At Present." The same is true for the future. Hence, it is the case not only that present events are present at present but also that past and future events are past or future at present. The correct explication of "*E* is past" is

(10a) *E* is past, and the being past of *E* is present, and the being present of the being past of *E* is present, and so on, infinitely.

An analogous complete explication is given to "*E* is future."[8]

Critique of Alternate Tenser Theories

It is the standard view of tensers that the elimination of the tenseless copulas in McTaggart's regress also eliminates the regress itself. This belief was first expressed by C. D. Broad,[9] and was later adopted or developed by G. J. Whitrow, A. N. Prior, F. Christensen, G. Lloyd, and others.[10] The idea is essentially that tensed sentences imply no regress, and that a regress ensues only if an attempt is made to analyze these sentences in terms of tenseless predications of presentness, pastness, and futurity.

I have already argued that tensed sentences like "*E* is present" entail an infinite regress, a regress of temporal properties inhering in their own inherences. In this section, I will defend this argument against a theory espoused by many of the above-mentioned tensers, that a view such as mine is based on the mistaken assumption that the grammatical predicates of sentences like "*E* is present," "*E* is past," and "*E* is future" refer to properties of presentness, pastness, and futurity. If this assumption is rejected, the theory implies, it will be seen that tensed sentences do not refer to temporal properties and, *a fortiori,* that they do not refer to infinitely regressive properties. The infinite regress is eliminated with the elimination of the temporal properties.

The theory in question maintains that "present," "past," and "future" are merely grammatical predicates and that their role in a sentence is most clearly brought out by translating the sentences with the temporal predicates into sentences without them. "The leaf's falling is past," according to this view, is translated as "The leaf was falling," thereby eliminating the apparent reference to a property of pastness.

An analysis of this sort raises more questions than it answers. In particular, one is left baffled about the status of the referent of the *analysans*. In the above example it is clear that "the leaf" refers to a *thing* and "falling" to a *property* possessed by this thing; but what about the referent of "was"? The "was" in part serves as a copula and refers to the *inherence* of the property of falling in the leaf. But surely "was" as a *past-tensed* copula has an additional function, that of indicating the *pastness* of the inherence of the property of falling in the leaf? And what could this pastness be, if not a property of the inherence of falling in the leaf?

Perhaps "was" does not refer to pastness at all. In that case, all that is referred to by "The leaf was falling" is *the leaf, falling,* and the *inherence* of falling in the leaf. Likewise, the present and future-tensed sentences "The leaf is falling" and "The leaf will be falling" do not refer to presentness or futurity but simply to *the leaf, falling,* and the *inherence* of falling in the leaf. But this implies that the past-, present-, and future-tensed sentences have the *same referent,* a referent that is devoid of temporal status. And such an implication is incompatible with a tenser's view of the world.

No illumination regarding the reference of "was" is achieved if "The leaf was falling" is analyzed as "The leaf is falling" prefixed by the sentential operator "It was the case that," so that we have "It was the case that the leaf is falling."[11] Depending upon how the sentential operator is interpreted, it is either about the sentence upon which it operates, "The leaf is falling," or about the referent of this sentence. If the former, the operator indicates that the sentence "The leaf is falling" *was true,* that it *used to refer.* This entails that there *was* a referent of the sentence, there *was* a falling of the leaf. But what could this mean except that the property of falling *formerly inhered* in the leaf? And if this is not interpreted as the inherence of the property of pastness in the inherence of falling in the leaf, how is it to be interpreted?

If the operator is instead interpreted as being about the referent of the sentence upon which it operates, the state of affairs of the leaf falling,[12] it indicates that this state of affairs *was.* An analogous question arises here about the pastness of this state of affairs; if not a property of the inherence of falling in the leaf, what is it?

Ferrel Christensen's version of this theory provides some sort of answer to these questions. He begins by indicating that tenses

> perform a very different sort of task from telling what the nature or properties of an individual are: they tell *when it has* its properties. That is to say, they tell whether it once did, or does now, or will yet possess such-and-such characteristics (or bear such-and-such relations), and also simply whether it did, does, or will exist. Indeed, it is precisely because the kind of information they carry is so different from that which is conveyed by predicates that the tenses have such a different logical form: to tell when the individual named in a sentence has the property (or relation) predicated of it, the tense operator acts in a special way upon both terms together.[13]

According to Christensen, terms standing for present, past, and future positions are "A-terms"; A-names ("the past") and A-predicates ("is past") are reducible to A-adverbs, the tensed element of copulas (such as the past-tensed aspect of the "was" in "The leaf was falling"). He continues: "To assert that A-names and A-predicates are reducible to the A-adverbs is to say that there are no such *individuals* or particulars as the past, the present, and the future, and no such *properties* as pastness, presentness, and futurity."[14] The referents of A-adverbs are not items in "an ontology of individuals and their properties and relations."[15] Christensen does not give a name to these referents, but it will be convenient to call them "A-adverbs" in the ontological sense, that is, referents of "A-adverbs" in the grammatical sense of this term.

I think it can be proved that these ontological A-adverbs are really properties. Christensen rightly says that grammatical A-adverbs "tell when the individual named in a sentence has the property (or relation) predicated of it."[16] If so, there is a distinction between the ontological A-adverb *per se* and its being *of* some individual's possession of its properties. In a tensed predication, it is said when *this* individual possesses its properties, implying that the ontological A-adverb designated is an adverb *of* this individual's possession of its properties and not *of* some other individual's possession of its properties. If the A-adverb's being *of* some individual's possession of its properties is not distinguishable in some sense from the A-adverb itself and the A-adverb itself is all that is designated by the tense of the copula, then it would be false that the past tense of the copula in "The leaf was falling" referred to an A-adverb *of* the leaf's falling. It would instead be the case that this past tense designated an A-adverb *simpliciter,* an A-adverb that stood by itself, unattached to anything, an A-adverb that is not *of* the leaf's falling or anything else.

But what of this *of*? What else could it be but the A-adverb's *modification of, characterization of, qualification of, inherence in* the leaf's falling? And so the ontological A-adverb after all is a property, for a property by definition is just that which *inheres in* (modifies, characterizes, qualifies) some item.

If Christensen nevertheless insisted that these were not *properties* that inhered in an individual's possession of its properties (perhaps because he has decided to use the word "properties" in some especially restricted sense) but ontological A-adverbs that inhered in (or modified) the individual's possession of its properties, I could for the sake of argument concede this point and prove that these A-adverbs nonetheless entail an infinite regress.

Take the ontological A-adverb designated by the past tense of the copula in "The leaf was falling." When does this ontological A-adverb modify the leaf's falling? It cannot modify it in the future, for if it did, it would not yet be true that "The leaf was falling." It cannot modify it in the past, for then it would no longer be true that "The leaf was falling." It cannot modify it permanently, for if the leaf *was* falling, then at some earlier time it is the case that the leaf is falling. So the ontological A-adverb must modify it *at present.* If all A-positions are ontological A-adverbs, then "at present" here refers to an ontological A-adverb referred to by the past tense of the copula in "The leaf was falling." And when does the present A-adverb modify the past A-adverb's modification of the leaf's falling? By the same reasoning, it must be concluded that it modifies it *at present*—and so on, infinitely.

Perhaps a way to eliminate such infinite regresses is to conceive of A-positions as individuals in their own right, as individual events or entities. Instead of saying "The leaf was falling" or "The leaf's falling is past," we can say "The leaf's falling occurred in the past," where "the past" stands for an event or entity of a special sort.

Such a view has been developed by Ian Hinckfuss, although he believes it to be valid only if simultaneity is not relative to a reference frame. "The Present," Hinckfuss suggests, is an "event" or "entity" the coincidence with which explains the presentness of other events or entities.[17] This implies that each present event E is related to "The Present" through the relation of coincidence and, by virtue of being so related, acquires the relational property of *standing in a relation of coincidence with The Present*. The question naturally arises, When does this relational property inhere in E? If E *is* coincident with The Present, the answer must be "At present": that is, the inherence of the relational property in E itself stands in a relation of coincidence with The Present and by virtue of so standing acquires the relational property of standing in this relation. And this second relational property must inhere in the inherence of the first relational property in E *at present,* and so on, infinitely. Consequently, the theory of The Present, The Past, and The Future as individual events or entities fails to eliminate the regress. Moreover, this theory is less desirable than the theory that A-positions are properties, for it is less parsimonious; it postulates A-events or A-entities of a special sort, an infinite number of relations to these A-events or A-entities, an infinite number of relational properties, and an infinite number of inherences. By contrast, the theory that A-positions are properties postulates only these properties and an infinite number of inherences.

I have not in this section discussed detenser theories claiming that tensed sentences are analyzable into, or at least have the same reference as, certain tenseless sentences and that time consists merely of relations of simultaneity, earlier than, and later than, and not of A-positions. If these theories were correct, there would of course be no regressive A-positions. However, I have argued elsewhere that detenser theories of time are mistaken, that tensed sentences possess an ineliminable reference to A-positions, and that A-positions are essential and mind-independent elements of time.[18]

Reflexive Properties

Properties are reflexive or nonreflexive; reflexive properties inhere both in terms and in their own inherence in terms; moreover, they also inhere in their inherence in their inherence in terms, and so on, infinitely. Presentness is reflexive, because if some event is present, then the *being* present of that event is also present, and so on.

Identity and *difference,* understood as first-order properties, are also reflexive. If the property of self-identity inheres in some thing, it also inheres in its own inherence in that thing. For the inherence of self-identity in the thing *is* (identical with) the inherence of self-identity in the thing. If this inherence were not identical with itself, then it would be true that the inherence of self-identity in the thing is different from the inherence of self-identity in the thing, which is contradictory. By the same token, it is also the case that the inherence of self-identity in its own inherence in the thing *is* (identical with) the inherence of self-identity in its own inherence in the thing, and so on, *ad infinitum.* Likewise, if a thing T_1 is different from another thing T_2, the inherence of the dyadic property of *difference* in T_1 is also different from its inherence in T_2, and so is the inherence of *difference* in the inherence of *difference* in T_1, and so on, without end.

If *oneness* is understood as a first-order property of individuals rather than as a property of concepts, then it too must be understood as reflexive. Some thing or event is *one* thing or event, and the inherence of oneness in the thing or event is *one* inherence, and the inherence of oneness in this inherence is itself *one* inherence, and so on, endlessly.

Individuality is also a reflexive property. To be an individual is to be a bearer of properties, to be that in which properties inhere.[19] Not only is a thing or event individual, but the inherence of individuality in the thing or event is individual, for this inherence, like every inherence, is one, self-identical, and different from other items. For any individual, *I,* it is true that

(11) *I* is individual, and the *being* individual of *I* is individual, and the *being* individual of the being individual of *I* is individual, and so on, without end.

Truistic properties, like *round or nonround* or *extended or unextended* are also reflexive. A thing is round or nonround, and its *being* round or nonround is round or nonround. Indeed, the inherence of *round or nonround* in a thing is *round or nonround,* for this inherence is nonround, and if it is nonround, then it is *round or nonround.* And the same applies to the inherence of *round or nonround* in this inherence, and so on, to infinity.

By no means all properties are reflexive, however. John is walking, but the inherence of walking in John is not walking; the sun is yellow, but the sun's *being* yellow is not yellow.

The above account aims to show that presentness belongs to the class of reflexive properties, along with difference, self-identity, oneness, individuality, and the truistic properties, to name a few. The regresses entailed by these properties are benign. A "benign" regress in the sense relevant here is one described as a regress of *analysandum* and *analysans,* an *analysans* being a sentence that makes explicit something implicit in the *analysandum.* For each regress, there is a *complete analysans,* a sentence which explicitly refers (by means of a phrase like "and so on, infinitely") to all the stages of regress. This *complete analysans* is the *complete explication* (in the sense defined earlier) of the original sentence (for example, "*E* is present" or "*I* is individual") that generates the regress; that is, it explicitly refers not only to what the original sentence implicitly refers to, but also to what it explicitly refers to. The complete explication is logically equivalent to, but really referentially distinct from, the original sentence.

There is no reason why these benign regresses cannot exist in reality. The concept of such a regress is not self-contradictory and hence is able to have real instances. The belief that the notion of an actual infinity is self-contradictory or somehow inapplicable to reality I have criticized elsewhere.[20]

Notes

1. There are some exceptions to this among the tensers: e.g., Richard Gale, who believes that pastness, presentness, and futurity are nonsensible properties of events. See his *Language of Time,* esp. chap. 5. However, Gale does not realize the implications of this view, that there must be an infinity of temporal attributions.
2. Cf. McTaggart, *Nature of Existence,* vol. 2; the relevant portions of McTaggart's theory are reprinted in R. M. Gale, ed., *The Philosophy of Time* (Garden City, N.Y., 1967), pp. 86–97.
3. Gale, ed., *Philosophy of Time,* p. 96.
4. Smith, *Felt Meanings,* chap. 4.
5. See, e.g., Michael Friedman, *Foundations of Space–Time Theories* (Princeton, N.J., 1983).
6. I am using "complete analysis" in this essay to mean *complete explication* (as defined later), and therefore I can allow the *analysandum* "*E* is present" to appear in the *complete analysans* (6c).
7. Note that I am using "refers" in a broad sense, to indicate any relation of a word to the item it stands for. Species of reference in this broad sense include *reference in the narrow sense* (as when it is said that names and definite descriptions "refer," but predicates and copulas do not), *expression* (predicates express properties), and *predication* (copulas predicate properties of things or events by standing for the inherence of the property in the thing or event).
8. I do not have the space to show this here, but my arguments in this section suffice to refute not only McTaggart's arguments for the thesis that attributions of presentness, pastness, and futurity entail vicious infinite regresses, but also the recent defenses of McTaggart's thesis by Mellor, Oaklander, and Shorter. See Mellor, *Real Time,* chap. 6; Oaklander, *Temporal Relations,* chap. 2; J. M. Shorter, "The Reality of Time," *Philosophia* 14, nos. 3–4 (Dec. 1984): 321–

339. For a very different but nonetheless sound criticism of Mellor's argument, see David Sanford, "Infinite Regress Arguments," in James Fetzer, ed., *Principles of Philosophical Reasoning* (London, 1984), pp. 93–117.

9. Broad, *Examination of McTaggart's Philosophy,* vol. 2, pt. 1. The relevant portions are reprinted in Gale, ed., *Philosophy of Time,* pp. 117–142; see esp. pp. 139–142.
10. G. J. Whitrow, "Becoming and the Nature of Time," in R. S. Chen and M. W. Wartofsky, eds., *Boston Studies in the Philosophy of Science,* vol. 22 (Dordrecht, 1976), pp. 525–532; Prior, *Past, Present and Future,* chap. 1; Christensen, "McTaggart's Paradox and the Nature of Time"; G. Lloyd, "Time and Existence," *Philosophy* 53 (1978): 215–228; and idem, "Tense and Predication," *Mind* 86, no. 343 (July 1977): 433–438.
11. The theory that ordinary tensed sentences are to be analyzed into tensed sentences prefixed by temporal sentential operators is developed by A. N. Prior in *Time and Modality* (Oxford, 1957); idem, *Past, Present and Future;* idem, *Papers on Time and Tense.*
12. By a "state of affairs" I do not mean a proposition or an abstract object, but a concrete thing or event *qua* possessing some property; e.g., the leaf *qua* possessing the property of falling.
13. Christensen, "McTaggart's Paradox and the Nature of Time," p. 297.
14. Ibid.
15. Ibid., p. 299.
16. Ibid., p. 297.
17. Hinckfuss, *Existence of Space and Time,* p. 98.
18. Cf. Smith, "Mind-Independence of Temporal Becoming"; idem, *Felt Meanings,* chap. 4; idem, "Sentences about Time."
19. I am here using "individual" in the wide sense. In the narrow sense of the term, an individual is an entity (substance) or event, like a cloud or a storm.
20. Cf. Smith, "Infinity and the Past," *Philosophy of Science* 54 (March 1987): 63–75.

ESSAY 16

McTaggart's Paradox and the Infinite Regress of Temporal Attributions: A Reply to Smith

L. NATHAN OAKLANDER

In Essay 15 Quentin Smith attempts to demonstrate "that the idea that presentness, pastness, and futurity are properties does indeed entail an infinite regress, but that this regress is neither vicious nor constituted of tenseless predications." Although I have no quarrel with his thesis that the regress is not constituted by tenseless predications, it is my main purpose here to show that the regress is in fact vicious. I shall do this by arguing that Smith's way out of McTaggart's paradox involves precisely the vicious infinite regress of temporal attributions that his analysis sought to avoid.

According to Smith, the paradox involved in McTaggart's argument for the unreality of time does not involve any contradiction in temporal attribution. Rather, it is McTaggart's own remarks about the infinite regress that are paradoxical, not the infinite regress! Why does he say that? McTaggart claims, correctly according to Smith, that the attribution of presentness, pastness, and futurity leads to a contradiction unless they are attributed successively. Smith continues: "However, in each case it *is* specified that the terms to which they are attributed have them successively. . . . How could it [the infinite levels of predication] *never escape* from contradiction if it *never was* contradictory? The first set of terms, the events, are contradictory only if it is *not* specified that these terms have presentness, pastness, and futurity successively. But it *is* specified that they have them successively!" (p. 182). Smith simply claims that an appeal to succession avoids the contradiction that McTaggart finds in temporal attributions, but he does not offer an argument. As we shall see, however, he needs to give an argument, since McTaggart's point is that an appeal to succession does not suffice to avoid the contradiction.

Smith does not deal directly with McTaggart's argument against the obvious reply to something being past, present, and future. Instead, he proceeds to give an explanation of what, for him, are McTaggart's paradoxical remarks. Smith claims that McTaggart mistakenly believes that the original set of terms is contradictory because he infers from

(1) The attribution of the characteristics past, present, and future to the terms of any series leads to a contradiction, unless it is specified that they have them successively

to

(2) The attribution of the characteristics past, present, and future to the terms of any series leads to a contradiction, which is subsequently resolved by specifying that they have them successively.

Smith then remarks: "This is an invalid inference, for a statement of the form *A unless B* does not entail a statement of the form *A and B*." Clearly, if McTaggart's paradox is based upon that inference, then it is invalid, but there is textual evidence to suggest that McTaggart's belief that the first set of terms never escapes contradiction is not based on that elementary logical blunder.

Smith claims that an appeal to succession nips the problem in the bud or, rather, demonstrates that there is no problem to begin with. Smith makes this point as if McTaggart had never thought of it, which of course he did, treating it as the most obvious apparent solution.[1] Virtually all the points that Smith makes in the first section of his paper depend on the assumption that there is no contradiction in something being past, present, and future. But McTaggart thinks that there is a contradiction in temporal attributions. Therefore, if we are to understand McTaggart's paradox, we must come to see why he thought that the appeal to succession is futile, involving either a vicious circularity or a vicious infinite regress.

The problem of time and change may be briefly stated as follows: what is the proper analysis of the fact that, say, an apple is green at one time and red at a later time or, synonymously, that an apple is green before it is red? In the facts so stated, the apple is first green and then red, not first red and then green; such is the point of saying that it is green before it is red and that there is change in a given direction. For detensers like myself, the direction of time is based on the unanalyzable temporal relation of succession; but for McTaggart, temporal relations are analyzable in terms of the flow of time, or the moving NOW.[2] As McTaggart argues in *The Nature of Existence,* "the series of earlier and later is a time series. We cannot have time without change, and the only possible change is from future to present, and from present to past. Thus, until the terms are taken as passing from future to present, and from present to past, they cannot be taken as in time, or as earlier or later; and not only the conception of presentness, but those of pastness and futurity must be reached before the conception of earlier and later and not *vice versa*."[3] For McTaggart, temporal relations are not there from the outset but are generated by the moving of the NOW along a nontemporal, but ordered, C-series. A C-series is any series whose generating relation is transitive and asymmetrical and which can be misperceived as a B-series.

Further evidence that McTaggart thought of temporal relations as reducible can be gleaned from his remarks in "The Unreality of Time," where he argues that time and change in a particular direction depend upon the A-series and the C-series. As he puts it:

> We can now see that the A series, together with the C series, is sufficient to give us time. For in order to get change, and change in a given direction, it is sufficient that one position in the C series should be Present, to the exclusion of all others, and that this characteristic of presentness should pass along the series in such a way that all positions on the one side of the Present have been present, and all positions on the other side of it will be present. . . . no other elements are required to constitute a time-series except an A series and a C series. . . . It is only when the A series, which gives change and direction, is combined with the C series, which gives permanence, that the B series can arise.[4]

This point cannot be overemphasized: for McTaggart, temporal relations between events are not ultimate but are analyzable in terms of the moving NOW.

With this background, we can begin to understand McTaggart's claim that the appeal to succession to avoid the contradiction contained in temporal attributions involves either a vicious circle or a vicious infinite regress. Appealing to succession involves a vicious circle because "it assumes the existence of time in order to account for the way in which moments are past, present, and future. Time then must be pre-supposed to account for the A series. But we have already seen that the A series has to be assumed in order to account for time. Accordingly, the A series has to be pre-supposed in order to account for the A series. And this is clearly a vicious circle."[5] In short, in order to account for something having incompatible temporal properties, the defender of passage must assume that the term in question has those properties in succession; but in order for a term to be first future, then present, and then past, we must assume that it has incompatible temporal properties. Thus, one cannot appeal to succession in order to explain how time and change are possible without falling into a vicious circle.

To develop this last point further, recall that an account of time must provide an account of, say, a poker's first being cold and then being hot or, synonymously, its being cold before it is hot. McTaggart's account of change involves the claim that every event in the poker's history changes with respect to the properties of pastness, presentness, and futurity. However, A-changes in events can account for time and avoid the incompatibilities problem only if events acquire and shed A-properties successively. Unfortunately, given McTaggart's positive conception of time, this can only mean that *first* the poker's being cold is present and the poker's being hot is future, and *then* the poker's being cold is past and the poker's being hot is present; or, more simply, that the cold poker is present *before* the hot poker is present. As the italicized words indicate, however, time or, more specifically, the temporal relation of *earlier than,* must be assumed in order to account for A-changes in events, "but we have already seen that the A series has to be assumed in order to account for time"[6] (since the B-series is defined in terms of the application of the A-series to the C-series). Consequently, the contradiction involved in the original A-series cannot be avoided by appealing to the relation of succession, since the A-series must be assumed in order to account for succession, and therefore, since the A-series is involved in paradox, succession is too.

McTaggart's difficulty with temporal prediction can be put in another way, in which the fallacy will exhibit itself as a vicious infinite series rather than a vicious circle. If we avoid the original contradiction by claiming that the terms have the incompatible A-properties at different times; that is E is future at t_1, present at t_2, and past at t_3, then the problem is avoided only if t_1, t_2, and t_3 refer to different moments of time. For if the events do not have their A-properties at different times, then they are either timelessly or simultaneously past, present, and future, and the paradox is unavoidable. What, then, is the basis for t_1 being *earlier than* t_2 and t_2 being *earlier than* t_3? Given McTaggart's analysis, it can only be that presentness moves along the series of moments in such a way that each moment is past, present, and future. But then the contradiction in the (first) level of events re-arises at the (second) level of moments. Nor is the resolution to be found in postulating another level of moments at which the preceding level can have its temporal properties. For this new series is genuinely temporal only if its terms occur in a given direction, but the direction of a series is generated by temporal attributions which have not and, I submit, cannot be freed from contradiction.[7]

It should be clear, therefore, that Smith can hardly be thought to have undermined McTaggart's paradox simply by claiming that it does not arise because incompatible temporal properties are had by events, moments, or whatever, successively. Nor has Smith argued, rather than merely claimed, that McTaggart's remarks concerning temporal attributions are paradoxical. What remains to be demonstrated is that Smith's own account of temporal attribution falls prey to precisely the vicious circularity or vicious infinite regress that McTaggart argued was inherent in the nature of time conceived of as involving the moving NOW, or passage.

McTaggart claims that since pastness, presentness, and futurity are incompatible properties, they cannot belong to the same event simultaneously, but must belong to it successively. Notice that this suggests that McTaggart is not claiming, as Smith maintains (Essay 15), that *"temporal predications are predicated by tenseless copulas,"* for if the copula is tenseless, then the temporal properties would not be exemplified at the same time but would simply be exemplified. Of course, there is a difficulty in the tenseless predication of temporal attributes, even if this is not understood as the simultaneous attribution of incompatible properties. But it is equally plausible, if not more so, to treat McTaggart as supposing that temporal attributions are predicated by tensed copulas. Thus, we begin with the three statements "*E* is now present," "*E* is now past," and "*E* is now future," which are mutually contradictory unless it is specified that *E* has these incompatible properties successively. "In tensed language," Smith writes, "this means that the event *is* present, *will be* past, and *has been* future, or that it *is past* and *has been* future and present, or that it *is* future and *will be* present and past" (p. 181). The crucial issue is whether this is anything more than a verbal solution to a metaphysical problem. I shall argue that, given Smith's analysis of temporal predication, the answer is that it is not, since the original contradiction is not avoided but just transferred to different terms.

According to Smith, the reality of temporal attributes implies an infinite regress of inherences of presentness inhering in their own inherences. That is, the correct analysis of "*E* is present" is

> *E* is present, and the being present of *E* is present, and the being present of the being present of *E* is present, and so on, infinitely (Smith's 6c).

He explains this by saying that "the first conjunct of (6c) predicates presentness of the event *E,* and each of the remaining conjuncts predicates presentness of a *different* inherence of presentness; the second conjunct predicates presentness of the $inherence_1$ of presentness in *E;* the third conjunct predicates presentness of the $inherence_2$ of presentness in its inherence in *E;* and so on" (p. 187). This passage makes clear that, for Smith, inherence exemplifies the temporal attribute of presentness. However, if inherence is present, then it must be past and future as well. This does not come out very clearly, since in his analysis of "*E* is past" and "*E* is future," all the inherence relations are present. Smith says that "the correct explication of '*E* is past' is (10a) *E* is past, and the being past of *E* is present, and the being present of the being past of *E* is present, and so on, infinitely. An analogous complete explication is given to '*E* is future'" (p. 187). Nevertheless, the inherence relation is past, present, and future, and, as I shall argue, the appeal to succession or higher-level inherence relations does not enable him to avoid the original contradiction involved in the first set of terms.

Suppose we resolve the difficulty of claiming that *E* is past, present, and future by claiming that it has those attributes successively. *E* is present, was future, and will be past; or it is past and was future and present; or it is future and will be past and present. The copula in each case is tensed, and that leads to a difficulty. Consider the first disjunct of conjuncts. If *E* is present, was future, and will be past, then the inherence of presentness *is now present,* and the inherence of futurity is now *past,* and the inherence of pastness *is now future*. Thus, in order to avoid the difficulty of *E*'s being simultaneously past, present, and future, Smith is forced to claim that *the inherence of a temporal property in E is simultaneously past, present, and future.*

Analogous remarks can be made about the other two disjunctions. For example, if *E* is now past and has been present and future, then the inherence of pastness is now present, and the inherence of presentness and futurity is now past. And, finally, if *E* is future and will be present and past, then the inherence of futurity is now present, and the inherence of presentness and pastness is now future. In either case, the first-order inherence relation has incompatible temporal properties simultaneously. Thus, the notion of succession, analyzed in terms of tensed inherence relations, does not really avoid the contradiction of something being past, present, and future, since it re-arises at the level of inherence.

It would be of no avail to attempt to avoid the contradiction at the first level of inherence by saying that inherence$_1$ has its temporal attributes successively. In tensed language this would mean that inherence$_1$ is present, will be past, and has been future or that it is past and has been future and present or that it is future and will be present and past. However, if, to consider just the first disjunction, inherence$_1$ is present, will be past, and has been future, then, given Smith's analysis of the copula, it follows that the inherence$_2$ of presentness is now present, the inherence$_2$ of pastness is now future, and the inherence$_2$ of futurity is now past. Thus, McTaggart's problem is simply transferred from the first level of inherence to the second level of inherence. Consequently, the resulting infinite regress is in fact vicious, since the original set of terms never escapes from contradiction at all.

One final point. In the second section of his paper Smith argues convincingly that "the various attempts to show that presentness, pastness, and futurity are real but are neither attributes nor regressive are unsuccessful." If his arguments are sound, and if my argument against his "way out" of McTaggart's paradox is successful, then we have provided further evidence for the view that the tensers' account of time rationally ought not to be accepted.[8]

Notes

1. McTaggart, *Nature of Existence,* vol. 2, sec. 330; idem, "Unreality of Time," p. 124.
2. Oaklander, *Temporal Relations.*
3. McTaggart, *Nature of Existence,* vol. 2, p. 271.
4. McTaggart, "Unreality of Time," p. 118.
5. Ibid., p. 124.
6. Ibid.
7. For a detailed examination of McTaggart's paradox and various responses to it see Oaklander, *Temporal Relations,* chaps. 2 and 3.
8. I wish to thank Melvin M. Schuster for his helpful comments on an earlier version of this essay.

ESSAY 17

The Logical Structure of the Debate about McTaggart's Paradox

QUENTIN SMITH

A central issue in the philosophy of time is whether the tensed theory of time is true or false. According to this theory (in the version of it I prefer), events possess not only relations of earlier than, later than, and simultaneity but also monadic and mind-independent properties of pastness, presentness, and futurity. If the tensed theory of time is false and the tenseless theory true, then time consists merely of the relations of simultaneity, earlier than, and later than.

A major area in which the debate between tensers and detensers has been conducted involves the argument known as McTaggart's paradox. McTaggart's paradox, simply put, is that each event possesses incompatible properties of futurity, presentness, and pastness and that appeals to different times at which the event possesses these properties merely reintroduce the incompatible possession of these properties at a higher level. The debate between tensers and detensers about the soundness of McTaggart's paradox has been going on for more than fifty years now, and there is no sign that a resolution is in sight. Indeed, a reader of the literature on McTaggart's paradox might well come away with an impression of futility, a sense that the debate repeatedly ends in the same impasse, with the tensers predictably making a certain move and the detensers predictably responding with a certain counter-move. Tensers typically respond to McTaggart's argument by claiming that there is a certain logical maneuver *M* that enables the contradiction deduced by McTaggart to be avoided. Detensers typically respond to the tensers' maneuver by claiming that *M* merely reintroduces the contradiction in a new guise. Tensers then respond that the same maneuver *M* can be applied to the detensers' response so as to render their imputation of the reintroduced contradiction ineffective. The detensers then deny this, and the seesaw continues, with no apparent resolution in sight.

My aim here is to uncover the logical structure of this seesaw between the tensers' maneuver *M* and the detensers' counter-move and to suggest a way out of the seeming impasse. I shall illustrate this logical structure and possible resolution in terms of a recent debate between Oaklander (on the detensers' side; see Essay 16) and myself (on the tensers' side; see Essay 15). I shall formulate the logical structure and resolution of the debate in terms drawn from our debate; but it should be kept in mind that this structure and resolution could be formulated in suitably different ways so as to apply to other discussions of McTaggart's paradox; for example, the recent discussions by Le Poidevin, Mellor, Lowe, Christensen, Schlesinger, Shorter, Horwich, and others.[1]

The first move in the debate about the soundness of McTaggart's argument is made by the tenser and is a criticism of one of the premises of McTaggart's argument. The premise is that if the tensed theory of time is true, then for each event *E* it is true that

(1) Event *E* is past, present, and future.

According to McTaggart, (1) is self-contradictory, since the three properties are incompatible. Note that this evaluation requires (1) to be understood as predicating the three properties nonsuccessively, since only in this case would (1) be self-contradictory. This is admitted by McTaggart: "the attribution of the characteristics past, present and future to the terms of any series leads to a contradiction, *unless it is specified that they have them successively.*"[2] But there is some ambiguity about (1)'s nonsuccessive predication of the three properties. If (1) predicates the three properties nonsuccessively, then it predicates them either *simultaneously* or *timelessly.* If it predicates them simultaneously, it is self-contradictory, since *E* cannot be past, present, and future at one and the same time. If it attributes the three properties to *E* both nonsuccessively and nonsimultaneously, then it attributes them to *E* timelessly, which is self-contradictory since *E* cannot be timelessly past, present, and future. McTaggart is not perfectly clear as to which of these two interpretations of (1) he adopts, and I will leave this issue undecided, since it does not affect the argument here.[3] It suffices to say that McTaggart regards (1) as self-contradictory, since it predicates the three properties *nonsuccessively* of *E*.

McTaggart proceeds to argue that the contradiction in (1) cannot be eliminated by moving to a higher level at which the properties are possessed successively. The expansion of (1) into

(2) *E* is past at a future moment, present at a present moment, and future at a past moment

is unsuccessful, since it is true of each of these moments *m* that

(3) Moment *m* is past, present, and future.

(3) is no less self-contradictory than (1), since *m* cannot be nonsuccessively past, present, and future.

The response by the tenser to this argument is to question McTaggart's original assumption that the tensed theory of time implies that (1) is true of each event *E*. This response seems warranted, since McTaggart provides no justification for this assumption. He simply asserts it. As I have stated, McTaggart admits that "the attribution of the characteristics past, present, and future to the terms of any series leads to a contradiction, *unless it is specified that they have them successively.*" But he goes on to claim that the tensed theory of time is incoherent, since the contradiction expressed in (1) cannot be eliminated by (2), since (2) entails the contradiction (3). Due to the reappearance of the contradiction, "the first set of terms [the events] never escape from contradiction at all.[4] But this claim is warranted *only if it has already been assumed* that the tensed theory of time entails that the terms possess the three properties nonsuccessively. Without this initial assumption, there is no contradiction in the first set of terms that needs to be escaped. The tenser of course rejects this initial assumption; according to him, the tensed theory of time entails not (1) but

(4) *E* will be past, is now present, and was future; or *E* is now past and was present and was (still earlier) future; or *E* is now future and will be present and will (still later) be past.

This is consistent, since the properties of pastness, presentness, and futurity are ascribed to *E* at different times. For example, the first disjunct says that *E* possesses pastness in the future, presentness in the present, and futurity in the past. This implies an infinite regress, but the regress is not the vicious infinite regress that McTaggart assumes to obtain. Consider, for example, the statement "*E* is now present." According to my analysis, this asserts that the property of presentness inheres in the event *E* and also in its own inherence in *E*. This analysis may be understood as a way of responding to the question "When does the property of presentness inhere in *E*?" Manifestly, if *E* is present, this property does not inhere in *E* in the past or the future but in the present. This means that presentness, not pastness or futurity, inheres in the inherence of presentness in *E*. A similar question can be raised about the latter inherence. When does presentness inhere in its own inherence in *E*? The answer is "At present." This means that presentness inheres not only in *E* and its inherence in *E,* but also in its inherence in its inherence in *E*. This regress continues infinitely, but in a benign manner, since at no stage in the regress is there a contradiction. This regress is also compatible with the other regresses that obtain if *E* is present. We need only consider one of them, that implied by "*E* will be past." If *E is now* present, then it *will be* past. This latter clause means that futurity inheres in the pastness of *E*. But when does futurity inhere in the pastness of *E*? The answer is already implicit in the present tense of "inheres" in the statement that "futurity inheres in the pastness of *E*." Futurity *now* inheres in the pastness of *E*. "*E* will be past" implies that "*E is now* such that it will be past." In terms of property-inherences, this means that presentness inheres in the inherence of futurity in the inherence of pastness in *E*. But when does presentness inhere in the inherence of futurity in the inherence of pastness in *E*? At present, since it is *right now* that the pastness of *E* is now future. The rest of this regress, it is apparent, is a regress of the inherences of presentness in its own inherences. This bespeaks the predominance of presentness in the regresses, which is to be expected, since if anything is past or future, it is *presently* past or future, and if anything is present, it is present *at present*. Despite the complexity of these regresses, I think it is intuitively clear that the two regresses described are compatible with each other and can obtain simultaneously:

First Regress (implied by "*E* is now present"): Presentness inheres in *E* and in its own inherence in *E* and in its own inherence in its inherence in *E* and so on, infinitely.

Second Regress (implied by "*E* will be past"): Pastness will inhere in *E,* and futurity inheres in the inherence of pastness in *E,* and presentness inheres in the inherence of futurity in the inherence of pastness in *E,* and presentness inheres in its own inherence in the inherence of futurity in the inherence of pastness in *E,* and presentness inheres in its own inherence in its own inherence in the inherence of futurity in the inherence of pastness in *E,* and so on, infinitely.

Let us pause at this juncture and recall that it is my intent to uncover the underlying logic of the tenser/detenser debate about McTaggart's paradox. I have illustrated above the first move in the debate, which is the maneuver *M* that the tenser makes to avoid the contradiction imputed by McTaggart. The second move is that of the detenser, which is to claim that the maneuver *M* reintroduces the very problem it was designed to avoid. We find that this, in fact, is the move made by Oaklander in his response to my maneuver *M* outlined above. Oaklander writes: "In order to avoid the difficulty of *E*'s being simultaneously past, present, and future, Smith is forced to claim that *the inherence of a temporal property in E is simultaneously past, present, and future*. . . . Thus, the notion of succession, analyzed in terms of tensed inherence relations, does not really avoid the contradiction of something being past, present, and future, since it re-arises at the level of inherence" (p. 200). Note that Oaklander regards the contradictory and nonsuccessive possession of the three temporal properties to be a simultaneous, rather than a timeless, possession. This interpretation may be adopted for the sake of responding to Oaklander's argument.

Let me point out that this passage is misleading insofar as it suggests that I *claimed* that the inherence of a temporal property is simultaneously past, present, and future, since a perusal of my essay will show that this claim is nowhere made. This passage can be more charitably interpreted as meaning merely that my theory *implies* that the inherence is simultaneously past, present, and future, even though I do not make this claim myself. Interpreted in this way, this passage may be taken as illustrating the second move made in the tenser detenser debate, the assertion that the tenser's maneuver *M* implies the original contradiction, albeit in a new guise.

Let me pass immediately to the third move in the debate, the counter-assertion made by the tenser to the detenser's assertion that maneuver *M* reintroduces the contradiction. The third move is that the detenser's assertion is itself invalidated by the very maneuver *M* that the tenser had introduced. This is, in fact, my response to the passage quoted from Oaklander's essay. Oaklander's assertion that the contradiction re-arises at the level of inherence is itself based on the unjustified assumption originally made by McTaggart that the tensed theory of time entails that pastness, presentness, and futurity are possessed nonsuccessively (simultaneously, in Oaklander's reading). Oaklander assumes that

(5) The inherence of presentness in *E* is now present

entails

(6) The inherence of presentness in *E* is simultaneously past, present, and future.

The fact that Oaklander assumes this is evinced by his assertion (Essay 16) that "if inherence is present, then it must be past and future as well," which Oaklander regards as implying that "the first-order inherence relation has incompatible temporal properties simultaneously." *But Oaklander gives no justification for this assumption*. Like McTaggart he simply *asserts* that whatever possesses the three temporal properties must possess them nonsuccessively. By importing this unjustified and foreign assumption into the tensed theory of time, Oaklander, like McTaggart, proceeds to deduce the incoherence of the tensed theory. But this assumption is not part of the tenser's theory of time. According to this theory, something possesses the three incompatible properties only successively. If the inherence of presentness in *E* is present, that does not imply the self-contradictory sentence (6) but the self-consistent sentence

(7) The inherence of presentness in *E* is *now* present, *was* future, and *will be* past.

If my account of the logical structure of the tenser/detenser debate about McTaggart's paradox is sound, then I should be able to predict Oaklander's future response to this criticism, which would amount to the fourth move in the debate. I predict that his response will be that I have merely asserted, without argument, that (5) does not entail (6) and that (7) is self-consistent. If he makes this response, he will be right. For I have not *argued* that (5) does not entail (6); I have merely claimed, without argument, that (5) does not entail (6) and that Oaklander has merely claimed, without argument, that (5) entails (6). Are we then at an impasse, each of us asserting without argument the opposite of what the other asserts without argument?

Russell once wrote that sometimes "one is confronted by one of those difficulties that occur constantly in philosophy, where you have two ultimate prejudices conflicting and where argument ceases."[5] But I believe that the point we have reached in the debate does not represent one of those difficulties and that there is a way out of the impasse. We can resolve the debate about McTaggart's paradox in one of two ways. In the first way, we can *stipulate* that tensed copulas and the predicates "past," "present," and "future" have a certain meaning and draw the consequences of this stipulation. This would resolve the conflict, for the detenser could stipulate that the meaning of sentence (5) is such that (5) entails (6), and the tenser could stipulate that the meaning of (5) is such that (5) does not entail (6) but does entail (7). If this sentence is stipulated by the two different parties to have two different meanings, there can be no disagreement, for stipulations cannot be true or false.

The second way to resolve the conflict is to argue that the empirical evidence about the use of (5) in natural languages is evidence that (5) possesses a certain meaning and that this meaning corresponds, as the case may be, to the tensers' or the detensers' theory. For example, the tenser would argue that tensed sentences as they are used in natural language do not entail that the temporal predicates are possessed nonsuccessively, and that this fact counts in favor of the tensers' theory of time. This way of resolving the conflict is of considerably more philosophical interest than the first way, since it enables a decision to be made between tensers' and detensers' theories of time. If the tenser could successfully make his case about these temporal predicates, this would contribute to establishing that the following argument is sound:

(8) Tensed sentences in their ordinary uses ascribe to events properties of futurity, presentness, and pastness.

(9) It is a rule of the ordinary uses of these sentences that these properties are ascribed to events in a compatible manner, that is, successively rather than nonsuccessively.

(10) Some utterances of these tensed sentences are true.

(11) Therefore, events possess, in a compatible manner, properties of pastness, presentness, and futurity.

Clearly (11) is a philosophically significant claim, since it is just the claim that the tensed theory of time is true. Thus, if we pursue this second way of resolving the conflict, we may be in a position to accomplish much more than we could ever hope to accomplish by pursuing the first, stipulative manner of resolution.

Let me begin by presenting some linguistic evidence that supports (9). Suppose that a person *P* uttered the sentence "The lightning stroke is now past" a few seconds after a stroke hit the ground. If the detensers' thesis endorsed by Oaklander were true of ordinary uses of tensed sentences, then it would be a rule of use of this sentence that if the sentence is true, then it is also true that the stroke is simultaneously present and future. But imagine a person *Q* responding to *P:* "I agree that the stroke is now past. This implies, naturally, that the stroke is also present. And this very stroke, the one that just passed away, is in addition in the future. As we all know, the stroke is *simultaneously* past, present, and future." This response is, of course, an outrage. *Q* would be regarded by *P* as either joking or insane. It is *obvious* that the rule of the ordinary use of this sentence does not imply that the stroke, if past, is *simultaneously* present and future. It is equally obvious that this rule does not imply that the storm, if past, is *timelessly* past, present, and future. It is a rule of the ordinary use of temporal predicates that they are satisfied by events successively, not simultaneously or timelessly.

But this conclusion is far from entailing that the tensed theory of time is true. The detenser could admit that the three predicates as ordinarily used obey the rule that they are satisfied by events successively but deny that these predicates are ordinarily used to ascribe properties of presentness, pastness, and futurity. The detenser could argue that these predicates, and tensed locutions in general, are ordinarily used to ascribe dyadic properties of earlier than, later than, and simultaneity. For example, he could claim that "*E* is past" means "*E* is earlier than this utterance." Thus, the next step that the tenser needs to take is to argue that the linguistic data about the ordinary uses of these predicates and other tensed locutions are consistent only with the hypothesis that they are used to ascribe properties of futurity, presentness, and pastness. I have presented some arguments on behalf of this thesis elsewhere[6] and believe, on the basis of these arguments, that the tensers' explanation of the use of these locutions is preferable. But I will acknowledge that even if this thesis is true, it does not entail that the tensed theory of time is true. For the ordinary uses of tensed sentences might all be false. Thus, I need the additional premise that "some utterances of tensed sentences (in their ordinary use) are true." I have not directly presented any arguments for this premise[7] and will not offer any here; I will simply point out the consequences of denying it. If it is denied, then it can never be uttered with truth that "I am now thinking," "The storm is past," or "I will go to the library tomorrow." Given the choice between this extreme skepticism and the tensed theory of time, I think that even the most diehard detenser might embrace the temporal phenomenon he most fears, the moving NOW.[8]

Notes

1. Le Poidevin and Mellor, "Time, Change, and the 'Indexical Fallacy'"; Lowe, "Indexical Fallacy"; idem, "Reply to Le Poidevin and Mellor"; Christensen, "McTaggart's Paradox and the Nature of Time"; Schlesinger, *Aspects of Time,* chap. 3; Shorter, "Reality of Time"; P. Horwich, *Asymmetries in Time* (Cambridge, Mass., 1987), chap. 2.
2. McTaggart, "Time," in Gale, ed., *Philosophy of Time,* p. 96; my italics.
3. It is arguable that he implicitly adopts the simultaneous interpretation, for he mentions this one (e.g., "the characteristics are only incompatible when they are simultaneous": ibid., p. 95) and not the timeless one. In Essay 15, I adopted the simultaneous interpretation, and Oaklander also adopted it in Essay 16. However, the choice between the simultaneous and timeless interpretations does not affect the soundness of McTaggart's argument, and I will leave both alternatives open in this paper.
4. Ibid. For an analysis of the invalid inference that seems to motivate McTaggart's assumption, see Essay 15. This analysis assumes the simultaneity interpretation, but a similar analysis can be constructed that uses the timeless interpretation.
5. Bertrand Russell, *Logic and Knowledge* (New York, 1956), p. 181.
6. See Smith, "Mind-Independence of Temporal Becoming"; idem, "Impossibility of Token-Reflexive Analysis"; idem, "Sentences about Time"; Essays 2, 12, and

34; idem, "Tensed States of Affairs and Possible Worlds'; idem, "Multiple Uses of Indexicals."

7. In *Felt Meanings,* sec. 20, ii, 131–134, however, I implicitly argued for it.
8. I am grateful to William Vallicella for helpful comments on an earlier version of this essay.

ESSAY 18

McTaggart's Paradox Revisited

L. NATHAN OAKLANDER

In Essays 15–17 Quentin Smith and I debated the question of whether McTaggart's paradox demonstrates that the tensed theory of time (according to which there are monadic, mind-independent [A-]properties of *pastness, presentness,* and *futurity*) involves a contradiction. Smith argued that McTaggart's paradox is a nonstarter, since it makes the ***unwarranted assumption*** that the tensed theory entails that events possess the three monadic temporal properties *nonsuccessively.* What I shall attempt to show here is that, on Smith's analysis of the tenses, the assumption in question is indeed justified and that McTaggart's paradox is thus unavoidable.

First, a quick review of the opening moves. On the assumption that A-properties exist and that there is no first or last moment of time, it follows that every event is past, present, and future. Of course, no event has more than one of those properties simultaneously or timelessly; rather, it exemplifies each such property *successively.* This fact may be represented linguistically by the use of tenses. Thus, the facts stated by "Event *E* is past, present, and future" are (allegedly) adequately represented by the following disjunction: "*E is now* past, *was* present, and *was* future; or *E is now* present, *was* future, and *will be* past; or *E is now* future and *will be* present and *will be* past." So far, so good; but what do the tenses signify in these transcriptions?

Consider "*E* is now present" and "*E* will be past." According to Smith, the truth of each statement implies an infinite regress of inherence relations, but "the two regresses described are compatible with each other and can obtain simultaneously":

First Regress (implied by "*E* is now present"): Presentness inheres in *E* and in its own inherence in *E* and in its own inherence in its inherence in E and so on, infinitely.

Second Regress (implied by "*E* will be past"): Pastness will inhere in *E,* and

> futurity inheres in the inherence of pastness in *E*, and presentness inheres in the inherence of futurity in the inherence of pastness in *E*, and presentness inheres in its own inherence in the inherence of futurity in the inherence of pastness in *E*, and presentness inheres in its own inherence in its own inherence in the inherence of futurity in the inherence of pastness in *E*, and so on, infinitely. (p. 205)

Earlier, I argued that these two regresses are not compatible, since they imply that the first-order relation of inherence *is now* present and *is now* future. I therefore concluded that the inherence relation has incompatible properties *simultaneously*. In his response, Smith claimed that I provided no justification for that conclusion. In what follows I shall attempt to provide one.

Consider, once again, the sentence "Event *E* will be past." On Smith's analysis, the inherence of pastness in *E* is such that futurity *now* inheres in it. To state the same analysis somewhat differently, *E* exemplifies$_1$ pastness, and exemplification$_1$ exemplifies$_2$ futurity, and exemplification$_2$ exemplifies$_3$ presentness. As Smith himself puts it: "If *E is now* present, then it *will be* past. This latter clause means that futurity inheres in the pastness of *E*. But when does futurity inhere in the pastness of *E*? . . . Futurity *now* inheres in the pastness of *E*. . . . In terms of property-inherences. *this means that presentness inheres in the inherence of futurity in the inherence of pastness in* E" (p. 205, my emphasis in last sentence).

The crucial—and fatal—move in Smith's analysis is the claim that the inherence$_2$ of futurity in the inherence$_1$ of pastness in *E is present*. For if the second-order inherence, or exemplification$_2$, *is now* present, then it *exists now*. However, if exemplification$_2$ *exists now*, then the term, in this case exemplification$_1$, that exemplifies$_2$ futurity must also *exist now*. Admittedly, I am assuming that if an inherence or exemplification "relation" (of any level *i*) *exists now*, then there must also *exist now* a term that exemplifies$_i$ a property; but this assumption seems eminently reasonable, if not necessarily true. Can anyone intelligibly maintain that the exemplification between *a* and *F* exists now but that either *a* or *F* does not exist now? I suggest not. Consequently, if the "tie" (exemplification$_2$) between exemplification$_1$ and futurity exists now, then it must be the case that exemplification$_1$ exists now. However, if exemplification$_1$ exists now, then it must be present. Since, by hypothesis, exemplification$_1$ is future, it follows that exemplification$_1$ is both present and future, or does now exist and does not now exist, and that is a contradiction.

It is no use trying to avoid this contradiction by claiming (a) exemplification$_1$ is future and will be present, or (b) exemplification$_1$ is present and was future, or (c) exemplification$_1$ is past and was present and (still earlier) future. In the first place, (a) contradicts the original assumption that *E is now* present, and (b) contradicts the assumption that *E will be* past. For if exemplification$_1$ is future, then it cannot exemplify$_2$ presentness as it must if *E* is now present. On the other hand, if we suppose that exemplification$_1$ is now present, meaning that exemplification$_1$ exemplifies$_2$ presentness and exemplification$_2$ exemplifies$_3$ presentness, then exemplification$_1$ cannot now exemplify$_2$ the property of being future, and that contradicts the assumption that *E* will be past. Finally, (c) contradicts the assumption that *E* was future. For if exemplification$_1$ is past, then exemplification$_1$ exemplifies$_2$ pastness, and exemplification$_2$ exemplifies$_3$ presentness. However, if exemplification$_2$ exemplifies presentness, then not only exemplification$_2$, but also exemplification$_1$, must now exist. That, however, contradicts the assumption that *E was* future, that exemplification$_1$ is past. For if exemplification$_1$ now exists, then it must exemplify$_2$ presentness.

Furthermore, if exemplification$_1$ is future and will be present, then Smith's analysis implies (1) exemplification$_1$ exemplifies$_2$ futurity, exemplification$_2$ exemplifies$_3$ presentness, and exemplification$_3$ exemplifies$_4$ presentness; and (2) exemplification$_1$ exemplifies$_2$ presentness, exemplification$_2$ exemplifies$_3$ futurity, and exemplification$_3$ exemplifies$_4$ presentness. But then, if we assume, as surely we must, that if exemplification$_3$ is now present, then exemplification$_2$ now exists, (2) implies that exemplification$_2$ is both present and future, or both now and not now. An analogous paradox (exemplification$_2$ is both present and past, or now and not now) occurs if we maintain that exemplification$_1$ is present and was future or if we say that exemplification$_1$ is past and was present. Clearly, we cannot avoid these difficulties by appealing to the different tenses involved in exemplification$_2$ having incompatible temporal properties, since this would give rise to precisely the same contradictions at the third level of exemplification or inherence.

I conclude, therefore, that Smith's account of tensed exemplification does not explain how an event (or the exemplification of an A-property by an event) can have the properties of pastness, presentness, and futurity *successively.* However, if an event or inherence relation must have its monadic temporal properties *nonsuccessively,* then the contradiction and the ensuing vicious infinite regress contain reasonable grounds for inferring, as McTaggart did, that A-properties do not exist.

ESSAY 19

Temporal Becoming

GEORGE SCHLESINGER

The Controversy Concerning the Existence of A-Statements

Here we shall consider what may fairly be called the profoundest issue in the philosophy of time: the status of temporal becoming. Some philosophers have even regarded it as the profoundest issue in all of philosophy; C. D. Broad, for instance, thought so and kept returning to the problem of temporal becoming throughout his life.

According to a view deeply ingrained in all of us, a view explicitly championed by McTaggart, the NOW is something that moves relative to the series of points that constitute time. Temporal points from the future, together with the events that occur at those points, keep approaching the NOW and, after momentarily coinciding with it, recede further and further into the past. The NOW, of course, is conceived not as some sort of object, but rather as the point in time at which any individual who is temporally extended is alive, real, or Exists with a capital *E*. I may be occupying all the points between the year 1900, my date of birth, and 2000, the date of my departure from this world, but only one point along this 100-year chunk of time is of paramount importance at any given instant, namely, the point that is alive in the present, the point that exists not in my memory or in anticipation, but of which I am immediately aware as existing in the present.

A typical event, on this view, is in the distant future to begin with; then it becomes situated in the less distant future; it keeps approaching until it becomes an event occurring in the present. As soon as this happens, the event loses its presentness and acquires the property of being in the near past. The degree of its pastness continually increases. Thus, events approach us (by "us" I mean that temporal part of our temporally extended selves which is subject to our direct awareness) from the distant future, become present, and then recede further into the past.

According to Bertrand Russell and his followers, this is a completely false picture. No event has the monadic property of being in the future, as such, to begin with. Consequently, it can never shed this property. An event, E_1, may occur later than some other event, E_0, but if this is so at all, then it is true forever that E_1 occurs later than E_0. Neither can any event be in the past. E_1 may be earlier than E_2, but once more, if this is so, then the fact that E_1 occurs earlier than E_2 is an eternal fact. Indeed, all the temporal properties of events and moments are permanent. E_1 has the unchanging relationship of either before or after or simultaneous with every other temporal entity in the universe. Apart from moments and the events that occur at them, there is no extra entity such as the NOW to which E_1 may have a changing relationship. Also, E_1 is as real at t_1 as E_0 is at t_0 and E_2 at t_2; that is, all events are equally real and alive at the times at which they occur and not at others, and they do not come momentarily to life as they are embraced by the NOW.

Opponents of Russell regard his temporal universe, which admits no transient properties at all, as essentially impoverished, whereas Russellians hold that their opponents admit into their universe nonexistent properties. The fascinating thing about this controversy is that although it is by no means about some remote aspect of the world—on the contrary, it concerns a most immediate and constantly encountered feature of the empirical universe—it nevertheless cannot be resolved within the scope of ordinary observation or scientific experimentation. Only through philosophical analysis does there seem to be any hope of making some progress toward the resolution of this fundamental controversy affecting one of the most ubiquitous aspects of the universe in which we live. A preliminary requirement, of course, is to understand exactly the two views, both of which have been subject to serious misinterpretation.

The controversy concerning temporal relations expresses itself also in argument about what kinds of temporal statements exist. According to McTaggart,[1] there are two fundamentally different kinds of temporal statements: A-statements and B-statements. The latter are the more familiar kind, for B-statements, like all statements in general, have permanent truth-values. "E_1 is before E_2" is a typical B-statement, which, if true at any time, is true at all times, and if false at any time, is false at all times. A-statements, on the other hand, are statements whose truth-value is subject to change. "E_1 is in the future" is an example of an A-statement, since it is true if asserted at any time which is earlier than the occurrence of E_1 but false if asserted at any other time.

McTaggart considers it essential that there be A-statements, for in their absence there is no possibility for change, and time would not be real if it did not permit change. But change occurs only when a fact that obtained at one time ceases to obtain at another; or, to put it differently, when a given statement that was true at one time becomes false at another, or vice versa. Russell tried to argue that A-statements may be dispensed with, because changes may be expressed with the aid of B-statements alone as, for example, in the case of a poker that is hot at t_1 but cold at t_2 and thus undergoes a change which manifests itself in the fact that "The poker is hot at t_1" is true, whereas "The poker is hot at t_2" is false. To this, McTaggart objected that no genuine change in the properties of the poker has been expressed with the aid of these sentences, because the first statement is true and never ceases to be true, whereas the second statement is eternally false. In other words, it has been a fact and will always be a fact that at t_1 the poker is hot, and, similarly, it is an unchanging fact that at t_2 the poker is not hot. Only the truth-value of the A-statement "The poker is hot now" really undergoes a change, for the statement is true when asserted at t_1 but false when asserted at t_2.

According to McTaggart, A-statements are those that refer exclusively to temporal properties of events or moments, and in no other domain do we encounter any such peculiar statements. For example, "*O* is here" does not have the feature characteristic of A-statements that it changes its truth-value. At first glance this may not be clear, for it may seem that it is true when asserted at the place where *O* is but false when asserted elsewhere. This, however, is not really so. I am fairly certain that McTaggart would accept the analysis according to which, when asserted at two different places, "*O* is here" amounts to two different assertions. The correct analysis of "O is here" is "O is at the place where I am," so when I am at p_1 the proposition in effect asserts that *O* is at p_1; but when I am at p_2 then, through the same words, I assert that O is at p_2. Thus, if *O* is in fact at P_1 and so am I, and I utter "*O* is here," then I make a true assertion which remains always true. If at p_2 I again utter "*O* is here," then I make a false assertion, but one which is different from the assertion made at p_1; for now I am asserting in effect that *O* is at p_2, and this is false and was false in the first place.

According to Russell there are no A-statements.[2] All statements have permanent truth-values. To a sentence such as "*E* is in the future," Russell applies basically the same kind of analysis as everybody does to a sentence such as "O is here"; namely, that when uttered at different times it expresses a different proposition.[3] One variation of this kind of analysis is due to Reichenbach and is also embraced by several other philosophers, among them Smart. According to this, "*E* is in the future" is reduced to the B-statement "E_1 is after the event of the utterance of this token," where "this token" refers to the sentence-token just being uttered. Consequently, when this sentence is uttered on two different occasions, once before E_1 and the second time after E_1, the first time it is asserted, the proposition is true and is unalterably so. The second time the proposition is asserted, it is a different one, because, unlike the first proposition, which claimed that E_1 is later than the first token, it claims that E_1 is later than the second token. The second proposition is now and has always been *false.*

The Advantages and Disadvantages of the Two Positions on the Nature of Time

The strongest reason, however, for preferring McTaggart's view to Russell's is the deeply entrenched impression, shared by all, of the transience of time and the generally held belief that time is moving. According to Russell, there is no room for any transience, as all temporal relations between events themselves and events and moments are permanent, and no temporal particular changes its fixed position in the temporal series of moments. According to McTaggart, however, it is possible to look upon the NOW as a particular which shifts its position relative to the series of events in the direction of the future. This movement is manifested by the fact that at one stage it is a fact that E_1 is in the future, which means that E_1 is at a point in time which is later than the time at which the NOW is situated. Yet, at another stage, this ceases to be a fact, and the NOW reaches the position in time at which E_1 is situated; and the two are simultaneous when, of course, it becomes true that E_1 is in the present.

Now whereas nobody denies that a deeply felt impression that time indeed flows relative to the present is a part of our mental makeup, many philosophers have already cited very strong reasons for why this impression must be mistaken. After all, if there really was a relative movement between the NOW and the series of moments, it would make sense to ask how fast this movement takes place. A moment's reflection, however, reveals that it is not because we lack this or that information that we cannot provide an answer to this question, but because it is in principle impossible to measure the speed of this movement, which therefore makes it necessary to deem it nonexistent.

A second, even stronger argument consists in pointing out that a movement always essentially involves two series, so that points in one may be correlated to points in the other. For example, when a car is moving along a road, this motion is embodied in the fact that one position of the car in the series of spatial points corresponds to a given point in the series of moments, while a second position of the car in the same series of spatial points corresponds to another point in the series of temporal positions. But how could the movement of the NOW along the series of moments be realized? What other series is there in which two different points correspond to any two positions that the NOW occupies along the time series?

Another famous objection to the belief that time flows relative to the present is due to Broad. When a car reaches a given point in space, that is one event; when it reaches another point, that is another event. It is these kinds of events that form the elements of moments that constitute our time series. When the NOW reaches a given point in this series of moments, that must also be some kind of an event, but one that surely cannot be a member of the very set which constitutes that moment. Thus, unless we are prepared to introduce an additional meta-series of moments which are made up of these events, we must deny the reality of these events and resign ourselves to the fact that the NOW hitting moments in time is not something that really occurs.

It should be mentioned that Broad notes that the difficulties attending the notion of temporal becoming could be overcome if we were prepared to postulate a higher-order series of moments. We may see at once that this is so. The movement of the NOW in the standard series of time may be explicated by explaining that the NOW is at t_1 in the ordinary series when it is at T_1 in the super-series and at t_2 in the ordinary series when it is at T_2 in the super-series. We may even assign a value to the speed of the NOW; it moves from t_1 to t_2 at the average speed of

$$\frac{t_1 - t_2}{T_1 - T_2}.$$

As to the third objection, the event of the NOW reaching t_1 may, if we like, be looked on as taking place in super-time. Naturally, we should be very reluctant to postulate a whole new series of temporal points. But Broad warns us that the difficulty is far more serious than that. The introduction of one extra temporal series will not solve our difficulties. If we cannot make sense of the first series without postulating a moving NOW, which in turn requires that we postulate a second temporal series, then we shall inevitably find it conceptually necessary to postulate a moving NOW for this new series as well, which, in turn, will commit us to a third series, and so on.

But here Broad moves far too quickly. Not only does he not bother to show that the regress is vicious; he also fails to show that there is any regress at all. It is by no means clear that if we want to endow the new series with a moving NOW, we will have to postulate yet another series. For just as the second series could be instrumental in helping to make sense of temporal becoming in the first series, in the same manner the first series could serve as the extra series through the use of which temporal becoming in the second series makes sense. For example, it might be said that the NOW in the second series is at T_1' when it is at t_1' in the first series and at T_2' when it is t_2', leading to the claim that the average speed of the NOW in the second series from

$$T_1' \text{ to } T_2' \text{ is } \frac{T_1' - T_2'}{t_1' - t_2'}.$$

In order to refute McTaggart conclusively, therefore, three things must be done, things that to my knowledge have not been attempted to anyone so far. First, it would have to be shown that McTaggart cannot make sense of the changes going on in ordinary time unless he postulates a meta-time of equal richness. That is, a meta-time that admitted B-relations only would not be capable of performing its required function. Second, it would have to be shown that in order to explicate the movement of the NOW in meta-time, we could not employ standard time in the same manner that we employed meta-time to explicate the movement of the NOW in standard time, and therefore we would be forced to introduce a third temporal sense. Third, it would have to be shown why the regress that results would have to be regarded as vicious.

. . .

By making use of Broad's somewhat fanciful idea that beside the regular time series there exists a second-order time series in which every point in the first-order series has an extended history, McTaggart's difficulty could be resolved. The difficulty arose out of the fact that we assigned incompatible properties to the selfsame moment m_1. We regarded this as unacceptable, because m_1 is devoid of temporal extension and thus has no room to accommodate incompatible properties. However, we face no difficulty once we are permitted to entertain the possibility of a higher-order time series in which m_1 endures indefinitely. All the moments of our regular time series coexist at each moment in super-time, and the position of the NOW in regular time varies from moment to moment in super-time. The moment m_1 and every other moment in regular time can assume different properties at different moments in super-time. In particular, m_1 may have the property of futurity, while m_2 has the property of presentness and m_3 the property of pastness at m_1^2, where m_1^2 is a moment in the second-order time series while later at m_2^2, m_1 acquires the property of presentness, m_2 of pastness, and m_3 of distant pastness. Thus, the problem of the extensionlessness of the moments in the continuum of which they form a part is resolved with the introduction of a higher-order time continuum in which they have unlimited duration.

The reason why this point is of interest is because through it we see that

McTaggart's difficulty would disappear under the same circumstances under which the various objections of Russellians to the moving NOW disappear. This confirms our view that McTaggart and Russell are ultimately referring to the same difficulty.

Notes

1. McTaggart, *Nature of Existence,* vol. 2, chap. 33.
2. Russell, *Principles of Mathematics,* sec. 442.
3. Russell's analysis is eliminative and not reductive, as A-statements are eliminated, and we are left with B-statements only.

ESSAY 20

McTaggart, Schlesinger, and the Two-Dimensional Time Hypothesis

L. NATHAN OAKLANDER

In "The Unreality of Time" and *The Nature of Existence,* McTaggart offers a positive conception of time and then provides arguments that purport to establish that such a concept is contradictory and therefore cannot be applied to reality. Although there are few, if any, philosophers who agree with McTaggart's conclusion that time is unreal, there are many who accept the general position that time involves passage or tense.[1] George Schlesinger is one philosopher sympathetic to McTaggart's positive views on time, and in his recent book, *Aspects of Time,* he attempts to defend McTaggart's account of the passage of time (see Essay 19) by resuscitating a gambit suggested some time ago by C. D. Broad:[2] namely, the notion that time has two dimensions. My purpose here is to argue that Schlesinger does not vindicate McTaggart's positive conception of time, since the two-dimensional time hypothesis that it allegedly requires is as beset with difficulties as the conception of temporal becoming it is supposed to render intelligible.

It will be useful for us to begin by stating the main elements of Schlesinger's interpretation of McTaggart's positive account of time, since it is the view that he (hesitantly) intends to defend. According to McTaggart, the NOW is something that moves relative to the series of points that constitute time. Schlesinger continues

> Temporal points from the future, together with the events that occur at those points, keep approaching the NOW and, after momentarily coinciding with it, recede further and further into the past. The NOW, of course, is conceived not as some sort of object, but rather as the point in time at which any individual who is temporally extended is alive, real, or Exists with a capital *E*. . . . A typical event, on this view, is in the distant future to begin with; then it becomes situated in the less distant future; it keeps

approaching until it becomes an event occurring in the present. As soon as this happens, the event loses its presentness and acquires the property of being in the near past. (p. 214)

(For other recent interpretations of McTaggart on time see Christensen and Rankin.[3]) Schlesinger also mentions with approval Gale's characterization of the NOW as a moving spotlight which successively illuminates different moments along the series of time.[4] Several major objections have been made to this way of conceiving of the moving NOW, but Schlesinger believes that they can be answered.

One such objection derives from Smart, who argues that if we think of time as a river or some kind of particular thing that moves, then it must make sense to ask, How fast is it moving? Yet the question, How fast did time flow yesterday? seems to be a senseless question. Smart says: "We do not know how we ought to set about answering it. What sort of measurements ought we to make? We do not even know the sort of units in which our answer should be expressed."[5] Since we cannot measure the speed of the NOW, there is reason to suppose that it does not exist.

A second and stronger argument derives from Broad.[6] Suppose that the NOW moves along a series of events. When the NOW "hits" a given event E_1, that event acquires the property of presentness and then loses it without delay. Although the acquisition and subsequent loss of presentness is itself *an event; that* event—that is, E_1's acquirement of presentness—cannot be a member of the very set which constitutes the first series of events. Thus, Broad concludes that if we accept the moving NOW, we must postulate a second time dimension in which events of the first time dimension acquire and lose presentness. As Broad puts it: "If there is any sense in talking of presentness moving along a series of events, related by the relation of earlier-and-later, we must postulate a *second* time-dimension in addition to that in which the series is spread out. An event which has zero duration, and therefore no history, in the first time-dimension, will yet have an indefinitely long duration and a history in the second time-dimension."[7] If one finds the notion of a second time dimension unintelligible, then one will reject the conception of the moving NOW that leads to it.

Schlesinger claims that this objection to the moving NOW is virtually identical with McTaggart's main argument for the unreality of time. The difficulty arose because in a world of absolute becoming, one and the same instantaneous event has the incompatible characteristics of pastness, presentness, and futurity. This, however, is impossible, because incompatible properties can only be possessed by the same entity if it has them at different times. Yet, events do not have any temporal scope during which they could accommodate incompatible properties. The "way out" of this difficulty, and of the preceding one, is to postulate a second, higher-order series of meta-moments in which literally instantaneous events have an indefinite duration. Schlesinger calls this idea "fanciful," but he seriously considers it as a way of defending McTaggart's conception of the transient aspect of time and resolving McTaggart's paradox.

According to Schlesinger, McTaggart's conception of time can be made intelligible and its difficulties overcome if we agree to postulate a higher-order series of moments. He answers Smart's objection as follows (p. 218):

> The movement of the NOW in the standard series of time may be explicated by explaining that the NOW is at t_1 in the ordinary series when it is at T_1 in the super-series and at t_2 in the ordinary series when it is at T_2 in the super-series. We may even assign a value to the speed of the NOW; it moves from t_1 to t_2 at the average speed of
>
> $$\frac{t_1 - t_2}{T_1 - T_2}.$$

Concerning Broad's objection, Schlesinger says that "the event of the NOW reaching t_1 may, if we like, be looked on as taking place in super-time" (p. 218). He is, quite naturally, reluctant to postulate a whole new series of temporal points but is comforted by his belief that we need not continually postulate a new time series to help us make sense of temporal becoming in the preceding one (pp. 218–219):

> For just as the second series could be instrumental in helping to make sense of temporal becoming in the first series, in the same manner the first series could serve as the extra series through the use of which temporal becoming in the second series makes sense. For example, it might be said that the NOW in the second series is at T'_1 when it is at t'_1 in the first series and at T'_2 when it is at t'_2, leading to the claim that the average speed of the NOW in the second series from
>
> $$T'_1 \text{ to } T'_2 \text{ is } \frac{T'_1 - T'_2}{t'_1 - t'_2}.$$

Finally, Schlesinger uses a second-order time series to resolve McTaggart's difficulty concerning A-characteristics. On the two-dimensional time hypothesis,

> All the moments of our regular time series coexist at each moment in super-time, and the position of the NOW in regular time varies from moment to moment in super-time. m_1 and every other moment in regular

time can assume different properties at different moments in super-time. . . . Thus, the problem of the extensionlessness of the moments in the continuum of which they form a part is resolved with the introduction of a higher-order time continuum in which they have unlimited duration. (p. 219)

Can this account of the movement of time be accepted as a way of avoiding paradox and resolving traditional philosophical perplexities concerning time? Let us see.

Schlesinger offers a challenge to those who would attempt to refute McTaggart's claim that the NOW is a particular which shifts in position relative to the series of events in the direction of the future. He says:

First, it would have to be shown that McTaggart cannot make sense of the changes going on in ordinary time unless he postulates a meta-time of equal richness. That is, a meta-time that admitted B-relations only would not be capable of performing its required function. Second, it would have to be shown that in order to explicate the movement of the NOW in meta-time, we could not employ standard time in the same manner that we employed meta-time to explicate the movement of the NOW in standard time, and therefore we would be forced to introduce a third temporal series. Third, it would have to be shown why the regress that results would have to be regarded as vicious. (p. 219)

It seems to me that three things that would have to be shown in order to refute McTaggart's conception of the transient aspect of time can indeed be shown, and in what follows I shall attempt to explain why.

Turning to the first issue, is it possible for the terms of the original series to undergo genuine change, in McTaggart's sense, if the terms of meta-time form a series that has B-relations (*earlier than* and *later than*) but not A-determinations (*pastness, presentness,* and *futurity*)? The answer to this question is emphatically *no.* McTaggart is very explicit in his belief that the B-series has no independent and separate reality but is entirely dependent and inseparable from the application of the A-series to the nontemporal C-series. According to McTaggart, you cannot have B-relations between events unless those events have A-determinations and change with respect to them. As further evidence, consider the following passage:

If there is any change, it must be looked for in the A-series . . . If there is no real A-series, there is no real change. The B-series, therefore, is not by itself sufficient to constitute time, since time involves change. The B-series, however, cannot exist except as temporal since earlier and later, which are the relations which connect its terms, are clearly time-relations.

So it follows that there can be no B-series when there is no A-series, since without an A-series there is no time.[8]

If it were possible for there to be genuinely temporal relations between terms without those terms having A-determinations, then it would be possible for there to be succession (a genuinely temporal series) without becoming. Clearly, McTaggart would not accept that conclusion, and since Schlesinger claims to be in complete agreement with McTaggart's positive conception of time and change, he could not accept it either.

Thus, a second series that admitted B-relations only would not be capable of performing the required function, because it would not be a temporal series at all. Consequently, an event could not possibly endure indefinitely in the second series. Indeed, it could not endure at all, since endurance requires time, and in a B-series without an A-series there is no time. To put the point still differently, since a B-series without an A-series is not a temporal series, the second series without becoming does not enable the first series to avoid the incompatible properties problem. Thus it would appear that McTaggart cannot make sense of the changes going on in ordinary time unless he postulates a meta-time of equal richness.

Turning to the second issue, the pertinent question is as follows: Can the movement of the NOW in meta-time be explained by reference to standard time in the same manner in which reference to meta-time is employed to explicate the movement of the NOW in standard time? Again, it seems to me that the answer is no, and to see why, let us first note that the movement of the NOW in standard time is never made clear. According to Schlesinger the moving NOW in the first series is explicated by saying that "the NOW is at t_1 in the ordinary series when it is at T_1 in the super-series and at t_2 in the ordinary series when it is at T_2 in the super-series" (p. 218). There are, however, serious problems with this first step. Since Schlesinger maintains that the events that are at the point in time at which the NOW is situated are those that are real and alive, or Exist with a capital *E,* and since he also maintains that when the NOW is situated at t_1, it is also situated at T_1, it follows that when the NOW is at t_1, not only do the events at t_1 Exist with a capital E, but so do the events that are at T_1. Furthermore, since Schlesinger claims that "all the moments of our regular time series coexist at each moment in super-time," it follows that when the events at t_1 are NOW, the events at T_1, that is, the events at t_1, t_2, . . . t_n, are also NOW, and that is absurd. For if the events in the original series are all NOW at T_1 in the second series, then they exist *simultaneously* and not successively. In other words, this account of the moving NOW does not make sense of time and change in the first series. It eliminates it!

My objection to Schlesinger's account of the moving NOW in the first series can be approached from a different direction. When the NOW is at t_1 in the original series, neither the past events at t_{1-n} nor the future ones at t_{1+n} exist. For events of the past do not exist *now* but only in the past, and "the future has no reality by means of which to reach out toward us and make an impact on the present."[9] On the other hand, when the NOW is at t_1, it is also at T_1, and when it is at T_1, all the moments and events in the first time series coexist. If, however, they all coexist, then they cannot be distinguished as past, present, and future, but must all be present; and that is unacceptable because (1) it entails that past and future moments both do not exist (at the first level) and do exist (at the second level); (2) since past and future events are NOW, they exist simultaneously with those that are present, and hence there is no ground for the original series being temporal; (3) since the moments that are NOW at T_1 cannot have different A-determinations—some being past, others present, and still others future—and then change those determinations when the NOW is at T_2, the appeal to a second time dimension does not, despite initial appearances, help to resolve McTaggart's difficulties.

Having said this much, we can deal briefly with the question of whether we can understand the moving NOW in the second series by employing standard time. Recall that meta-time consists of a temporal series of moments the contents of which are whole series of moments of original time. If the second series is to be a *genuinely temporal* series, then its terms must change with respect to their A-determinations, and the question arises as to how this is possible. Could it be that each term T_i which has a momentary existence in the second series has an indefinite history in the first series and thus changes properties at different t_i's in the first series? This would make no sense whatsoever, for if it were true, then each t_i would contain the contents of every term in the second series. Consequently, all events in the first series would exist at each t_i, and that contradicts the original presupposition that the terms of standard time do not all exist at the same t_i. Furthermore, if *all* the terms in the second series coexist in each moment of the first, and if we say, as Schlesinger does, that the NOW is at T_1 when it is at t_1, then it would follow that if one of the terms of the second series is NOW, then all the terms of the second series are NOW. But then the second series, like the first, no longer deserves to be called temporal. It would appear, then, that the movement of the NOW in the second time dimension cannot be understood in terms of the movement of the NOW in the original time dimension but would require the postulation of a third time series, a fourth, and so on.

We are thus led to the third and final question: Does the resulting regress have to be regarded as vicious? Here I think that the answer is yes, for at no stage along the regress of time series can we stop and say that the questions and problems for which they were introduced have been answered or solved. In other words, the infinite regress of time dimensions is vicious, because the notion that we are attempting to understand by an appeal to a higher-order time series arises in exactly the same form in that higher-order series, and consequently, regardless of how many time dimensions we introduce, we never manage to answer the problem for which they were introduced. More specifically, the problem of time and change centers around the following question: How are we to understand the commonsense belief that an apple is green at one time and red at a later time or, equivalently, that an apple is green before it is red? What must time be in order for it to be possible for a single entity to have a property and then lose it? Presumably, the account of time developed by Schlesinger and McTaggart is intended to answer that question. Their answer involves the notion of the NOW moving along a series of events such that it is at the point, t_1 at which the green temporal slice of the apple exists at one time, T_1, and is at the point, t_2, at which the red temporal slice of the apple exists at another time, T_2. Alternatively, they could say that the NOW is simultaneous with the green slice *before* it is simultaneous with the red slice of the apple. As Schlesinger says, it is useful to think of the NOW as a moving spotlight which illuminates *successively* different moments along the series of time. Unfortunately, this account does not help us to understand the nature of time and change because it presupposes it. For Schlesinger is treating the NOW as a substance that has a property at one time then loses it at another time (it is *first* simultaneous with the green section and *then* it is later than the green section), and we want to know how this is possible. Thus, the moving NOW involves precisely the same notion that we hoped to make intelligible, thereby involving us in a vicious circle.

The circularity of Schlesinger's solution to the problem of time can be seen in still a different way. For Schlesinger, the account of an ordinary substance changing from green to red is explicated in terms of qualitatively different temporal cross-sections of events (for example, the green section and the red section) which comprise that substance changing from future to present to past. Such change, he claims, is unintelligible unless temporal slices of substances have an indefinite duration in a second time dimension and exemplify different properties at different points in that second series. But then we are faced with the original question at the second level: How can a single thing (say, the red section) of an indefinitely long duration have a property (say, futurity) at one time and then not have it at another time? By parity of reasoning, Schlesinger must admit that change in the red section of the second series is to be understood in terms of temporal slices (for example, the red sections being present, the red sections being past) that undergo becoming. Thus, to "solve" the problem of change in the second series, we must appeal to a third time dimension in which an instantaneous event in the second series, say, the red section's being present, has an unlimited duration and exemplifies different A-properties at different times. It should by now be evident, however, that, and why, such a move is futile. Thus, the infinite regress of time dimensions is a vicious one, because at each level the concept which it was introduced to comprehend remains incomprehensible.

The three things which Schlesinger claims would have to be done in order to refute McTaggart's account of the moving NOW in terms of a meta-time can be done, as I have shown. Thus, the hypothesis that there exists a second time dimension neither aids our understanding of the passage of time nor resolves McTaggart's difficulties concerning that notion. I conclude, therefore, that Schlesinger has not provided an adequate defense of McTaggart's positive conception of time.[10]

Notes

1. Cf. Broad, *Examination of McTaggart's Philosophy;* R. D. Chisholm, *The First Person* (Minneapolis, 1981); Gale, *Language of Time;* Prior, *Past, Present, and Future;* idem, *Time and Tense* (Oxford, 1968).
2. C. D. Broad, "The Philosophical Implications of Foreknowledge," *Aristotelian Society Supplement* 16 (1937): 177–209; idem, "A Reply to my Critics," in P. A. Schilpp, ed., *The Philosophy of C. D. Broad* (New York, 1959); cf. H. H. Price, "The Philosophical Implications of Precognition," *Aristotelian Society Supplement* 16 (1937): 211–228.
3. Christensen, "McTaggart's Paradox and the Nature of Time"; K. Rankin, "McTaggart's Paradox: Two Parodies," *Philosophy* 56 (1981): 333–348.
4. Schlesinger, *Aspects of Time,* p. 132.
5. Smart, "River of Time," p. 485.
6. Broad, *Examination of McTaggart's Philosophy,* pp. 277–280.
7. Ibid., p. 278.
8. McTaggart, *Nature of Existence,* vol. 2, p. 13.
9. Schlesinger, *Aspects of Time,* p. 60.
10. This essay was funded by a Faculty Development Grant, University of Michigan–Flint.

ESSAY 21

How to Navigate the River of Time

GEORGE SCHLESINGER

In my initial efforts to defend McTaggart's theory, I too strictly followed the letter of Broad's suggestion and attempted to explicate time's motion with the aid of a logically possible higher-order time system, thereby unnecessarily complicating matters. More recently I have realized that it is possible to keep to the spirit of Broad's idea and at the same time simplify the exposition considerably by employing, instead of a super-time, a second time system that is symmetrically related to ours. After a brief description of the nature and purpose of such a system, I am confident that little room is left for misunderstanding and that the objections of Oaklander (Essay 20) will seem irrelevant.

Let me then postulate two physical systems, *X* and *Y*—for example, two very similar solar systems—in which all the laws of nature are identical to ours with the exception that light travels at infinite speed. As a result, inhabitants of *X* can receive immediate replies from *Y* to their inquiries, and of course no problems arise in connection with determining the simultaneity of distant events. Let us now suppose that with the aid of the vast number of similar mechanical, electrical, chemical, biological clocks the inhabitants have in both systems, they determine that t_0 in *X* coincides with t_0 in *Y*. Subsequently they find, however, that t_1 in *X* is simultaneous with t_2 in *Y* and t_2^x with t_4^y, t_3^x with t_6^y, and so on. This presents us with a straightforward case in which the assertion that the NOW moves twice as fast in *Y* as in *X* makes good sense and can be fully explicated in terms of observations. The inhabitants of both systems, who are in regular contact with one another, would claim to have observed that when the NOW in *X* is at t_1, the NOW in *Y* is at t_2 (according to *Y*-clocks) and that during the period in which NOW^x merely traveled from t_1^x to t_2^x, NOW^y moved from t_2^y to t_4^y.

It should be pointed out that alternative hypotheses would also be available. Most likely, the simplest among these would be one which postulates that the intervals in Y are shorter than those in X. To be precise, the length of the temporal interval $t_0 - t_1$ in X could, for instance, be claimed to equal the length of $t_0 - t_2$ in Y. These two intervals being of exactly the same magnitude implies that their extremities coincide, which accounts for the fact that, given that $t^x{}_0$ and $t^y{}_0$ are simultaneous, $t^x{}_1$ and $t^y{}_2$ are also simultaneous. Presumably therefore, a Russellian would, in the situation we have described, insist that the second hypothesis, which is no more complicated than the first and is compatible with his view of the nature of time, be adopted.

It is vital to realize that the admissibility or even preferability of the Russellian hypothesis under the circumstances just depicted does not defeat the point of my story. We must recall what it is we are trying to establish: namely, that the charge made against those who believe in the flow of time that their position is downright incoherent, or at least is incapable of being spelled out clearly, is unfounded. While practically all agree that the passage of time intuitively appears to be one of the most central features of reality, when it comes to the formulation of this idea, there seems to be no intelligible way of doing it. After all, to partake in a real motion, an individual must cover a certain amount of space in a certain amount of time; are we thus driven to maintaining the absurdity that when time itself is moving, then a certain amount of time is covered in a certain amount of time? In the context of our story, however, the Russellian would no longer claim that our hypothesis is incoherent or unintelligible.

Russellians would have to admit that they fully understand our account of what is actually taking place. Furthermore, they would even have to admit that in the context of our story we could assign a definite value to the magnitude of the NOW's speed in different systems, without having to establish how much what is being covered in how much time in one and the same system. Adopting as our unit the speed of the NOW in X, the NOW in Y may be said to move at speed 2, while in some other systems it moves at speed 3, 4, and so on. All that a Russellian would insist on is refusing to accept our hypothesis. But then there is no need for us to find a situation in which our claim that time moves is adopted by all; for our purposes it is fully sufficient to have a logically possible one in which our claim would have to be admitted to make perfectly good sense.

In order to achieve clarity, it is important to note the following brief points: (1) The moving NOW is obviously not a material object; indeed, it is not a particular of any sort. Being present is a property which every event assumes and then sheds. A property or a universal can of course be exemplified by different events in different places and at different times. Thus, none of the difficulties mentioned by Oaklander concerning the question of *where* exactly the NOW is, arise. (2) Just before that, Oaklander asks: Can the movement of the NOW in meta-time be explained by reference to standard time in the same manner in which reference to meta-time is employed to explicate the movement of the NOW in standard time? To make this question relevant to the current, simplified version, where we have perfectly symmetrical systems, substitute X for "meta-time" and Y for "standard time." Obviously, our answer will then be stronger than merely "Yes, it can be explained," since as soon as the first explanation is given, the need for any further explanation vanishes. Saying that in the situation described we would regard the speed of the NOW in Y to be double the speed of the NOW in X *amounts* to saying that we regard the speed of the NOW in X to be half the speed of the NOW in Y. (3) In order to account for a change from red to green, nothing nearly as complicated as Oaklander indicates is required (p. 227). For that, we do not even need to make use of the transient aspect of time. By merely saying that i is red at t_1 and green at t_2, we have already made clear that i changes from red at t_1 to green at t_2. The transient aspect is required only in an explication of "i being red at t_1" itself changing from being in the present to being in the past.

There is no doubt that the transient theory of time is consistent and intelligible. But is it true? I do not believe that this is the kind of question to which a final, conclusive answer is possible. As with all genuinely metaphysical theories, what we may reasonably expect is further clarification concerning its precise presuppositions and implications and an increasingly more detailed list of its advantages and disadvantages. The progress that is likely to be made in the future is the construction of new arguments showing that one theory is ultimately simpler, accords more smoothly with common sense, or fits more easily certain acknowledged facts, without arriving at an incontestable conclusion that either of the two theories is false or unintelligible.

ESSAY 22

A Reply to Schlesinger

L. NATHAN OAKLANDER

In his response to my Essay 20, Schlesinger claims to show that "there is no doubt that the transient theory of time is consistent and intelligible" (p. 231). He supports his claim by appealing to the observations of the inhabitants of two physical systems *X* and *Y* in which the *X*-clocks are running only half as fast as the *Y*-clocks. In my reply I shall argue that the situation he describes is irrelevant to the task he wants to accomplish and that, therefore, doubt does indeed arise concerning the coherence of the theory he propounds.

To get to the heart of the matter, consider the following passage: "The moving NOW is obviously not a material object; indeed, it is not a particular of any sort. Being present is a property which every event assumes and then sheds. A property or a universal can of course be exemplified by different events in different places and at different times. Thus, none of the difficulties mentioned by Oaklander concerning the question of *where* exactly the NOW is arise" (pp. 230–231).

But they do arise. Schlesinger claims that being in the present, or NOWNESS, is a universal. A universal is a timeless entity that is one and the same or wholly contained in each particular that exemplifies it. Therefore, on Schlesinger's view, literally the same NOW is at every time, and this impales him on the horns of a dilemma: his position is either incoherent or circular. To see what is involved, note first that moments must exemplify NOWNESS timelessly or temporally. However, if NOWNESS is exemplified timelessly at all times, then there is only one time; or rather, since time requires succession, there is no time. The "temporal" series is a *totum simul* in which all moments exist NOW. The alternative, according to which moments are NOW at different times, renders his position circular. For, if there is a second "time" series, then the problem for which it is introduced, namely, to account for succession in the first series, re-arises in it. The problem is that since the NOW is a universal, it is literally and wholly at each moment in the second series, thus making all moments NOW. It is no use saying, as Schlesinger does, that the NOW in X is at t_1 when the NOW in Y is at t_2 and that the NOW in X is at t_2 when the NOW in Y is at t_4, and so on, since the NOW in Y is the *same* as the NOW in X, and hence all "times" in X exemplify the same NOW as all "times" in Y. Thus, the existence of temporal succession in both X and Y is still unaccounted for. Nor does it help to say that t_1 in X is NOW at t_1, and t_2 in Y is NOW at t_2, and so on, for those statements have an unchanging truth-value and so do not reflect the transient aspect of time.

The difficulties with Schlesinger's reformulated theory of time can also be seen by considering what he says about change: "In order to account for a change from red to green, nothing nearly as complicated as Oaklander indicates is required (p. 227). For that, we do not even need to make use of the transient aspect of time. By merely saying that i is red at t_1 and green at t_2, we have already made clear that i changes from red at t_1 to green at t_2" (p. 231).

Can the problem of change really be handled that simply? I do not think so. Schlesinger's account of change is adequate if and only if t_1 and t_2 name successive times. What, then, accounts for t_1 and t_2 being members of a temporal series? Schlesinger claims that i's changing does not require the transient aspect of time. However, since i can change only if there exists a temporal series, and since i can change without the passage of time, it follows that, on Schlesinger's view, there can be a temporal series without temporal passage. But it is followers of Russell who claim that there can be a temporal series without passage. In what sense, then, is Schlesinger's view a defense of the intelligibility of the transient theory of time? On the other hand, if he treats moments as themselves subject to becoming, then (1) contrary to what he claims, a thing's changing does require the moving NOW, and (2) the incoherence or circularity involved in claiming that NOWNESS is a universal that all moments exemplify reemerges.

Schlesinger believes that the question of the truth of the transient theory of time, like all metaphysical disputes, is not one "to which a final, conclusive answer is possible." I would certainly agree that the question of the flow of time will always have proponents arguing for and against its reality. However, it does not follow that neither position is true and neither false. I have argued that Schlesinger's explication of the statement "The NOW moves" is neither consistent nor intelligible, and so a theory of time based on it cannot possibly be true. (The inconsistency involved in treating NOWNESS as a property that events acquire and shed is also argued for by Mellor in his book *Real Time*.)

ESSAY 23

Temporal Becoming Minus the Moving Now

DAVID ZEILICOVICI

Being later than is an order relation which, ranging over moments and events, generates the common static time series. This much is, I think, generally accepted, the only open question being whether this series is *all* there is to time. McTaggart thought it was not.[1] He believed that each moment is not just later than or simultaneous with or earlier than some other moments but is also absolutely and objectively past, present, or future. Absolutely, yet not permanently, for the future necessarily becomes first present and then past. It is this "movement," the dynamic aspect of time, the phenomenon of temporal becoming, which, if real, is not accounted for by the static time series. In order to correct this failure, McTaggart introduced the (in)famous moving (upon the ordered points of the static series) NOW and, with it, the need to explain, *inter alia,* the strange meaning of the movement or what is involved in the definition of "movement." For how, indeed, can the event of the NOW reaching *t* be a member of the same series that *t* belongs to? This led Broad to consider the rather wild possibility of two-dimensional time.[2] The idea was to have the NOW of common time move by occupying different common moments at different meta-moments. But, according to McTaggart, meta-time would not be full time unless it possessed a moving (meta-) NOW of its own. Hence the charge of infinite vicious regress in time dimensions.[3]

A recent instalment of the multidimensional fairytale is a debate between professors Oaklander and Schlesinger. In 1980 Schlesinger (Essay 19) attempted to save the older approach by suggesting that meta-time does not have to be full McTaggartian time and that two-dimensional time could do the job, the NOW of each series moving symmetrically during temporal intervals of the other series. He also held that the regress, if any, has not been proved vicious. Three years later Oaklander (Essay 20) strongly refuted all these alternatives, but Schlesinger (Essay 21) then countered with a *one-dimensional* way out, which, rather misleadingly, he presented as a mere simplification in the exposition of his earlier attempt. Within a single, common time dimension, two time-measuring physical systems, in instant communication with each other, are envisioned. At some given moment all clocks are synchronized. Subsequently it is observed that whenever all clocks in one system show any time T, all clocks in the other system show $2 \times T$. Schlesinger's thesis is that among other legitimate explanations of this phenomenon is the theory that the NOW of the first system moves twice as fast as the NOW of the second system. This, he argues, shows that the term "moving NOW" is empirically meaningful.

I am at odds to decide whether Oaklander is taken in by Schlesinger's strange mode of presenting what is essentially a new argument. For although in his reply (Essay 22) he does refer to the new physical systems, rather than to the old time dimensions, he continues to speak of the second time series in "Schlesinger's reformulated theory." This is important, because Oaklander tries to refute Schlesinger's new theory mostly by the same objections he raised against the old one, a stratagem which is only partially successful precisely because it willfully ignores the new and powerful hypothetico-observational part of Schlesinger's later argument.

What, in my opinion, Oaklander should have said is that the possible observations referred to by Schlesinger fail to make the moving NOW intelligible, because the complaint against this concept is not that it is *not clear* but that it is self-contradictory (or at least hopelessly confused) and, as such, *in principle unclarifiable*. Schlesinger claims to have demonstrated the possibility of "a straightforward case in which the assertion that the NOW moves twice as fast in Y as in X makes good sense and can be fully explicated in terms of observations" (p. 229).

But for something to move faster than something else, it must belong to the category of things to which the concept of movement is applicable. Schlesinger not only fails to show that the NOW belongs to this category; he also leaves unaddressed the argument which purports to prove that it makes no sense to say it belongs to it! Moreover, Schlesinger's case is not that "straightforward." His representatives in the two physical systems observe clock dials; they do not observe any moving NOW. Therefore Schlesinger can hold that the "moving NOW" is "explicated in terms of observations" only indirectly; for this term figures in a theory which is a candidate for explaining what is observed. There are, however, many restrictions on theory candidature. For example, no legitimate candidate may attempt to convey meaning to "square circle" or, in general, to any term which, for whatever reason, is in principle meaningless. And if a theory is held to be the sole direct conferer of meaning on its own theoretical terms, then it is not the ordinary meaning of some (similarly sounding) nontheoretical terms that has been explained.

The upshot of our discussion is that we must agree with Oaklander's evaluation (Essays 20 and 22) to the effect that neither by improved multidimensional theories nor by imaginary empirical evidence can the moving NOW be saved. It is crucial to realize, however, that Oaklander is under the false impression that by crushing the moving NOW, he has also refuted Schlesinger's claim (Essay 21) that "the transient theory of time is consistent and intelligible." This mistaken belief clearly stems from the identification of the general transient theory with its particular moving NOW versions. And unless the word "change" is strictly re-defined for use in such sentences as "An event changes from future to past," a version *is* basically a moving NOW version even if "movement" is understood, in an extended sense, to mean change (as is common in Greek philosophy). For if "change" is understood in the ordinary way, you still "represent temporal becoming as if it were some sort of motion, . . . [that of] the A-series moving up the B-series."[4]

Now, while it is true that the *intuition* of temporal becoming does involve the *image* of something (a river, a NOW, a spotlight) moving, this intuition should not be confused with a *metaphysical theory* of dynamic time. On the other hand, it should not be prejudicially taken for granted that all transient theory terms (such as "change," when re-defined) are just other names for the moving NOW. The task of the "philosopher sympathetic to McTaggart's . . . account of the passage of time" (Essay 20) is not to remain with a vague intuition of transience while attempting to clarify (or rename) the unclarifiable; it is rather to devise a solid and clear theory that will articulate in its own manner and terms what is essential, vital, and redeemable in what was intuitively, albeit dimly and not very consistently, grasped.

As a matter of fact, this is exactly what McTaggart tried to accomplish by substituting for the imagery of motion such (semi-) theoretical concepts as his A-series and the special sense in which events change: transition from one A-series to the next. By this theory, A-determination places moments and events objectively (that is, in a manner not depending on an observer's temporal position) in the past, present, or future, and such A-determinations are conveyed by A-statements. Since, intuitively, events approach us from the future, realize themselves in the present, and then recede into the past, A-statements change their truth-value with the passage of time. For any fixed event (*E*), all tokens of "*E* now" express the same proposition, sometimes truly, at other times falsely. These logically problematic changes in truth-value correspond to actual changes in the world. McTaggart conceives a *sui generis* kind of change (*change of time*) which applies not to objects but to moments, periods, events, and processes and concerns their status in the A-series. It reflects something elementary and very difficult to describe, which happens (*sic*) to time itself.

That McTaggart failed in defining nonmetaphorically, clearly, and consistently his A-determination as "nonordinary properties" of events does not demonstrate that this cannot be accomplished. In the rest of this essay I shall attempt to develop (and also interpret and try to defend) an idea of Broad's into a nonmoving NOW-including transient time theory which supplies such definitions and which, as I shall try to show in the last section, is immune to Oaklander's criticism.[5]

Broad's idea is that "the sum total of existence is always increasing."[6] I take this to mean that, at any present moment, *future time* does not exist.[7] Thus, what has to be "increased" is the time series, and this feat is accomplished by the creation of future *moments*. But although Broad develops this position and even uses it in order to establish that the time series has an intrinsic direction, which should help explain changes in A-determinations of events,[8] thirteen years later he writes: "I do not suppose that so simple and fundamental a notion as that of absolute becoming can be analysed."[9] To me, however, it seems that Broad's idea (in my interpretation) is tailored to serve as the pivot of a rather extremist open-future theory of dynamic time. It is a theory cast in the old-fashioned empiricist mold, because it presupposes that the question of what there is cannot be separated from the question of what can be known, taking "to be known" to mean "to be observed or to be (somehow and somewhat) judged favorably by the observer."[10]

Unlike present events, future events are not, at present, objects of direct knowledge ("knowledge by acquaintance"[11]). And unlike the case of past events, there are, at present, no traces, records, or memories of future events. Moreover, Russell has argued convincingly that (recent) remembered events belong, together with perceived events, to our sensory data.[12] It follows that the difference between past and present facts, with respect to the possibility of direct and not overly inferential present knowledge, is often a matter of degree, whereas present knowledge of future facts is always exclusively and strongly inferential, in that future events supply, at present, neither direct nor indirect data.[13] But time is the ordered set of moments, and moments are defined as equivalence sets of events under the relation of simultaneity. So future time can be no less synthetic and no more a priori than future events; in both cases we may predict existence but not take it for granted. It follows that the (existing-) time axis is, at any moment, finite in the positive direction. Beyond the bound lie mere conjectures, which may, of course, be used to conjecturally map the conjectured future, thereby conjecturing an infinite time axis.[14]

One result of having at each present moment an infinite time axis composed in part of existing moments and in part of predicted moments is that if $t < t'$, t-tokens of such statements as "t' exists at t'" are not always as uninformative as they look. For, if t happens to be the present, t' may remain forever just an empty placeholder on the time axis, whereas the above t-token conveys the informative prediction that this is not the case. The nonemptiness of the t-token is important, because what we are proposing is a "creationist" theory about moments and not, like Broad perhaps, about events. And although we believe that moments are defined by classes of simultaneous events, we hold that the two kinds of theories are not exactly the same. Thus if E is an event, t is the present moment, and t' is a latter (future) moment, there is still a difference, in our view, between a theory which proclaims the truth of the two tenseless statements:

(1) E does not exist at t

and

(2) E exists at t'

and a theory that proclaims the truth of the two tenseless statements

(3) t' does not exist at t

and

(4) t' exists at t'.

In order to see the difference between the two kinds of theories, consider Schuster's claim that no creationist (or annihilationist) theory, in either its tensed or tenseless version, can be coherently formulated.[15] Schuster argues that

> the use of expressions such as "the sum total of existence" is particularly well-designed to blur the distinction between one time and another . . . [while it] is illegitimate to qualify an expression and then to claim it is unqualified. And that is just what the creationist would like to do. . . . [Because] for a creationist at t' (1) and (2) would be true, . . . [but] these statements only succeed in placing the existence of E at one time (t') instead of another (t); they do not establish the unqualified non-existence of E which is a necessary condition for its creation at t'. That can only be done by admitting the truth of the statement "E does not exist at t',"

although to do so would involve the creationist in a contradiction because he also accepts (2). He therefore settles for (1) and (2), believing that he says more than he does, because he confuses "*E* does not exist at *t*" with "*E* does not exist at *t′*" by allowing the former to carry out the function of the latter.[16]

Now, even if Schuster's general argument, of which I have presented only a small part, is sound (which I doubt, but have no room to argue here), it is clear, even from the passage quoted above, that it is directed only against the creation of events at (*already*) existing moments and therefore can hurt only an event-creationist. A moment-creationist, on the other hand, begins by holding that (3) is true at *t* (not (1) at *t′*). And while, being a creationist, he holds also that (4), which is *not* the negation of (3), is true at *t* (not (2) at *t′*), he does not contradict himself. Moreover—and this is the main difference—he does not succeed merely in placing the existence of *t′* at one time, *t′*, instead of another, *t,* because *if not for the truth of (informative) (4), he does not have "the one time (t′)" to place anything at*! In other words, even if some creationists do confuse unqualified statements with qualified ones, they do not need to do so. For as long as future moments, even if defined by events, are not given as existing prior (logically) to future events, *the qualified statements will succeed in conveying the creationist position.* Given that future moments are not a priori existent delivery rooms in which events are to be born, it is simply *not* true that "'The sum total of existence at *t* does not include *E*' is no more a denial of the existence of *E* at *t′* than is 'The sum total of existence in this room does not include John' is a denial of the existence of John in another room."[17] The other room may very well be there to contain John, but *E* has no ready-made cot waiting for its birth at *t′*.

Having defended the distinction between existing and conjectured time, we shall now use it to define the A-series. We start by defining a present-at-*t* event as any event occurring at the upper bound of until-*t* existing time. If some predictions are fulfilled, there will be a next moment; *t* will lose the frontier characteristic, and our event will lose presentness while retaining intact all its ordinary properties and all its temporal-order B-relations to its previous partners. But, because it is now related to new members, this event (and all others) will belong to a different time series, the series which includes the previous one but whose "membership is increased."[18] We call the different time series emerging in this manner A-series. The single B-series differs from any of the many A-series precisely in failing to distinguish between existing moments and predicted moments and by being, as a consequence, unbounded. B-sentences *report* B-relations between existing temporal particulars and *predict* them for nonexisting ones, without registering any ontological difference between the two kinds of particulars or any epistemological difference between the two kinds of acts. It is just this failure of discrimination which permits us:

1. To speak of a *single* B-series.
2. To speak of a *multitude* of A-series, if permitted to speak of A-series at all.

And, what turns out also to be required for (2),

3. To remain an A-theorist while not feeling obliged to deny that the B-series (by itself) *is* a time series.

Let me explain and try to justify each of these points.

1. The addition of a new moment to existing time fails to create a new B-series, because the new moment appears in the old B-series anyway. Thus the two series are indistinguishable from each other and so must, a fortiori, be counted as one and the same series. It is true, of course, that what appears in the new series is a "real," existing moment, while what figures in the old series is a mere predicted shadow of a moment. But such a feature cannot be used as a springboard for making a distinction between two *B-series* because *being later than,* the generating relation of any B-series, signally fails to mark any such distinction between the two kinds of moments. Interestingly enough, the uniqueness of the B-series holds true not only for the B-theorist, who impartially makes the reality of all moments an article of his faith,[19] but also for the A-theorist, who castigates *being later than* for just this sin, by accusing it of being either a nontemporal relation or a nonexhaustive temporal relation.

2. One thing *our* version of the A-theory does not do is to ignore the difference between existing and predicted moments. It even defines an A-series as that part of the B-series which contains only existing moments. The result of this definition is that if a prediction of a new moment turns out to be true, we stay with the same B-series but get a new A-series which is nonidentical with the old one. In order to maintain this position, we have, of course, to admit that the B-series is a time series independently of the phenomenon of temporal becoming as represented by the A-series. For if the B-series is not a time series, then none of its segments can be a time series; while if the B-series is a time series but borrows its temporality from the A-series, then our position is circular.

It is precisely this kind of objection that Oaklander, in his greatest offensive against the dynamic view of time, raises to other concepts of the A-series and also to the standard concept of an unstable-in-truth-value A-statement.[20] He points out that a *single* A-series won't do because, in it, events possess fixed, unchanging A-determinations.[21] But, he argues, a *sequence* of A-series must, in order to avoid circularity, presuppose the relation of succession *earlier than,* and even the B-series itself, to be temporal, independently of the phenomenon of temporal becoming.[22] Unfortunately, continues Oaklander,[23] this presupposition is not part of the *traditional* (and Oaklander envisages no other) A-theory. That Oaklander is right in this last claim may be seen from Gale's (correct) summary of the position of leading traditional A-theorists, such as McTaggart[24] and Broad.[25]

> It is because and only because the events comprising a B-series undergo temporal becoming that we can distinguish a B-series from a non-temporal series. We cannot do this solely in terms of the logical properties of generating relations of the B-series—its being asymmetric, transitive and irreflexive—since there are non-temporal relations, such as "larger than," which have the same logical properties.[26]

Often this attitude towards the B-series manifests itself in additional claims with which we are not immediately concerned here.[27]

3. Nevertheless, the idea that the B-series is, even without the A-series, a time series should not be dismissed lightly. For it must be kept in mind that the concept of a sequence of A-series can escape circularity within a theory which grants that the B-series by itself is, "at the ontological level,"[28] a time series. And while it is true that *traditional* A-theories refuse to make this concession, the question arises as to whether Oaklander is fair in holding this refusal against the transient view in general. In order to answer this question, we must consider the rationale of the traditional position. Broad explains that, as time involves change and change is made possible only by the A-series, the sole supplier of the much needed commodity called "*intrinsic* direction," then "the relation of *earlier than* can hold only between terms which have A-characteristics; just as harmonic relations can hold only between terms which have pitch."[29] But while this way of looking at things is only too natural within the context, aims, and quality of traditional A-theories (see next section), it does not follow that it applies outside these versions of the dynamic theory of time. Every honest A-theorist must, by definition, believe that change is essential to time; but it is an open question whether he must also believe that it is essential to the B-series. Must all A-theorists be, like all B-theorists, monists about time? After all, the basic transiency intuition is dualistic: there seem to be two aspects to temporal facts,[30] or even

two different types of temporal facts. First there are facts about [static] temporal relations of precedence and subsequence between events, and, second, there are [dynamic temporal] facts about pastness, presentness and futurity of these same events. Corresponding to the first type of temporal facts there is a series of events called the "B-series" . . . its generating being *earlier* [*or later*] *than;* corresponding to the second, is a series of events called the "A-series."[31]

Now, it is obvious that the two kinds of facts need to be correlated to each other, but this does not necessarily mean unilaterally preferring one kind to the other. Why are we not free to grant that *earlier than* is "an unanalyzable temporal relation,"[32] precisely because it ranges over the unanalyzable temporal particulars known as moments? This would not, of course, be the whole A-story of time, since there is the other type of temporal fact. But even without the second instalment, there is no more danger of confusing moments with physical bodies than there is of confusing physical bodies with places, for the sole reason that *earlier than, larger than,* and *to the right of* are all order relations which do not induce *intrinsic* order in their respective domains. Therefore, instead of rejecting the concept of a sequence of A-series on the grounds of its being either contradictory or circular, we should blame some of the traditional A-theorists for not availing themselves of what appears to be an open and attractive course of action: that of embracing a B-without-A-time-series.

An even more damaging case against *traditional* A-theorists may be built in two stages. We first charge them with gross maltreatment of the B-series. After finding them guilty on that charge, we proceed to expose their motives and find these motives to be even more objectionable than the original crime. The facts pertaining to the first stage may be summed up in the following manner. On the phenomenological level, traditional A-theorists are certainly aware of two independent kinds of temporal facts,[33] but when it comes to theory construction, they reject the notion that the B-series may express the static theme of time while the song of change may be relegated to other voices. They try to support this position by arguing from the premise "Time requires change" to the conclusion "The B-series requires the A-series." But the great originators of the classic versions of A-theory must surely realize that this argument is a *non sequitur.* And so we are led to speculate that their real motives are to be found elsewhere: perhaps in a tendency to invent over-ambitious theories in which the inventors themselves have less than full confidence.

On the one hand, traditional A-theorists wish their theories to explain not only changes in A-determinations of events, such as the change of the death of Queen Anne from future to past,[34] but also changes in properties of objects, such as the change in the poker from being hot at t to being cold at t'.[35] And as they do so by holding that the last kind of changes are strongly dependent on the first kind, because t cannot be earlier than t' unless t and t' already possess changing A-determinations, the result of their ambition is tantamount to a denial that the B-series without the A-series is a time series.[36]

On the other hand, the same traditional A-theorists seem to lack sufficient confidence in the ability of their own theories to explain the first kind of change if they permit the B-series to join unsponsored the time-series club. And as their method of defining change in A-determinations of events is to first take A-determinations as primitive (static) terms and only then use some shabby device, such as the NOW moving along the B-series, in order to have them change—they have good reason to feel insecure. For, given this method, the admission of an independently temporal B-series has two effects: it makes the road leading to McTaggart's paradox[37] even straighter than it would otherwise be,[38] and it opens the door to the charge that the very concept of an A-determination is an empty one. For the second effect, consider a t-token of "*E* now." B-theorists, who believe that only the B-series is objectively real, claim that this means no more than "*E* occurs at t" or "*E* is simultaneous with the token."[39] They criticize the spurious distinction between an event and its realization, which, they claim, is a main muddle of the dynamic intuition.

It seems to me quite possible that, cowered by this criticism, traditional A-theorists do not believe (and, I think, they are right) that their own ambitious versions of the dynamic theory can support a better method of defining change in A-determinations of events. And thus, instead of changing their definition, so that they might face their critics on better ground, these A-theorists prefer to give up the (very intuitive) notion that the B-series is, by itself, a time series. Sadly, they end up by meeting the same adversaries, armed with the same weapons, on the worst possible terrain.

In contrast to traditional versions, our version of the transience theory can accommodate a by-itself-temporal B-series. We judge that the task of explaining changes in A-determinations of events is ambitious enough for any version, and in ours we do not even try to explain changes in properties of physical bodies. And as we think that the idea of novelty, perhaps the main attraction of the dynamic intuition, is connected with change in the A-determinations of events and not with change in the properties of objects, we do not feel uneasy about the importance of our proper task. Being willing to accept the existence of two equally basic kinds of temporal facts, we can also afford to entertain two basic components of meaning for one and the same A-sentence. This, in turn, enables us to meet the objections which traditional A-theorists, the sacrifice of B-series notwithstanding, have left unanswered. Thus, while we grant that a *t*-token of "*E* now" does convey that *E* occurs at *t,* we hold that this is just one (the B-) component of its meaning. The *t*-token is that of an A-sentence only because it has an additional (A-) component of meaning to the effect that *t* is the limit of existing time. Furthermore, the claim made in the A-sense concerns, on our theory, the total state of the time series at *t* and thus the state of affairs in the whole physical world of which *E* is only a minute part. It cannot, therefore, be true that the transformation of *E* from possibility to reality, its actualization or determination, are simple-mindedly and emptily doubled for *E*. For, unlike the B-component, the A-component does not specifically apply to *E*. Moreover, the term "now" may be dispensed with altogether, for we shall next use our theory to introduce a sense of the term "change" which will serve as a theoretical substitute for the metaphorical "moving NOW."

The only concept of change properly defined so far is the ordinary concept of *change in time*. Change in this sense consists in a subject having some property at one moment and not having it at another.[40] It is clearly a concept devised for *things* and precludes any possibility of *events* changing because, by definition, events *are* changes, and changes do not change. Moreover, in spite of Smart's devastating attack in 1949,[41] when McTaggart spoke of changes in events, he did not have in mind second-order changes in things.

Like McTaggart, we are badly in need of some concept of change that would allow for an event changing without a change in any of the properties of the object involved in the event.[42] Otherwise, we would have a second event, not something which an event undergoes. This is made even plainer by noting that while things change in time by acquiring or by shedding properties, events must undergo changes as they are, that is, as completely ordinary (and we have so far defined no other) property-closed. And it is worthwhile stressing that this is true even of future events if they are to stay self-identical. The dynamic conception of time (even when cast in less radical versions than ours) does, indeed, go well with the idea of an important difference between the open future and the closed past. But to hold the future open just is to view different future events as different, but not yet realized (or not yet determined), possibilities, which means only that it is yet undecided which of the many possible events will occur. It surely does not mean that well-defined future events change their properties. On the contrary, since distinct future particulars are nothing more than distinct aggregates of compatible properties, it follows that an unrealized possible event must be property-closed if it is not to be confused with *another* unrealized possible event.

So, if it is change in events that we want, we must resurrect, and also try to improve, McTaggart's other concept of change, that of *change of time*. The improvement may begin by postulating that the two senses of "change" must denote concepts close enough to each other to warrant the use of the same word, followed by "in time" in one case and "of time" in the second. Otherwise the intuition that events undergo "changes" when they turn from being future to being past fails to be captured by the theory. This clearly requires that pastness, presentness, and futurity be established as nonordinary properties of events; that is, as properties not subject to the Leibniz identity principle.[43]

Now, our theory can claim to have made it plausible that something nontrivial ("is at the limit of existing time") that could have been truly predicated of an event at t can no longer be truly predicated of it at t'. And if this something, which constitutes the A-determination of the event, can be shown not to be a relation, it could, since it is not an ordinary property, be defined as a nonordinary property.

It may, at first sight, appear that just because we have made so much of the fact that A-determinations concern the entire time series, we will find it embarrassing to argue that they are not used, B-wise, as vehicles for expressing relations between events. The difficulty seems to be particularly acute in the case of past events. On the dynamic view, these have no static A-determination, of course; in distinct A-series past events are characterized as being of different degrees of remoteness in the past. And this seems to be the sheer result of their increasing their distance (a relation!) from the events which define the instantaneous upper bound of existing time. Strictly speaking, however, the temporal interval between any two (existing or predicted) events is constant in all A-series, the magnitude being allocated in the B-series. What happens is that a prediction turns out to be true so that a new event has the A-determination of presentness in a *new* A-series, and the distance of any given past event from this event is greater than it was from different events which enjoyed this distinction in *another* A-series. Thus the case of past events introduces no novel threats to our position. Still, the main conundrum remains: how can participation in any given A-series fail to be anything but a relation to the moment which is the unique bound of that series?

In his *Scientific Thought*,[44] Broad entertains the suggestion that when Tom becomes taller than his father, there occurs a change in (the value of) a certain existing relation; but that when a new male child is born, so that

Tom is no longer the youngest son, the change is in Tom and not in any relation. For, Broad argues, no previous relation which could change previously obtained between Tom and his nonexisting brother. But, as C. W. K. Mundle has pointed out,[45] Broad tends to treat Tom and his brothers as events.[46] To this I may add another objection: that Broad fails to consider *being the youngest brother* as a relation not between Tom and his nonexisting brother, but between Tom and his ever existing family.

Both objections fail, however, in the case we are considering. The entity whose A-determination is stated and the entity which is being added to existing time *are,* in our case, both events, or those classes of events which we call moments. Furthermore, while it is true that in both cases there can be no relation between old (existing) and new (not yet existing) particulars, there is a difference; for while Tom is related to his family (which has undergone transformation by increase but is still his family) both before and after the happy event, there is no relation between the old moment and a permanent A-series. For, by both the transience intuition and our theory, the new A-series is not a transformation of an old A-series; it is a strictly new entity which did not exist (not even as an incomplete series) before that which makes the new A-determination true (that is, the addition) is there. And this is really the crux of the issue: an increase in a family may literally be spoken of in terms of expectations, but it is always incorporated within the solid boundaries of an existing background and unquestioned (though questionable) future. The increase of time turns out, when questioned, to be unadulterated expectation and, as such, clearly no relatum.[47]

I do not, of course, argue that the above is the only reasonable way of viewing phenomena whose very reality is debatable. Indeed, B-theorists strongly deny that the transient aspect of time is an objective physical phenomenon and are therefore perfectly free (or willingly forced) to try to devise linguistic or psychological reductions that tend to represent A-statements as incomplete relation statements.[48] What I do maintain is that no A-theorist may adopt such a view. Anyone who embraces "*E* is past" as empirical is inconsistent unless he holds fast to the notion that pastness is no more a relation than it is an ordinary property. He must, therefore, either recant or admit some extraordinary devices, such as irreducible tenses (see Essay 1) or nonordinary properties, in order to make sense of the linguistic category to which "is past" belongs.

It remains to show that the change of time is not our old discredited moving NOW in fancy disguise. Otherwise we cannot boast of having a moving-NOW-free theory and must live in constant fear that the new concept will be unable to survive the objections leveled at the old one. The claim that our theory is faulty in just this respect may be made in two alternative ways, one rather simplistic, the other more sophisticated.

1. It might be argued that the moving NOW haunts our theory because it is definable within it. It is indeed so easy, natural, and intuitive to use the term "now" for the upper limit of existing time that our omission to have done so seems artificial, if not plainly hypocritical. The objector then proceeds to spring the trap by insisting that the property of being now belongs at different moments to different moments ("is not unique"[49]) so that what we have defined is the moving NOW.

The answer to this kind of formulation is that nowness is not an ordinary property. In fact, its nonordinariness consists precisely in its being used to make the SAME claim at nonsimultaneous moments by tokens of a single sentence, *tokens which differ in truth-values but not in the statements they convey* (see p. 215). Furthermore, even if "now" is defined as suggested, what is being defined is *not* a moving NOW. For while the transition of the now from movement to moment is change of time, it is certainly not movement. Movement is always upon the same (B-) time series, while the different moments to which nowness applies belong to different (A-) time series. It was in order to avoid such confusions that the usage of the word "now" was forgone altogether.

2. An astute philosopher like Oaklander might argue as follows. An object has some property at t which it does not have at $t'(\neq t)$. This is change in time. An event has an A-determination at t which it does not have at t'. This is change of time. Both changes consume time—just as movement does. It seems that change of time is alarmingly close to the movement of the NOW, in that it makes use of precisely that which it is supposed to account for: the transition from t to t'. The theory under attack answers that there is (at t) no t', so "change of time" cannot mean transferring anything to t'. Instead, it envisages something incomparably more elementary, which concerns t' but does not move to it, does not happen to it, and certainly does not occur in it. But, aware of the impropriety of resting its case on negatives, the theory goes further and (rather inconsistently) states that t' is added to the time series. This still means that t' was not at t and is at t'. So the increase of time satisfies the definition of "change in time" and thus is, after all, an event. It is, moreover, an incoherent event, because events consume time, while the theory insists that no (existing) time is available for the process of increasing the time series.

The mistake in this argument is similar to that made by trying to apply ordinary arithmetic to transfinite numbers ("How can a set be of the same cardinality as that of one of its own proper subsets?"). Our ordinary intuitions fail and mislead us when we think of creation, just as they do when we think of infinity. There simply seems to be no valid reason for holding that creation in general and creation of new time in particular are events.

There are, on the other hand, compelling reasons for not counting the increase of the time series as a change in time. By definition, such a change must have a subject, a thing that acquires and discards ordinary properties. But the moment that is being added is no such subject. It is not a thing, and it is not there. Were its addition an event, which it certainly is not, it still could not serve as the subject; events also are not things. Furthermore, they are changes, and changes do not change. And if the critic tries to offer the time series as the subject, we ask him to specify the kinds of time series he condones. If he is a thorough going B-theorist, he can come up with nothing better than the (one and only) B-series. But this already contains indiscriminately all existing and all predicted moments. Nothing is being added to this series; nor is it a subject of change in *any* sense. While if the critic condescends to entertain A-series, he should understand them in the strict A-sense in which they are absolutely momentary affairs which admit no extension. The addition of a new moment means a new A-series, of which the new moment is the instantaneous upper bound. It emphatically does not mean change in time occurring to the older A-series, which, far from being increased, is being replaced. And the replacement itself, of course, is the very change of time which is the whole point of A-theory; it is certainly not a change in time. Also, even if time series were things, in the sense of the change-in-time definition, which they are not, the presence of a new moment is surely not an ordinary property as required by that definition.

The whole situation is characteristic of what happens whenever the B-theorist attempts to come down on the transient view. Like Dummett's observer, who can be either in time, uncommunicatively aware of the dynamic process, or out of time, in principle able to master no more than temporal relations,[50] the B-theorist, clearly the outsider, can only grasp A-theory in his *own* terms. But then he understands and slays some already comatose half-breed, not A-theory. He can understand the transient view via the absurd moving NOW just because it is supposed to move upon his own unbounded B-series. As this series presupposes no objective distinction between (existing) past time and (conjectured) future time, the B-theorist is complacently able to say that he has proved there is no such distinction. He then adds that nonexisting distinctions cannot be drawn even by the momentary ghost position of the (nonexisting) moving NOW. When presented with the strange idea of change of time, part of a radically different conception, all the B-theorist can do is to pull it by the nose, convinced that its mask will fall to reveal the comfortable funny features of old moving NOW.

While this block in communication is the source of much sorrow, including the McTaggart paradox,[51] it ought not to deter us from considering the question of the possibility of a theory of dynamic time on its own merits. And if we do this impartially, we cannot fail to realize that while Oaklander, Smart, and other philosophers have raised powerful objections to the moving NOW, they have not shown that a transient theory is necessarily wed to this concept or to any of its direct equivalents.

Notes

1. McTaggart, *Nature of Existence,* vol. 2, chap. 33.
2. Broad, *Examination of McTaggart's Philosophy,* pp. 277–278; idem, "A Reply to My Critics."
3. See Smart, *Philosophy and Scientific Realism,* p. 136.
4. Gale, ed., *Philosophy of Time,* p. 83.
5. In parts of secs. 7 and 8 of my paper "A (Dis)Solution of McTaggart's Paradox," *Ratio* 28, no. 2 (Dec. 1986): 175–195, I briefly introduced somewhat similar ideas. However, the rudimental theory I described there is quite different from the one I am developing here. This is, in part, a result of my having had second thoughts about the role for which an A-theorist should cast the B-series in his theory and even about the kinds of facts this theory should see itself as committed to explain.
6. C. D. Broad, *Scientific Thought* (London, 1925), p. 66.
7. However, some of Broad's own formulations of his positions—e.g., "Future events are non-entities" (ibid., p. 68), tend to show that he believed that the sum total of existence has to be increased by addition of future events and not by addition of future moments. This is also M. M. Schuster's explicit interpretation of Broad's position in "On the Denial of Past and Future Existence," *Review of Metaphysics* 21 (1968): 462. This, if true, makes Broad an event-creationist, whereas in my view he is, or at least ought to be, a moment-creationist.
8. Broad, *Scientific Thought,* pp. 57–70.
9. Broad, *Examination of McTaggart's Philosophy,* p. 281.
10. Bertrand Russell, *Human Knowledge: Its Scope and Limits* (New York, 1948), pp. 496–499.
11. Bertrand Russell, *The Problems of Philosophy* (London, 1912), chap. 5.
12. Cf. ibid.; idem, *Human Knowledge,* pp. 96–97, 188–189, 422–424.
13. This, of course, is only contingently true, so that if, as is logically possible, some of us have or develop precognitive powers which could not be explained away as covertly inferential, our version of A-theory might have to be abandoned. But, as refutability is not too high a price to pay for a narrowly empiricist version which does not go beyond presently acknowledged facts, I refrain from using stronger arguments which support the thesis of there being a radical difference between the future and past (cf. Gale, ed., *Philosophy of Time,* pt. 3).
14. Zeilicovici, "(Dis)Solution of McTaggart's Paradox," sec. 7.
15. Schuster, "On the Denial of Past and Future Existence."
16. Ibid., pp. 454–455.
17. Ibid., p. 454.

18. Broad, *Scientific Thought,* p. 66.
19. See Gale's characterization of the B-theory credo in *Philosophy of Time,* p. 70.
20. Oaklander, *Temporal Relations,* pp. 59–61.
21. Ibid., pp. 48, 52.
22. Ibid., pp. 52–54.
23. Ibid., pp. 37, 40, 43, 51–52.
24. See McTaggart, *Nature of Existence,* vol. 2, pp. 30, 271. But this is McTaggart as interpreted by Oaklander in *Temporal Relations,* chap. 2, sec. 1. I think that on the issue of McTaggart's evaluation of the status of the B-series, Oaklander's interpretation is closer to the letter and spirit of McTaggart than the interpretation which Oaklander (ibid.) attributes to Schlesinger. On the other hand, Schlesinger's interpretation represents much better the position that, as I argue a little later, a modest A-theorist should adopt.
25. Broad, *Examination of McTaggart's Philosophy,* pp. 295–300.
26. Gale, ed., *Philosophy of Time,* p. 79.
27. E.g., most A-theorists argue that B-relations must and can be defined in terms of A-properties, if the B-series is to acquire the status of a temporal series (cf. Gale, ed., *Philosophy of Time,* pp. 72–77; McTaggart, *Nature of Existence,* vol. 2, p. 271). But, rather unexpectedly, the task of defining B-statements in terms of A-statements turns out to be quite a formidable one (cf. Schlesinger, *Aspects of Time,* pp. 42–47). And as reduction, even when successful, does not mean elimination, I believe, as does Broad (cf. *Examination of McTaggart's Philosophy,* p. 302), that it is not essential to the main and stronger argument.
28. Oaklander, *Temporal Relations,* p. 31.
29. Broad, *Examination of McTaggart's Philosophy,* p. 303.
30. Ibid., pp. 265–272.
31. Gale, ed., *Philosophy of Time,* p. 67.
32. Oaklander, *Temporal Relations,* p. 51.
33. Cf. Broad, *Examination of McTaggart's Philosophy,* pp. 265–267; McTaggart, *Nature of Existence,* vol. 2, pp. 9–11.
34. See McTaggart, *Nature of Existence,* vol. 2, p. 13.
35. Ibid., p. 14.
36. I agree with Oaklander (*Temporal Relations,* p. 43), who makes the connection between McTaggart's rejection of Russell's account of change in any subject (which is that the subject has something at t which it does not have at $t' > t$) and McTaggart's rejection of the B-series as, by itself, a temporal series. I think that had McTaggart not willfully chosen to involve temporal becoming in the explanation of change in properties of physical objects, he would not have misguided himself into a rejection of the B-series.

    In contradistinction, the view I recommend throughout this essay is that an A-theorist should not take upon himself the burden of accounting for changes in properties of physical objects. He should grant the status of temporal series to the B-without-the-A-series and concede that Russell succeeds in explaining changes in properties of physical objects. The A-theories can (and should) still hold that Russell either fails to explain changes in A-determinations of events or is wrong in claiming that change in a property of an object is the only observer-independent kind of change.
37. Cf. McTaggart, *Nature of Existence,* vol. 2, chap. 33; M. Dummett, "A Defence

of McTaggart's Proof of the Unreality of Time," *Philosophical Review* 69 (1960): 497–504; Schlesinger, *Aspects of Time,* pp. 41, 47–52; Gale, ed., *Philosophy of Time,* pp. 65–70.

38. For this effect see Zeilicovici, "(Dis)Solution of McTaggart's Paradox"; Oaklander, *Temporal Relations.*
39. Cf. Smart, *Philosophy and Scientific Realism,* pp. 131–135; Hans Reichenbach, *Elements of Symbolic Logic* (New York, 1947), pp. 50–51.
40. Cf. Russell, *Principles of Mathematics,* 2nd ed. (Cambridge, 1903), p. 469; Newton-Smith, *Structure of Time,* pp. 13–17; Mellor, *Real Time,* chap. 7.
41. Cf. Smart, "River of Time."
42. See Mellor, *Real Time,* pp. 104–107.
43. See Zeilicovici, "(Dis)Solution of McTaggart's Paradox."
44. Broad. *Scientific Thought,* pp. 66–67.
45. C. W. K. Mundle, "Broad's Views about Time," in P. A. Schlipp, ed., *The Philosophy of C. D. Broad* (New York, 1959), p. 358.
46. Schuster ("On the Denial of Past and Future Existence," pp. 461–452) also points out that Broad treats A-determinations of relational predicates of events. He then proceeds to reprove Broad for giving the event of the change in Tom's status a question-begging creationist interpretation and then jumping, without any reason or proof, to a general creationist theory in which creation itself is treated as an event. Schuster concludes, therefore, that the Tom analogy does not help Broad to present a coherent creationist theory.

    While I think that Schuster's criticism of Broad's maneuver is correct, I do not think it counts so much against the role of the Tom analogy in elucidating what is meant by "creation of events" as against three of Broad's other premises: holding A-determinations to be relations, holding their changes to be events, and holding creation to be an event.
47. Notice, however, that if the relation which generates the B-series is accepted as temporal, despite the fact that it (rightly or wrongly) ignores altogether the phenomenon of increases of time, then, while it may be deemed inexhaustive of time because it does so, it cannot be classified among the nonrelations just because it obtains in some cases between members which exist at a given moment and members which are, at that moment, still within the scope of the increase.

    In contrast to the B-series, the A-series is founded wholly around the concept that at any given moment some members of the B-series do not exist. In other words, the objection of the nonexistent second member may not (under these conditions) be raised against the notion of *earlier than* being a relational predicate. The outcome is that only A-determinations, not B-relations also, are nonordinary properties.
48. Cf. Gale, ed., *Philosophy of Time,* pp. 70–77; Adolf Grünbaum, "The Status of Temporal Becoming," in Gale, ed., *Philosophy of Time,* pp. 322–355; Reichenbach, *Elements of Symbolic Logic,* pp. 284–298.
49. See Grünbaum, "Status of Temporal Becoming," sec. 5.
50. Dummett, "Defence of McTaggart's Proof of the Unreality of Time."
51. For this diagnosis see Zeilicovici, "(Dis)Solution of McTaggart's Paradox."

ESSAY 24

Zeilicovici on Temporal Becoming

L. NATHAN OAKLANDER

The aim of David Zeilicovici's essay is clear and admirable. He wants to develop a theory of temporal becoming that (1) gives full ontological status to B-relations, (2) gives full ontological status to the transitory aspect of time, and (3) avoids commitment to the moving NOW and the subsequent (McTaggart's) paradox. But it does not seem to me that he succeeds in accomplishing these difficult undertakings, and in what follows I shall attempt to explain why.

A useful place to begin is by briefly considering the theory that he wants to avoid. On the *traditional* tensed or A-theory of time, the NOW is a particular or property that moves along an ordered, but as yet nontemporal, C-series. The terms of the C-series exist (tenselessly) in unchanging relations to each other, and these unchanging relations become temporal relations as the NOW moves across them so that one term, *E,* is NOW *when* another term, *E'*, is future, and *then, when* the NOW "hits" *E'*—that is, *when E'* is present—*E* is past. The problem is that this account of time presupposes time. As the words "and *then*" and "*when*" indicate, in order for there to be temporal relations among terms in the C-series, there have to be temporal relations among the "events" of *E's being* NOW and *E's not being* NOW, otherwise we have a contradiction. But, given the traditional view, in order for there to be temporal relations between terms, there must be a NOW moving across them. Thus, we are caught in either a vicious circle (of trying to define temporal relations in terms of the moving NOW and the moving NOW in terms of temporal relations) or a vicious infinite regress (where the contradiction of every term being past, present, and future, or NOW and not NOW, is passed on from one series of terms to another).

At this point, Zeilicovici offers a theory that attempts to avoid the pitfalls of the traditional view by maintaining both that (1) *at any present moment future time does not exist* and that (2) the B-series is a *single time series*. On this view, temporal becoming involves a *change of time,* and such a change involves the creation of a "*new entity* [an instantaneous A-series] which did not exist (not even as an incomplete series) before that which makes the new A-determination true (that is, the addition) is there" (p. 246, my emphasis).

Thus, change of time consists in the "replacement" of an old A-series by a new A-series. What is crucial to note about "replacement" is that "the increase in the sum total of existence" is not an increase in a *previously existing* A-series, for if it was, then we would have a change in time and not a change *of* time. Rather, as moments are created, a totally new A-series comes into existence, and the old A-series ceases to exist, that is, is replaced.

Zeilicovici's view is intriguing, but we must clarify certain points in it before we can detect its strengths and weaknesses. For example, what does he mean by "moments" of the "B-series" and the "A-series"? Zeilicovici's official view of "moments" is relational. He says that "time is the ordered set of moments, and moments are *defined as equivalence sets of events under the relation of simultaneity*" (p. 237, my emphasis), and his response to Schuster's argument against the creationist view of time seems to imply a commitment to the relational theory. For, on the event-creationist view, events come into being "*at already existing moments*" (so this view is vulnerable to the moving NOW critique); but on the moment-creationist view that Zeilicovici adopts, "future moments are not a priori existent delivery rooms in which events are to be born" (p. 239) (and so, allegedly, this view is not open to Schuster's criticism). Thus, on Zeilicovici's official view, there do not exist (tenselessly) absolute moments when events come into existence (or are created), but rather, what come into existence are moments relationally understood—that is, sets of simultaneous events.[1]

Nonetheless, other remarks that Zeilicovici makes suggest a commitment to absolute time. He claims that A-theorists, no less than B-theorists, can maintain that the B-series (by itself) is a time series. After all, he exclaims: "Why are we not free to grant that *earlier than* is 'an analyzable temporal relation' . . . just because it ranges over the unanalyzable temporal particulars known as moments?" (p. 242).

Soon we shall see why an A-theorist is not free to grant this, but at present we should note that no relationalist would countenance "unanalyzable temporal particulars known as moments." One might say that Zeilicovici just misspoke here; but then what are we to make of his claim that "if $t < t'$, t-tokens of such statements as 't' exists at t'' are not always as uninformative as they look. For, if t happens to be the present, t' *may remain forever just an empty placeholder on the time axis,* while the above *t*-token conveys the informative prediction that this is not the case" (p. 238, my emphasis).

What can he mean by saying that "*t'* can remain an empty placeholder on the time axis" if not that *t'* may exist as a term in the B-series even if no event (or substance) comes to occupy it? And that is to be committed to absolute time. The ambiguity reflected in his talk about "moments" has an analogue in his account of the B-series.

The B-series, according to Zeilicovici, is a genuinely *temporal* series whose generating relation is *being later than*. Furthermore, the B-series is "static," or unchanging, in that neither the relations between the terms in the series nor the series as a whole change any of its properties. Most important, there *is one and only one B-series*. Zeilicovici's reasoning for this thesis is important and worth quoting at length.

> The single B-series differs from any of the many A-series just in failing to distinguish between existing moments and predicted moments and by being, as a consequence, unbounded. B-sentences *report* B-relations between existing temporal particulars and *predict* them for nonexisting ones without registering any ontological difference between the two kinds of particulars or any epistemological difference between the two kinds of acts. . . . (pp. 239–240)
>
> The addition of a new moment to existing time fails to create a new B-series, because *the new moment appears in the old B-series anyway* [my emphasis]. Thus the two series are indistinguishable from each other and so must, a fortiori, be counted as one and the same series. It is true of course, that what appears in the new series is a "real," existing moment, while what figures in the old series is a mere predicted shadow of a moment. But such a feature cannot be used as a springboard for making a distinction between two B-*series* because *being later than,* the generating relation of any B-series, signally fails to mark any such distinction between the two kinds of moments. (p. 240)

His position seems to be this: that all the terms that are ever in the B-series are always in the B-series. There is a single B-series composed of *the same terms* at every (present) moment, and that is so even though at every present moment some terms of the B-series exist and some terms of the B-series (those that are predicted) *do not exist;* that is, they are terms beyond the upper bound. But what does it mean to say that an existing moment in the "new" B-series "figures in the *old* series [as] a mere *predicted shadow* of a moment"? (p. 240, my emphasis).

Either the predicted shadow is a member of the B-series or it is not. If it is *not* a member of the B-series, then the new B-series is not the same as the old B-series. On the other hand, if the predicted shadow *is* a member of the B-series, then it would appear that it is nothing other than a "placeholder" on the (absolute) time axis "waiting" to be filled by newly created events. In the first case, he must abandon his claim to be treating the B-series (by itself) as a temporal series independently of the phenomenon of temporal becoming as represented by the A-series. In the second case, he must abandon his claim to be analyzing temporal becoming *minus* the moving NOW.

To connect this point with the previous one about the ambiguity concerning "moments," note that if moments are understood relationally, then the "predicted shadow" cannot exist as part of a *single* B-series. For if the predicted moments are sets of simultaneous events, then at any present moment there are an infinite, or at least an indefinite, number of (future) sets of simultaneous events. After all, on this view the *future is open*. But to say that the future is open is just to say that there are many possible events that will follow the present moment. Each possibility constitutes a different B-series. Admittedly, there is only one *actual* B-series, but the actual B-series, the single B-series, is composed entirely of terms that *exist*. Thus, if he adopts a relational view of moments, and if he maintains that the future does not exist but is only predicted or conjectured, Zeilicovici cannot speak of a *single* B-series; that is, he cannot say that the old B-series is the same as the "new" B-series. Indeed, he cannot say that the B-series is a time series independently of temporal becoming at all, because it turns out that the B-series simply is the A-series. If, however, moments are understood as absolute, that is, as intrinsically temporal particulars, then predicted moments are part of a *single* B-series, and the becoming of events, or the transitory aspect of time, must reintroduce the moving NOW. By playing on the ambiguity of "moments," Zeilicovici is able to think that a *change of time* preserves a single B-series (where moments are absolute) within which there exists the creation of different (new) A-series (where moments are construed relationally). Perhaps we can understand this last point more clearly by considering his conception of the A-series.

Simply put, the A-series is that part of the B-series that exists. As he puts it:

> A present-at-t event [is] any event occurring at the upper bound of until-t existing time. If some predictions are fulfilled, there will be a next moment; t will lose the frontier characteristic, and our event will lose presentness while retaining intact all its ordinary properties and all its temporal order B-relations to its previous partners. But, because it is now

related to new members, this event (and all others) will belong to a different time series, the series which includes the previous one but whose "membership is increased." (p. 239)

Hence the passage of time is the creation of new A-series, each containing more members than the previous one that it has replaced. Thus, he thinks that he has a single B-series, and within that single B-series a multitude of A-series. In this way there exists a change *of* time (A-series) without a change *in* time, that is, without any single thing (like the NOW) moving along the B- (or C-) series. But, given Zeilicovici's problems with the B-series, his attempt to distinguish a *change of time* from a *change in time* fails.

If the B-series is composed of absolute moments "waiting" to be occupied, then the creation of each new A-series at a certain time is itself an *event* that must take place in time and so must undergo temporal becoming. On the other hand, if the B-series is composed of moments construed as sets of simultaneous events, and future moments do not *exist* as members of the B-series, then we cannot speak of a single B-series or of an "old" B-series being the same as a "new" B-series. Rather, as each moment is created, we get a different A-series which itself is a B-series, and that is precisely the part of the traditional A-theory that Zeilicovici so strenuously wants to avoid. Finally, if he maintains the relational view of moments and insists that there is only one B-series (which I have argued all of whose terms must exist), then obviously Broad's idea that "The sum total of existence is always increasing" must be abandoned. Indeed, it would appear that the only way to reintroduce transience at this point is by appealing to the moving NOW. On the hopelessness of that way of trying to make sense of temporal becoming, both Zeilicovici and I agree.

Note

1. Note that Zeilicovici does emphasize that each A-series is instantaneous; but this is troublesome because each A-series, in order to have its membership increased, must contain *past* as well as present members. However, if a newly created A-series contains members that are both past and present, then how can what is created be a moment—a set of *simultaneous* events?

ESSAY 25

The Stream of Time

GEORGE SCHLESINGER

As we have seen, there are some very fundamental similarities between space and time; there are also some dissimilarities. There exists, however, a unique aspect which, according to some, constitutes the most fundamental difference between space and time and which, according to others, does not exist at all. I am referring to the belief that time—unlike space—has a dynamic aspect as well as a static one.

Those philosophers who insist that the passage of time is not what their opponents have called a "myth" claim that, on the contrary, there is hardly any experience that seems more persistently and immediately given to us than the relentless flow of time. We are all greatly concerned by the swift passage of time and regret things that have *ceased* to exist and whose memory *keeps fading*. It is quite common for people to feel that life is *passing them by* and to fear that time is *running out,* as well as hoping for the *arrival* of better times and joining Prior in thanking goodness that certain painful experiences are over.[1]

To approach the matter from a different angle, many of us regard the existence of two radically distinct types of temporal properties a central feature of experience. We think of any given moment *m* as being later than an indefinitely large set of other moments and of *m*'s relation of subsequence to these moments as a permanent fact of the universe. We also think of *m* as being earlier than another huge set of moments and of this, too, as an enduring fact. The relationship of being earlier than or later than (which McTaggart and his followers, the transientists, have labeled B-determinations) confers no privileged status upon *m,* since (disregarding now the possibility of time having a beginning or coming to an end), every other point in time also has these relations to two very large sets of moments.

On the other hand, we feel strongly that there is a property that confers upon *m* a unique privileged status, namely, that of being in the present. The property of "presentness" may be too basic to lend itself to a strict definition; however, our attitude to the present may be described as regarding it as distinct from every other temporal position. For, while the future is yet to be born and the past is fading rapidly, the present is palpably real. This characteristic of *m* (which McTaggart called one of its A-determinations) is a transient feature of that moment; *m* grows bright and comes to life for an instant, after which its presence or immediacy is passed on to the next moment.

Objections to Transientism

While the account just given seems to be an unquestionably correct description of what might rightly be claimed to be one of the most pervasive and compelling features of felt experience, on only brief reflection, it becomes apparent that it contains absurdities. Consequently, the majority of contemporary philosophers insist that the notion of a moving NOW, implying that future events approach us, reach us momentarily, and then keep receding further and further into the past, does not just fail to correspond to the facts; it is altogether devoid of sense. Smart and, more recently, philosophers like Paul Needham, Keith Seddon, and David Zeilicovici have referred to the dynamic view of time as "incoherent," "unintelligible," "self-contradictory," and "in principle meaningless."[2]

Among the reasons for the strong opposition to transientism, consider first the claim that the moment *m* which is in the present enjoys some special, privileged status that sets it uniquely apart from all other moments, which have already receded from or have not yet reached contemporaneousness. Surely the distinctive status that the present bestows upon *m* does not persist; in fact, it lasts only for the very brief duration of *m*. But then, every other moment too is granted this much; it resides in the present during its own duration. And if one moment enjoys its "nowness" no more enduringly than any other, given that each one is in the limelight of the present for exactly the amount of time it lasts, how can we ascribe a distinct privilege to any particular moment? Thus, to use Grünbaum's expression, "nowness" turns out to be "not unique."

This leads to the uncovering of another absurdity affecting the dynamic theory of time. We have indicated that the radical difference between an A-term and a B-term is that the latter is a two-place predicate, since it denotes the relative temporal order between various moments or events, whereas the former is a one-place predicate ascribing the property of presentness (or futurity or pastness) *simpliciter* to a given moment or event. But I have just pointed out that the property of "nowness" can not be correctly ascribed to the same moment all the time. Thus a brief event *E* (for example, the wall clock's hands showing exactly 12 P.M.) has the property of occurring in the present at a single moment, *m* (that is, when the time is 12 P.M.). Consequently, since *E* is in the present only at *m,* then by correctly ascribing the property of presentness to *E,* we are in fact attributing the existence of a certain *relation* between *E* and *m*. But if so, A-predicates are also two-place predicates. Moreover, the statement "*E* is present at *m*" expresses nothing more or less than the typical B-statement "*E* occurs at *m*."

These points, which represent only a few of the objections that have been made by philosophers, if unanswered, are sufficient to demonstrate not merely that transientism—the essence of which is the claim that there exist two radically different temporal properties—is false, but that it is devoid of content. For they indicate that one can lend substance neither to the claim of changing truth-values of A-statements nor to the claim that A-predicates are one-place predicates.

The Difficulty in Renouncing Transientism

Among the many metaphysical controversies that exist, the debate concerning the question of whether time itself is involved in any kind of flow, motion, or change, as well as whether time has any characteristics other than B-characteristics, may justifiably be regarded as a unique episode. For example, virtually all philosophers strongly believe in the validity of induction, in the existence of other minds, and in the existence of an external world. Yet there is hardly anyone who would insist that he or she is incapable of forming an intelligible picture of a universe in which induction failed to work most of the time, in which all apparent human bodies they came across were mere robots, or where most experiences were misleading, since nothing external to the senses existed at all. We tend to reject vigorously such ideas, not because of their conceptual incoherence but because they offend our intuitions and would force us to resign ourselves to living in a dreary universe, having an isolated, vacuous existence with a totally unpredictable future lying before us.

Furthermore, most philosophers will readily concede that, in spite of their firm belief in the overall validity of induction, there have been instances in which that method has failed, and that in the course of the history of science, various beliefs thought to have been solidly supported by the principle that the unobserved is going to be like the observed subsequently turned out to be downright false. Neither is our belief that a certain kind of behavior is a sure sign of the possession of a mind quite absolute or straightforward. In the last fifty years or so, we have come to witness physical systems that are capable of copying, and even outstripping, the most typical of human behavior, namely, intelligent reasoning, as is routinely done by advanced computers, and yet we do not believe them to be actual sentient beings, capable of pain, pleasure, joy, or discomfort. And lastly, although on the whole we do believe in an external world, experiences that completely fail to reflect external reality, instances in which our senses give us an entirely false picture of objective reality, are common to us all. For instance, many times we have vividly experienced partaking in extraordinary adventures, performing dramatic deeds in exotic surroundings, only to wake from our feverish dreams and discover that in reality we were confined to our bed.

In the context of temporal transience the situation is radically different. First of all, no transientist will admit, or indeed sees any reason for admitting, that at a given interval or under any specific circumstances time may lack its characteristic transient feature. He would not feel compelled to admit that at any point in the long history of the world, there was a moment which had B-features but none of the A-features of past, present, or future.

However, what is more significant, the transientist would insist that he is simply incapable of conceiving a temporal universe which consisted of the B-series alone. It is glaringly obvious to him that it is conceptually impossible that at any given moment there be present more than a single moment. Given that our universe endures longer than a single moment, he finds the conclusion inescapable that the remaining points in time are either no longer in the present—that is, they have receded into the past—or must be still ahead of us—that is, in the future, waiting to become present. The transientist may, of course, be well aware of the dissenting view according to which his argument fails to get off the ground. His initial assertion that a given moment can be present for no longer than a moment and that the rest must be either in the past or in the future is, according to his opponents, based on the fallacy that there are such things as past, present, and future. In reality, they (the Russellians) insist, all that can be said is that a given moment occurs precisely at the time of its occurrence, while all the rest occur either before or after that moment. Against this, the transientist will complain that his adversary treats all points in time as being equal, since what has just been said applies equally to all moments. However, such egalitarianism sweeps aside one of the most indispensable aspects of time, conceptually speaking: namely, that among the indefinitely long B-series of moments, only a single one enjoys the privileged status of having the spotlight of reality flash upon it and thus has the momentary capacity of being experienced directly (as opposed to being merely remembered or anticipated). The important point is this: the transientist is not merely more comfortable with this view than with the alternative; he rejects the alternative not because it is uninviting or threatening in its bleakness, but because he finds it altogether incoherent. Time bereft of its A-features is simply unintelligible.

It is significant that even veteran opponents of transientism like Smart admit that "certainly we *feel* that time flows." It is most instructive to read how Keith Seddon—who, in his very clearly written book, relentlessly opposes the dynamic theory of time—describes his own sentiments regarding the matter:

> Part of the answer as to why we feel time flows obviously lies in the fact that to talk about time at all involves talking with spatial metaphors and movement metaphors. The language we have all grown up with dictates patterns of thought. Even though I now feel convinced that time does not flow, if I think about the past or future I can do this only by thinking in terms of the river-of-time image; without the image I could not think about time at all. Since I cannot dispense with the image, all I can do is remind myself that as far as the metaphysical truth about time is concerned the image is false.[3]

I believe that this passage, written by an implacable anti-transientist, clearly shows that a belief in the dynamic feature of time, unlike a belief in the validity of induction, say, has been adopted by its adherents not because of considerations of convenience, but because they felt it to be conceptually imposed upon them. Seddon and others in the Russellian camp refuse to yield to their natural feelings in this matter, because they have convinced themselves that it would lead to insurmountable logical difficulties. They have not succeeded, however, in coming up with an equally manageable alternative way of thinking. They have only been able to devise a (undeniably ingenious) way of expressing, in a language that makes no reference to A-determinations (that is, in a tenseless language), most of the things we want to say and hitherto have been forced to say in tensed language. Thus, they were resourceful enough to circumvent the need to complain that "I am now having an excruciating toothache" by saying instead, when the circumstances warranted it, "My excruciating toothache is simultaneous with this utterance." They remained, of course, quite aware of the fact that it is of very little concern to the tormented individual with which utterance his racking pain happens to be simultaneous, or whether *any* utterance occurs at all. Indeed, it is in general of no relevance to the urgent quest for relief whether any particular event connected with it is occurring at the time of affliction. There are exceptions: for example, simultaneously suffering from another pain as well, where the two aggravate each other, or when the pain is simultaneous with arrival at the dentist, and thereby the prospect of alleviation comes into sight. But to assert the simultaneity of any of these events and the toothache would not even remotely resemble what the patient wishes to convey by "I now have an excruciating toothache." On the other hand, what is of overriding importance to the sufferer is that the toothache is not, unfortunately, *now over.* Thus its existence is not yet confined to memory only; nor is it an event *yet to come,* such as would allow one the opportunity of taking adequate preventive measures. In any case, the conventional cry "I am now having an excruciating toothache" conveys directly what is of great concern to the aggrieved: namely, that, regrettably, he is tormented by pain that is yet to be abated, that is not merely looming in the future, but is raging and demanding attention now.

A student of the philosophy of time, upon hearing "My excruciating toothache is simultaneous with this utterance" will of course be able to infer everything expressed by the conventional expression it replaces; but the two expressions are certainly not identical, and only "I am *now* . . ." conveys directly the urgent information the utterer wishes to make known.

It should be mentioned that recent defenders of the tenseless view have come to admit that tensed sentences cannot be translated into tenseless sentences without loss of meaning. They, the anti-transientists, are now ready to embrace the idea that tensed sentences may, under certain circumstances, be indispensable. However, they insist that the truth conditions of such sentences are never tensed facts; the proper truth conditions are tenseless. Thus, we are told "Any token *S* of 'It is now 1980' is true iff *S* occurs in 1980" (Essay 5). And undeniably the truth or falsity of *S* occurring in 1980 is a tenseless fact.

I am unable to examine this suggestion closely, since I am not sure I understand it well. For example, on the transientist's view, among the truth conditions of "It is now 1980" would be the fact that today's *perfectly reliable* newspapers printed 1980 as the current year. Now, according to most philosophers, to understand a sentence is to be able to recognize (if not to know, as Quine has claimed; or at least be able to describe, at a minimum) circumstances under which that sentence is true. Thus, if I were challenged to show that I understand the sentence in question, I could refer to the fact that today's newspaper named 1980 as the current year. According to the new tenseless view, however, I could not do so; I would have to say, "*S* occurs in 1980." But how am I going to accomplish this? What word or phrase would I have to use for "*S*"? Surely, it would be of no help if I said "A token of the sentence 'It is now 1980' occurred in 1980"; for it might be true that some such token occurred in 1980, yet the particular token (the meaning of which I was asked to produce) occurred in some other year. Clearly, therefore, what I am expected to do is to give a unique description of the token in question by mentioning some of its individuating characteristics. The precise spatiotemporal location of the token would amount to such a characteristic, but in the present context it would be unusable: the token is supposed to convey a contingent statement, whereas "The token occurring in 1980 occurs in 1980" is a tautology.

A detailed enough description of the auditory properties of the token would be adequate. Yet it would be absurd to insist that unless I can produce such a description I cannot be said to understand what was said. I may not be able to recall at all how the token sounded, whether it was said slowly or quickly, in an American, British, or foreign accent, and yet fully understand its meaning. Thus, I am unable to figure out how S is to be referred to.

A Coherent Picture of the Moving Now

As a preliminary to our attempt to tackle these problems, let us remind ourselves of some elementary points relating to the notion of "actual world." There are many different views about what exactly the "actuality" of the actual world amounts to, but most agree that there are infinitely many possible worlds and that one of them is unique—not merely from its own perspective, but absolutely so—in that it is unreservedly actual. We maintain that it is correct to regard the world in which we find ourselves as the actual world. What we mean by this is not that it is actual only relative to the world which we inhabit, since if this were so, all worlds would have the same status—each one actual from its own point of view. Neither do we mean that it is so from the standpoint of *every* possible world, because that would make the statement attributing actuality to this world *necessarily* true. Moreover, it would also be inadmissible to claim that the statement was true from an "objective" standpoint, which is a point located outside any universe. No such point is available, since all actual or possible points have already been included in some world. Thus, as a number of philosophers have concluded, this statement has a unique status; it is true *simpliciter*.

Now, of course, the actual world may last any amount of time; it is not hard to imagine ours lasting a short period only: a month, an hour, or even just a fraction of a second. We could imagine that it is universe W_n, the actual universe, in which I am writing these words, at this very moment, m_n (where m_n, though very short, is of a finite duration). The actuality of W_n lasts only for this interval at which m_n is experienced to be present, and every moment preceding it is regarded as past and those following it are anticipated as future moments.

Let us focus on an indefinitely large set, Σ, of worlds, namely, W_1, W_2,

. . ., W_n, W_{n+1}, . . ., which are all qualitatively identical. Every event that takes place at some moment m_i, in one member of Σ also takes place at the same moment in every other member of Σ. In other words, all these worlds harbor absolutely identical B-series. If so, what makes these worlds numerically distinct?

The answer is: the event that sets the members of Σ apart is not one that takes place *within* any of them, but a different kind of event, which we may call an A-event (but in a sense that differs quite substantially from McTaggart's), which happens *to* each world. The uniqueness of W_n, for instance, consists in its being actual at no other time but during the small interval m_n. It is at the beginning of m_n (which, to remind ourselves, is a term in the B-series) that actuality begins its visit at W_n, only to depart at the end of m_n. During actuality's brief stay at W_n, m_n springs to life and assumes immediacy or presentness. The actuality of the next (where the term is used in an A-sense—that is, the A-next) universe, W_{n+1}, also lasts a fraction of a second only, and we shall denote the brief interval during which W_{n+1} is in the privileged state of enjoying its actuality by m_{n+1}. In W_n, moment m_{n+1} is never palpably real; however, at moment m_n the statement "m_{n+1} is in the future," if actually asserted, is true *simpliciter.* Similarly in W_{n+1}, m_n is never present to be directly experienced; but the actual statement in W_{n+1} (which of course can only be made at m_{n+1}) "m_n is in the past" is true *simpliciter.*

There Exists No Meta-Time

In order to clarify what is involved, let us represent the transientist's view of reality (without introducing a meta-time) by adopting a two-dimensional picture of time (see Fig. 1). The sequence of elements constituting each horizontal line represents the successive moments in the B-series. The sequence of lines may be said to constitute the A-series. Thus the various lines follow one another in an A-sense. Each segment along the B-series axis contains precisely the same set of events, the only difference between the parallel lines being the different position of the thick line, representing the shifting location of the moment unique in any given world, the moment which is alive during that world's tenure in actuality.

The ordering relations along the B-axis are the B-determinations "before" and "after." Thus, since m_n occurs before m_{n+1} in the B-series, the segment representing the former is to the left of the latter.

It is crucially important to realize that the perpendicularity we observe in the diagram is not part of anything but the diagram. The series themselves

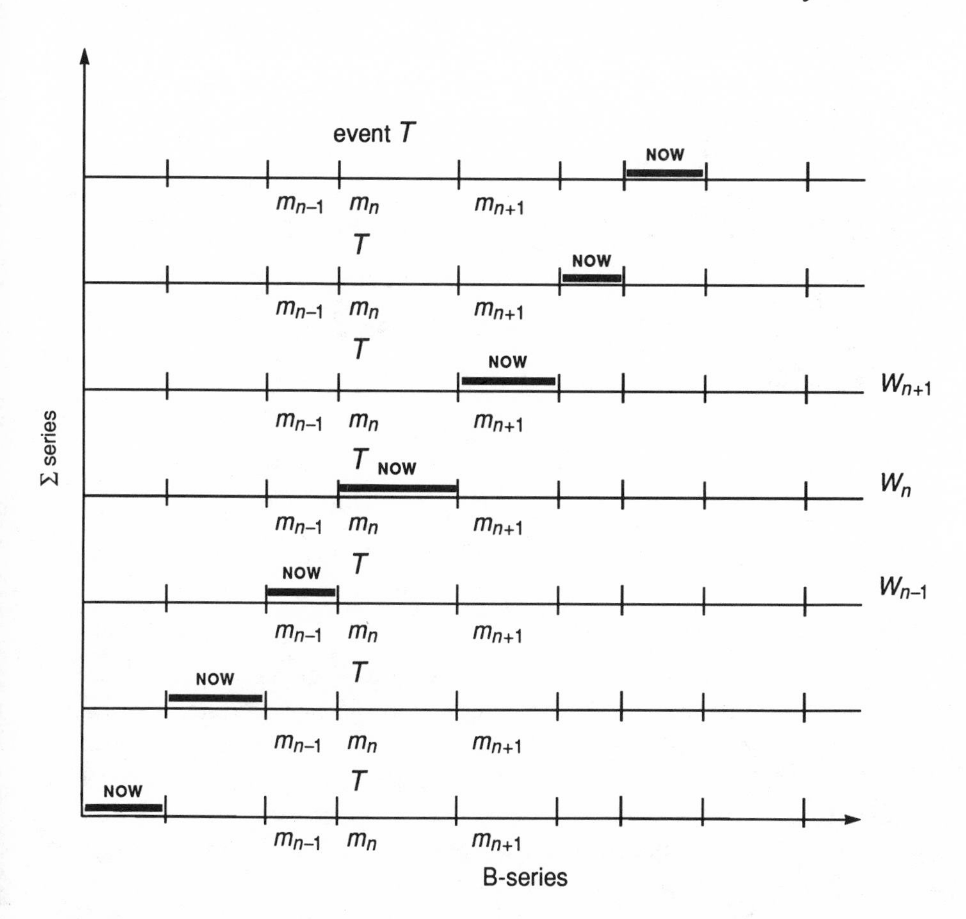

Figure 1

do not form a right angle, nor do they—or could they—possibly form any other angle. No geometrical features can be meaningfully ascribed to them.

It will be of considerable help if we compare time as seen by the transientist with the situation obtaining in the case of an ideal gas, where pressure varies directly with temperature (Fig. 2). It would never occur to any one that in a gas, temperature and pressure were perpendicular to one another. The same goes, of course, for the two time series. Also, in Figure 2 the ordering relations of the values represented along the horizontal axis, are successive magnitudes of temperature, whereas the ordering relations along the vertical axis are successive magnitudes of pressure.

Returning to Figure 1, the ordering relation along each horizontal axis is the increasing magnitude of the B-series as measured from some arbitrary

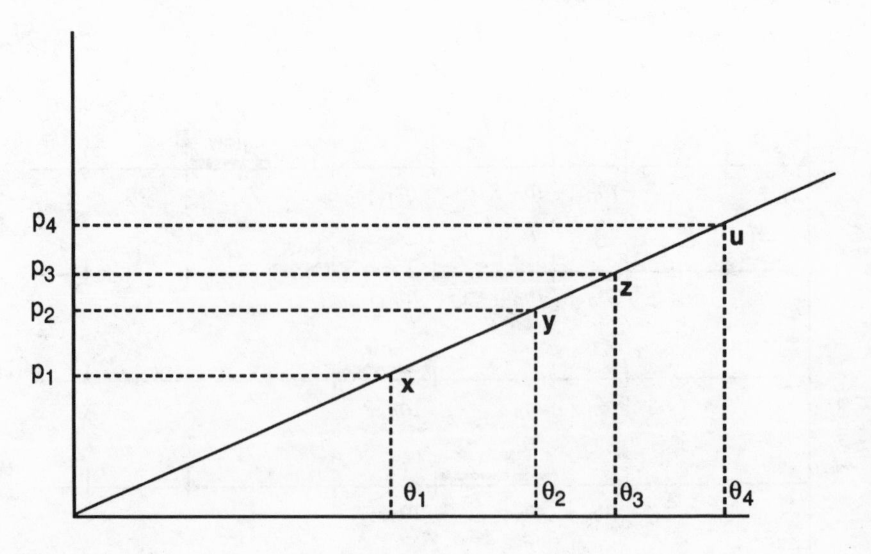

Figure 2

point on the left, which might be the origin (the point where the two coordinates intersect). The ordering relation along the vertical axis is the increasing number of worlds that have achieved actuality.

In Figure 2, X is the point where temperature θ_1 is seen to correspond to pressure p_1. Thus, X may be said to be a "temperature/pressure correspondence point." So, of course, are the points Y, Z, and U. We would have no difficulty in understanding if we were told that the "temperature/pressure correspondence point" was shifting its position; it partakes in a motion depicted by the diagonal line, which may be taken as representing a movement in the direction of increasing pressure as temperature increases.

It is possible to speak in a similar fashion about the "passage of time," which is represented by the different positions of the thick line on the successive horizontal lines. In Figure 1, the counterpart of the temperature/pressure correspondence point is the B-series/A-series correspondence point, and it too appears to be moving in an upward-rightward direction. This movement represents the shifting NOW (which is the "B-series/A-series correspondence point"), which is a movement toward later moments in the B-series as actuality shifts to higher and higher-indexed worlds.

Our A-series differs considerably from what McTaggart called A-series. The ordering relations of pastness and futurity generating his A-series consisted of the same elements as those constituting the B-series, that is, moments and events. In our account, the two series consist of different elements; the A-series is a sequence of worlds belonging to Σ, worlds of very brief duration, and it is ordered by the particular position that the NOW occupies in any given W. We, unlike McTaggart, obtain a dynamic picture of time not through a representation wherein points on adjacent lines slide relative to one another, but by a diagram in which corresponding points on perpendicular lines are plotted against each other. Nevertheless, as we shall see, our A-series retains those features of McTaggart's that are vital for the transientist.

We may also use our diagram to show how the transientist accounts for the notion of "temporal becoming." Many regard that notion as untenable, since it implies that there exist two radically different kinds of events. On the one hand, we have, for instance, event T, defined thus:

$$T = \text{the Titanic begins its maiden voyage on 12 April 1912,}$$

which is, was, and always will be a fact; hence the statement asserting this to be so has a permanent truth-value. On the other hand, there is, on the dynamic view, what seems to be a higher-order occurrence, T^B, which happens to T and which is conveyed by the statement "T ceases to be in the future and *becomes* a present event." According to the transientist, the last statement has no permanent truth-value. This, however, is hard to reconcile with the fact that T^B occurs precisely when T occurs and is of equal duration; for how can the statements referring to them be of such a fundamentally different nature?

The answer is shown by the diagram, in which the B-event, T, is marked on each line and stands for a member of Σ, at precisely the same point. Thus, the statement concerning T is equally true in every one of the worlds represented by these lines. The second event, T^B, is to be thought of as an A-event, as it does not take place *within* any of these worlds; T actually assuming presentness is brought about through W_n acquiring actuality, and this happens only to W_n. Thus, in whichever world actuality may happen to be, the statement ascribing moment m_n to the occurrence of T is true. On the other hand, the statement concerning T^B is true only when actually asserted in W_n at m_n; that is, at the point in time at which reality briefly favors W_n.

No Privileged Position in the B-Series

The picture that emerges from the previous section shows an important aspect of the two basic categories of temporal properties, and thus of radically different temporal statements. Any B-statement like "T is after E," or "T occurs at m_n," if true, is true in every member of set Σ. On the one hand, any A-statement like "m_n is now" can be true only in a single member of Σ. This will help to clear up the difficulties concerning the "privilege" a given moment may enjoy (difficulties which were at the root of the objections to the dynamic view mentioned in the second section). On the other hand, we seem to have been led to the unquestionable conclusion that it is absurd to insist on bestowing this privilege on any specific moment or moments to the exclusion of the rest, since all moments are present at the time of their occurrence. But we have conceded that a "privilege" shared equally by all moments is bereft of content. Now, however, it should be clear that we are in fact not facing such an intolerable dilemma. For we realize that there is, in fact, no privileged position; furthermore, it does not make sense to speak of such a position created by the resting point of the NOW in the B-series as such. To do so would be as absurd as referring to the particular position of the thick line in the sets of parallel horizontal lines. In fact, that position varies with each individual member of the set Σ. The thick line in our diagram does, however, occupy a peerless position on each particular line: it represents the moment at which each individual world (belonging to Σ) enjoys its short-lived actuality. Thus the privilege which the transientists are trying to bestow on a given moment consists in a favored position occupied by that moment (for example, by m_n in W_n) and only by that moment; it is the favored position at which the "light of actuality shines" upon the world it occupies. In W_{n+1}, it is moment m_{n+1} which has this privilege, and so on.

The Nature of A-Predicates

Earlier we cited the charge that the transientist holds inconsistently, that ascribing "presentness" *simpliciter* to an event like *T* attributes a monadic property to it while admitting that *T* has this property at one particular moment, m_n, only. He or she is therefore forced to concede that "*T* is now" ascribes a dyadic relation (namely, that of simultaneity) between the event of the Titanic's proud launching and the moment 12:14 P.M. on 12 April 1912.

On the suggestion advanced here, however, the transientist is in no such predicament. Consider, for instance, "Ripe tomatoes are red," which is, strictly speaking, true only relative to some worlds. Nevertheless, as a rule we regard the statement as containing a monadic predicate "red"; we do not relativize redness to the actual world, since the statement "Ripe tomatoes are red" is true *simpliciter*. Similarly, since "*T* is now" is true in W_n, if asserted in that world at the moment when that world enjoys actuality, "*T* is now" is true *simpliciter*.

It is more important to emphasize that it would not be disturbing at all if someone were to insist that presentness is a dyadic relation (between T and W_n). Earlier, on the traditional interpretation, it was essential that "T is now" be regarded as ascribing a monadic relation to T, for otherwise we would be forced to read it as "T is simultaneous with m_n." But in that interpretation, A-statements refer to monadic properties. To see it clearly, let me give a detailed description, according to the new version of transientism, of the way A-statements differ from B-statements, as well as of the way that both of them differ from statements that are neither A nor B.

A B-statement like "T is simultaneous with m_n" has the same truth-value regardless of at what time and in what world it is asserted. An A-statement like "T is now" is true *simpliciter* if it is *actually* asserted and at the moment at which T occurs. Indeed, "T is now" is equivalent to "T is occurring at the moment m_n," which is in a privileged position in the sense that W_n assumes actuality at that moment. Or, to use an expression used earlier, "T is now" is equivalent to "T is at the A-/B-series correspondence point." But note that the statement "T is now" is absolutely false (and not merely *fails to be* true *simpliciter*) in every Σ-world except W_n, and even in that world it is false except when actually asserted.

On the other hand, a statement like "T is in the past" when actually asserted can be true *simpliciter* and also false *simpliciter.* If asserted at any moment m_{n-x}, where $x > 0$, in any Σ-world which is actual at that moment, it is true *simpliciter,* otherwise it is not. The conditions which ensure that "W_n is such that it will become actual" is true *simpliciter* are the same.

A B-statement refers to a dyadic property; that is, it expresses a relation (of before, simultaneous with, or after) between particulars like events and moments *that* take place within every Σ-world. An A-statement ascribes a monadic property (of futurity, presentness, or pastness), to some event or moment which is a member of the B-series.

A statement like "'T is now' is true in W_n" is neither a B-statement nor an A-statement. It is not a B-statement, because it fails to relate the temporal particular T (that is, the event T) to any event that takes place *within* the Σ-worlds. And, of course, it is not an A-statement, because either it is true regardless of where and when it is asserted or it is false without any qualification.

Time Is Real

In this section we shall see why we are now in a position to reply to McTaggart's argument leading to his famous paradoxical claim that the concept of time is involved in a contradiction so serious that we must abandon the belief in time's reality.

Let me first state that virtually everyone who has dealt with the subject agrees that McTaggart has succeeded in raising an issue that touches the very foundations of the philosophy of time. His argument has certainly engendered an immense amount of discussion. Many have interpreted it in a way that seemed to them necessary to ensure a cogent argument; there have also been dozens of rebuttals. Notwithstanding, numerous philosophers have agreed that McTaggart has brought to light a genuine difficulty. However, we are not obliged to infer from it the intolerable contention that time is unreal. Instead, a variety of suggestions can be made as to the conclusion we may legitimately draw.

The core of McTaggart's argument is relatively easily stated. He has pointed out that the one-place predicates "is in the future," "is in the present," and "is in the past" are mutually incompatible. (When a friend tells me ruefully, "Our silver wedding anniversary has already passed," while his wife practically simultaneously avers, "We are busy with preparations for our silver wedding anniversary that is about to come," then at least one of these statements must be false.) Yet futurity, presentness, and pastness *may* legitimately apply to every individual event and moment. The reply that one is tempted to give appears so obvious that it is hard to grasp how anyone could fail to see it immediately: there is no hint of a contradiction here, since these predicates never apply jointly to the same thing. The same event may be seen in the future from the vantage point of the present, as well as in the past when viewed from a later temporal location. (In the case of the wedding anniversary, both wife and husband could be telling the truth if the former was speaking several months earlier than the latter.)

Some philosophers, probably bent upon rendering McTaggart's paradox very short and blunt, assume that one of his premises (premise 4) was "A proposition that can truly be asserted at one time can truly be asserted at any other time."[4] According to this view (which Farmer attributes to Pears, among others), if it is ever the case that "*E* is present" is true *simpliciter,* then it is always true *simpliciter* that *E* is present. However, this interpretation of McTaggart is very unlikely to be correct; McTaggart would clearly not entertain premise 4, since for him transience is the most essential feature of time; that is, he regards it as the central aspect of A-characteristics that they may be acquired and then shed by temporal particulars. Hence, even if it is a general principle that the truth-values of propositions are unchangeable, A-statements constitute an exception to this principle. Perhaps the reason why some philosophers have saddled McTaggart's analysis with this and similarly absurd interpretations is that his conclusion that time is altogether unreal is simply intolerable; therefore, he must have been led to such a conclusion by either confused arguments or blatantly unsound premises.

In fact, however, McTaggart's argument is much more sophisticated than that and is not vulnerable to obvious objections. On a superficial look, it seems that we can deal with McTaggart's contention that the employment of A-terms leads to inconsistency, as we have mentioned earlier, by making the natural and compelling point that we never apply incompatible predicates to the same term at the same time. This simple reply actually contains two different statements, neither of which, when examined, proves adequate. The first proposition amounts to an attempt to resolve the difficulty by pointing out that of one and the same event E it may be truly asserted that it *is* in the past at present, while it is also true to say that it *was* in the future in the past. This reply does nothing more than shift the difficulty, however. Now we may point out that "*is* past at present" is itself incompatible with "*is* future at present," yet both terms may be predicated (at different times, of course) of the same moment. Of course, this difficulty may be thought to be soluble along the lines attempted in the context of the first difficulty, by distinguishing the different tensed expressions resulting from qualifying "is future at present," but this again could be shown to be merely shifting the difficulty to the incompatibility of a somewhat more complex expression.

The reply avoids the very first inconsistency by claiming that there is no contradiction to begin with, since E may be in the past at m_n and in the future at m_{n-1}. This, simple though it is, would be fatal to the view that two radically different temporal characteristics exist; it would mean jettisoning A-determinations altogether. Statements like "E is in the past at m_n" imply no transience, have permanent truth-values, and thus are straightforward B-statements.

McTaggart's paradox boils down to this dilemma: either we retain the notion that a moment may exemplify the property of being, say, in the past *simpliciter,* in which case we are led to an inconsistency, or we deny the possibility of applying the term "past *simpliciter*" and apply a notion like "past" that is relativized only (to certain fixed moments in time), in which case we are forced to admit that A-determinations do not exist.

The problem just described is a severe one, which, I believe, does not lend itself to any solution as long as we are not prepared to drop some of McTaggart's assumptions. It can be resolved, however, through application of our version of transientism, which differs in a number of ways from McTaggart's yet preserves its essential features as it renders the transient feature of time intelligible. Thus, we should be willing to go along with

McTaggart that the ascription of two determinations like futurity and pastness to *E* amounts to inconsistency. We should also concede that to try to escape the inconsistency through relativization of different moments deprives us of A-determinations. The solution will emerge from our reply to an objection made to the thesis presented here.

Three Objections

The question of whether time has both static and dynamic features is among the most contested metaphysical issues, so it is not surprising that any new version of transientism that is suggested should meet with strong objections from the followers of Russell. Here I propose to cite three of the most commonly voiced objections.

First, our suggestion depicts time in a manner that does not in any way correspond to experience. When it is said that space is three-dimensional, everyone has a clear picture of what is meant and can verify that it corresponds to the concept he or she has formed of space. The average person is also capable, if required, of explaining why space, obviously, has three dimensions. Furthermore, we are able to point at planes within our universe, that is, objects that have merely two dimensions and even describe a universe of "flatlanders"; that is, one in which space itself has two dimensions only. In addition, we are able to handle conceptually four- or, five- or, indeed, any multi-dimensional space.

Nothing even remotely resembling this is true concerning time. Nobody has hitherto observed time to have two different dimensions; it is hard even to make sense of a two-dimensional "plane" that is supposed to constitute time. Furthermore, we are incapable of imagining what two-dimensional or multidimensional time would be like. In brief, our conception of time is one that does not admit the possibility of multiple dimensions.

The reply to this objection is that there is no intention whatever of ascribing to time itself the kind of dimensionality we ascribe to space. When we speak about the three-dimensionality of space, we think of the existence of three mutually perpendicular coordinates in space itself. As explained before, the A-series and the B-series are the two essential components of time, which of course have no geometrical relations whatever, just as neither the pressure nor the temperature of a gas is capable of entering into a geometrical relation with anything.

Thus, it is an erroneous belief that our temporal experiences contain nothing that would even remotely suggest two-dimensionality. To be sure, nobody has observed time itself to have two different dimensions similar to those of space, and indeed, it is hard even to make sense of a two-dimensional "plane" that constitutes the elements that constitute time. But even many of the most ardent adherents of the static view admit, however reluctantly, that we do have a vivid impression that there exist two different kinds of temporal determinations, an impression that some of them find impossible to escape in everyday discourse. My aim has been to provide a diagram, which, if properly interpreted, offers an authentic, visually translated display of the two series, the combination of which gives rise to our temporal experiences. The diagram was certainly not meant to be a true-to-life snapshot of the two series that constitute time. The series themselves are not perpendicular to one another, nor are they the constituents of any other angle. Indeed, no geometrical features can be meaningfully ascribed to them.

Second, a favorite objection has been to argue that the present suggestion falls into the very trap it was designed to avoid. In other words, our version, like McTaggart's, is thought to be forced into avoiding the problem of predicating contrary temporal terms of the same particulars by abandoning the possibility of A-determinations.

A clever variation on this theme has been advanced by David Buller and Thomas Foster in a recent paper.[5] They begin their argument by giving an account of McTaggart's paradox. They have judiciously chosen what I believe is the most plausible interpretation of McTaggart's argument. Subsequently they claim that my model succumbs to a comparable paradox: "For any world W_i in Schlesinger's model . . . W_i's being actual is incompatible with any other world's being actual; for to say that W_i is actual is simply to say that no world W_k, such that $k \neq i$, is actual. But in Schlesinger's model every world is actual. It follows that every world in the model is both actual and not actual."[6] If this were really true, then we could as well claim that an admittedly innocuous statement like "Ripe tomatoes are red" (an example used earlier in this chapter) is also involved in a paradox. The redness of tomatoes is a fact "relative" only to our kind of world; there are plenty of possible worlds (each claiming actuality for itself) where tomatoes are granite-grey or marine blue or whatever! However, what really counts is that no sentence but the one attributing redness to ripe tomatoes is true *simpliciter*. Sentences ascribing different colors to tomatoes are true in many other possible worlds, but we cannot assert them to be true without reserve (*simpliciter*). Similarly, given that "W_i is actual" is asserted at m_i, the assertion is true without reserve. Sentences denying this in W_k such that $k \neq i$ are not true without reserve.

Still, it is instructive to go along with the Buller–Foster criticism which assumes the existence of a contradiction from which there is no escape except by engaging in a series of steps culminating in the following:

(1) "*E* is present" is true when W_n is actual.
(2) "*E* is present" is true when W_n is actual and W_n is actual at m_n.
(3) "*E* is present" is true at m_n.

They then claim that (3) is a B-statement, and that this is incompatible with my thesis (the details are not needed for present purposes).

However, this part of their argument contains a subtle error (sufficient in itself to eliminate their objection). For clarity's sake, let us distinguish between A-present and B-present. "*E* is present at *m*" means nothing more than "*E* occurs at *m*" or "*E* and *m* are simultaneous"; thus the term "present" signifies B-present. However, when "*E* is present" is said, then of course A-present is what is meant. Now it should be obvious to the authors that (1) cannot refer to the B-present, for that would make (1) unintelligible. The B-presence of *E* has to be related to some event or moment within the B-series. The sentence (1b), "*E* is B-present" is true when W_n is actual, is equivalent to the unintelligible (1b′), "*E* is simultaneous with——" is true when W_n is actual.

Clearly, therefore, Buller and Foster intended to attribute to me the following:

(1a) "*E* is A-present" is true when W_n is actual,

which at best leads to:

(3a) "*E* is A-present" is true at m_n.

But (3a) is an A-statement. In plainer language, it amounts to:

(3a′) "*E* and m_n are both A-present" is true (provided conditions 1 and 2, mentioned earlier, obtain).

Third, according to this last objection, everything I have claimed may be correct; but if so, I have succeeded in defending the viability of the transient view of time only at an exorbitant cost. Only by granting an actual role to an infinite number of worlds, instead of the single world familiar to us, am I able to fend off all the alleged contradictions afflicting the dynamic theory of time. The objectors question whether it is reasonable to pay such a high price for salvaging the dynamic theory of time.

In reply, I should first remind the reader of the minimal view we have adopted concerning the status of possible worlds. Only the actual world has any substance. The present version of transience does not, however, increase the magnitude of what is actual, that is, of what has real existence. Thus the claim of extravagance might well be compared to Yogi Berra's request that the pizza he ordered be cut into four and not eight pieces, as he is not hungry enough to eat eight.

Furthermore, I should like to indicate, without going into all the details, that those who prefer to do so can formulate our account without making any reference to more than a single world. They could make use of the successive states of the actual universe as members of the A-series. Such an approach would not be objectionable as long as each state could be identified independently of the time of its occurrence, by, for example, having each state distinguished from all other states by size or by the density of the universe associated with that state.

How Fast Does Time Move?

We should also be able to see how perhaps the most famous objection to the dynamic view—that any form of movement or dynamic change requires us to speak of a rate—can be met. Indeed, we may assign a *rate* at which it takes place; moreover, this rate could conceivably increase or decrease. It has been pointed out, by Smart and Seddon among others, that it makes sense, for instance, to speak about a car moving, regarding the car to be covering a certain spatial distance during a given temporal interval. Should the first interval be shortened, the car will be said to be moving at a faster rate; should it lengthen, it will be said to be moving more slowly. Thus, if temporal movement really does take place, then it should be possible (at least in principle) to answer the question "How fast does the NOW move toward the future?," or "How fast are future events approaching us, and at what rate are the events of the past receding from us?" These kinds of questions must be possible to answer, "because movement is change of position (albeit in this case a change of position in time) and such change, to be change at all, must be supposed to occur at some rate or other. Since the river of time represents lapse of time and not extension in space, the only possible answer seems to be 'events pass by at the rate of one second per second' or 'We are moving into the future at the rate one second per second.'"[7] Undeniably, if we can do no better than offer such a pointless answer to the question of how fast time is moving, then we are doomed to speak utter nonsense in response to the next question, namely, how we are to envisage a change in the rate of time's flow. Are we perhaps to say that time could flow at a different rate than it has done so far, since it may gather speed and cover two or three seconds per second?!

Let us first note that in order to have movement or change or even a change in the rate of change, it is not necessary that one of the series involved consist of temporal intervals or of spatial intervals. It makes sense, for example, to say that in a given Scandinavian country taxation rises increasingly rapidly with increase of income. This would be so if, for instance, an individual earning $5,000 per annum had to pay $500 in taxes, while one earning twice that amount must pay three times the amount (from an income of $15,000, $5,000 is exacted, and so on). As long as we have series of two different kinds—in this case, one sequence consisting of different amounts of tax to be paid, the other a succession of incomes earned—we can talk about movement, and even about the acceleration of movement.

It is not hard to discern that the schema outlined above does provide two different sequences, consisting of radically different kinds of elements, scope for the changing position of the NOW. Anyone who wishes to do so may take advantage of our schema by positing that the various moments referred to so far are not of equal length; they vary at random within the range of 10^{-1} and 10^{-2} seconds. Consequently, the successive hours in the B-series do not contain the same number of moments either. This makes it possible that in the last hour actuality may have traveled through such and such number of Σ-worlds, while in the hour before it traveled through a smaller or larger number of such worlds. If we remind ourselves that each Σ-world's hold on actuality lasts as long as the duration of its own "privileged moment," we see at once that a change in the rate of time's flow is in principle possible.

What Is Being "Approached" by Future Events?

The ever resourceful anti-transientist J. J. C. Smart has recently advanced a new argument for why the idea of events that have already taken place receding into the past, as well as events in the future approaching us, is unintelligible. The core of his argument is that it is impossible to identify clearly who we are referring to as the particular, relative to which the distance of events keeps changing. He writes: "What is the 'us' or 'me'? It is not the whole person from birth to death, the total space–time entity. Nor is it any particular temporal stage of that person. A temporal stage for which an event *E* is future is a different temporal stage from one for which event *E* is present or past."[8] It is undeniably true that an individual who lives from say, 1850 to 1950, occupies all the temporal points lying between those extremities and none outside them; thus, he is confined permanently between these two boundary points. Smart is also right in saying that not one of that individual's temporal parts shifts from any of the positions it occupies. However, on the dynamic view, that person has a unique, privileged temporal phase; namely, the slice which is momentarily real or alive, that is, the slice that is present in the world which is actual. On the view that actuality keeps shifting, from the earlier to the later members of the sequence, it follows that the actual stage of every person keeps shifting from earlier to later positions in the B-series.

Determinism and Transientism

Unquestionably, the major reason why those who subscribe to the dynamic theory of time do so is that it reflects our most ubiquitous experiences of time. Yet there exist some additional, albeit less important advantages that the dynamic theory may be claimed to bestow upon its adherents. One of these involves the issue of determinism and free will.

The doctrine of determinism, as everyone knows, teaches that the universe is completely law-governed. Thus, given all the initial conditions prevailing everywhere at any given moment, and combining these conditions with those laws, every event occurring at any time can be accurately predicted. This doctrine has been seen by many as a threat to the idea, precious to us all, that we humans are able to choose and act according to the dictates of our own will. Most of us recoil from the idea that our conduct, in all its details, happens with absolute inevitability as determined by factors operating entirely outside our jurisdiction.

Hundreds of attempts have been made to deal with this problem, which, however, is not our concern here. The point I wish to raise is that any philosopher who has felt that human freedom was threatened by universal determinism felt so only when confronted with the possibility that the laws of physics coupled with past physical conditions imply what is happening now. Virtually no one has ever expressed misgivings over the fact that we may be bereft of all freedom because from the totality of facts at some future moment (and combining these with the laws of nature) all present events may be inferred; in other words, it does not seem to curtail our freedom of action if the state of the universe at some future time renders everything that happens now inevitable. The question I should like to consider is, why does it make such a radical difference from which direction the constraints on us originate; why should we feel robbed of our cherished autonomy in the case that our current conduct is imposed upon us by external factors prevailing in the past, whereas we would accept with equanimity being told that our current conduct was imposed upon us by conditions obtaining in the future?

It might seem natural to dismiss this question by simply pointing out that past events which are nomically connected with present events are regarded as causes, whereas future events with similar connections to the present are regarded as effects. The idea would be that it is reasonable to recognize a threat to our freedom if all our acts are caused by external factors, but not when our acts generate effects that are manifested in the future. Many would object to this, however, by saying that the substance of a causal connection between events consists of nothing more than one event being a sufficient or a necessary or a sufficient and necessary condition for the other. Now if *E* is a sufficient and necessary condition for *F,* then the two events are symmetrically related; hence *F,* too, is sufficient and necessary for *E*. Therefore, just as it is irrelevant whether *E* is to the north, south, above, or below *F,* so we should expect it to be irrelevant in what temporal direction *E* is relative to *F*.

A considerably more plausible explanation of why temporal order is a decisive factor can be given by those who subscribe to the dynamic theory of time as well as to the doctrine of the open future. On their view, reality consists of everything that has happened so far. Future events and moments do not yet exist. Holding the view of the openness of the future does not commit one to denying the doctrine of determinism. The course of the future may be fully determined, but as long as any event has not yet become a present event, it has not become part of reality. On this view, there is complete symmetry between the statements "*E* occurs at *t*" and "*F* occurs at $t + n$," as they mutually imply one another. There is a crucial difference, however, between the two events referred to when *E* is present while *F* is in the future, for *E* is already part of reality whereas *F* is not. A nonexistent event is incapable of any action, let along bringing about other events. Thus it is out of the question that F should, at a time when it does not yet exist, produce *E*. Hence *E* must be the progenitor of *F*. Thus the transientist has a fairly persuasive explanation as to why we are not concerned that our present actions might be imposed upon us by future conditions. Events yet unrealized cannot be seen as a real source of constraint upon present happenings.

It should be clear that those who deny the existence of any temporal series except the B-series are unable to explain in the way set out above why an earlier event can have the role of a producer and shaper of a later event, but not vice versa. The Russellians' position prevents them from embracing anything like the idea of an open future. The reason is that, were they to entertain that notion, time would be freed from all those features which they see as preventing them from admitting A-characteristics. Recall that one of their major objections to the dynamic theory of time is that A-determinations relate event and moment to a single special point, namely, the point in time which is the present. This point is different from any other point and is thus treated in the dynamic theory as a privileged point. This, however, is inconsistent with the view that every moment has precisely the same privilege, since every moment acquires and holds on to its presentness until it shifts to a later moment. But, on the doctrine of the open future, reality grows with the advance of time (or the realized part of the B-series gains in length by one moment with every moment that passes); that is, the terminal point of the actualized portion of the B-series at m_j occupies a point that is further into the future than the terminal point occupied by m_i. The unique, privileged point in time may be said to be the tip of the achieved portion of the B-series. It is thus no longer the case that any specific moment is present *simpliciter;* the privilege of being in the present is relativized to the length of the realized B-series. Thus, point m_i is privileged relative to an actualized B-series that ends at m_i, while the point m_j is privileged relative to a realized B-series that ends at m_j. It should be emphasized that neither the location of m_i nor that of m_j can, by definition, be derived from the length of the realized B-series. This is so not only for the reasons mentioned earlier (that the lengths of the moments constituting the elements of the B-series are determined by a random process), but also because the various moments need not be referred to through their indices but can be referred to through the description of their unique empirical features.

McTaggart's famous objection, that ultimately we have to give up A-determinations altogether, no longer applies either. McTaggart thought that since we are forced to relativize them to certain fixed moments—for example, "*E* is in the future" must be rendered "*E* is later than m_i"—then all A-determinations are in fact abolished and become B-determinations. On the open-future view, of course, we may escape McTaggart's paradox by rendering a statement like "*E* is in the future" as "*E* is in the future when the realized portion of the B-series is such and such a length." Thus we relativize A-characteristics to the length of actual time that has passed, but that does not reduce them to B-determinations, since the length of the series involved changes constantly.

Finally, let us consider the puzzlement that has been raised by Smart, Seddon, and others, that movement requires more than a single variable. To say "Events pass by at the rate of one second per second" makes no sense and certainly does not express a rate of change. This difficulty does not arise for those who maintain the openness of the future. They have two different temporal variables: the position of the terminus of the realized B-series and the length of that series. The motion of time consists in the covariation of these two.

Russellians, who have recourse to none of these, are nevertheless not bereft of all explanation of why we may be concerned about past events but not about future events determining our current actions. They will say that if *E* and *F* are mutually sufficient and necessary conditions, then it is an irreducible fact that *E* is the cause of *F,* which is its effect. It is part of the meaning of "*F* being caused by *E*" that *E* forces *F* to happen and determines the shape it takes.

Thus, it seems that the transientist has an advantage over those who subscribe to the static theory of time. The former is able to make use of a premise that few would wish to find fault with: namely, that nonexistent particulars are powerless to act. Subsequently, by maintaining the openness of the future, the transientist is able to offer a clear, vivid explanation of why future events cannot influence current happenings. The Russellian, by contrast, who can do no better than cite it just as a primitive, is incapable of explaining why position in time, which has nothing to do with existence, nevertheless plays a crucial role in determining what can be the cause of what.

Comparing and Contrasting the Various Versions of the Dynamic Theory

So far we have seen three versions of transientism: (1) McTaggarts's original view, (2) the view that time may be treated as two-dimensional, and (3) the view that time's motion consists in the constant growth of the realized B-series.[9] The following is a summary of the major features in which these resemble and differ from one another.

1(a) According to all three versions, time has dynamic A-characteristics as well as static B-characteristics. According to McTaggart, the only elements that play a role in the production of temporal phenomena are moments and the events which occur at them. His A-series and B-series consist of identical sequences of terms except for the temporal predicates which apply to these terms.

(b) According to the first new version considered here, set Σ, the sequence of qualitatively identical worlds (or states of affairs), plays an essential role in bringing forth the A-series.

(c) According to the second new version, which is based on the openness of the future, an indispensable role is played by the division between two different parts of the B-series, the realized and the potential parts. Figure 3 provides a picture how the idea of a constantly growing, realized B-series enables us to offer a coherent account of transientism. This figure is similar to Figure 1 in that each horizontal line represents the same B-series and differs only in the length of the realized part (depicted by a continuous line). Each unbroken line is longer than the one below it, illustrating how the achieved part of the B-series keeps growing.

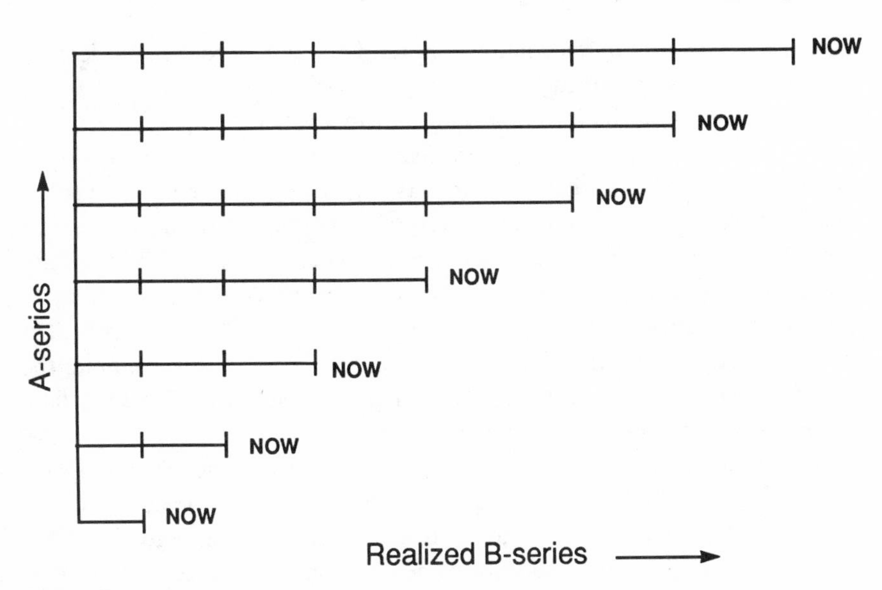

Figure 3

2(a) In McTaggart's version what gives rise to time's passage is the A-series sliding along the B-series.

(b) in our first revised version, all moments, all events, and indeed all other particulars have a fixed position in the B-series. Nor is the B-series itself in motion relative to anything. It is actuality which shifts its position and briefly vivifies the succession of Σ-worlds. Given that each world's coming to life coincides with a different term in the B-series, the A-series/B-series correspondence point keeps shifting from an earlier to a later position in the B-series. This constitutes the source of the phenomenon known as the passage of time.

(c) Our second revised version makes no use of any moving physical particulars either. What moves is the point that divides the two parts of the B-series. As we see in Figure 3, the realized portion of the B-series keeps increasing its length, and consequently the terminal point of each successive unbroken line (which represents what we experience as the present) keeps shifting from left to right (that is, from earlier to later moments).

3(a) According to McTaggart, A-statements are unique. There is a principle which says that every statement is either permanently true or permanently false. A-statements constitute the sole exception to this principle. For example, "*T* is in the past" is true if asserted later than the time of *T*'s occurrence but is false otherwise. To some philosophers who hold that different tokens of a statement are identical if and only if they have identical truth-values, this odd feature of A-statements represents one of the barriers against entertaining the idea of granting them a role in temporal discourse.

(*b*) Transientism as construed with the aid of the Σ-series clearly does not face any such problem. Like all statements in general, A-statements are either true or false. If at this moment I were to make the assertion "*T* is a past event" (recall that *T* = the launching of the Titanic), I would be speaking truthfully. As indicated earlier, nobody would claim that the same statement must have the same truth-value in every possible world. Consequently, it is not a source of concern that in some possible Σ-world (for example, the world which is actual in 1892) "*T* is a past event" would make a false assertion, as it would not be an actual assertion.

(*c*) On the open-future interpretation of transientism, one may also maintain the principle of invariance of truth-values. On this interpretation, when, for instance, someone asserts (ϕ) "*T* is now," the intention is to claim that *T* occurs at the tip of the realized B-series. If, indeed, the location of *T* is at the terminal point of the achieved B-series, then the assertion is true; but if it is asserted at any other point, it is false. Does this show that one and the same statement like ϕ may have different truth-values depending on the time of its assertion? We should notice that no two extremity points mark the end of the same realized B-series. Each realized B-series is unique in containing the number of moments (of finite length and counted from an arbitrary date designated as zero).

Our objection to Russell's theory of time could not be based on the conceptual priority of the A-series, since in our picture, while it is impossible to envisage the A-series on its own, it is possible to picture to ourselves the B-series standing by itself. Such a picture would of course satisfy Russellians, but it would be unacceptable to transientists. Transientists' rejection of such a picture is based on universal, omnipresent experience. The static theory is highly ingenious and considerably more parsimonious (as it eliminates a whole species of temporal properties) than its rival. Yet there are some who hold that while unexpected novel ways of looking at familiar things and surprising accounts of common phenomena add spice to the work of the philosopher, his main objective must be authenticity and the sophisticated rendering of common sense without gross distortion. Admittedly, ingenious inventions are often stimulating, but they do not constitute an end in themselves. The wise words of Dr. Samuel Johnson are relevant here: "The irregular combination of fanciful invention may delight awhile by that novelty of which the common satiety of life sends us all in quest; but the pleasures of sudden wonder are soon exhausted and the mind can only repose on the stability of the truth."

Now Johnson might be opposed, as he was criticizing *Tristram Shandy,* a work of imagination, and there may be any number of aesthetically equally valid, incompatible fictitious stories. However, in analytic philosophy, unless we think that its essence is cleverness for its own sake, there is but a single correct way of reasoning, and Johnson's strictures surely apply.

It is regrettable that more people are not aware of Einstein's view regarding our topic. Rudolph Carnap in his *Autobiography* wrote: "Once Einstein said that the experience of the Now means something special for man, something essentially different from the past and the future, but that this important difference does not and cannot occur within physics."[10] Thus Einstein would reject, for instance, Mellor's idea that B-theoretic time alone is real time. He would similarly disagree with Horwich that "the 'moving now' conception of time is indeed incoherent", and with all the many philosophers who insist on depriving time of its central feature.

Einstein's succinct statement is in harmony with the way in which time appears to us. The distinction he makes is rather conspicuous whenever we pick up any book in the physical sciences. Nowhere in a text dealing with the properties of mute systems are we going to come across a single A-statement. However, in the preface we are quite likely to find such statements, as when the author thanks his wife for having constantly encouraged him and expects in the future to do this, that, and the other.

It is worth noting that Einstein's theories of relativity seem to deviate so sharply from normal thinking that even some ninety years later many find themselves incapable of making the conceptual effort they demand. Why, then, in the context of time's passage does he refuse to depart from the way the phenomena are commonly perceived? The answer is that Einstein did not relish the production of startling, clever theories. He was not seeking to reconstruct experience so as to create astonishment. It is a misconception that relativity theory clashes with common sense. It is, after all, dealing with aspects of nature (for example, particles moving with velocities far greater than anything observed before), about which no commonsense view exists, since they are inaccessible to observation without the aid of highly sophisticated means.

Returning to the subject of definitions, we may note that B-statements can be defined in terms of non-B-statements. For example, "E is before F" may be defined as "There exists some Σ-world in which E is in the past and F is in the future." As pointed out earlier, this is not an A-statement but a sentence which makes essential use of A-concepts of future.

(d) On the open-future interpretation of transientism, it is also possible to define B-statements in terms of non-B-statements. Our previous example

"*E* is before *F*" may be translated as "There exists some B-series where *E* is in the realized part while *F* is in the unrealized part."

4(a) McTaggart advanced the famous claim that B-properties may be defined in terms of A-properties, but not vice versa. This has turned out to be a somewhat problematic claim, and I shall not repeat here the various difficulties involved in an attempt to carry out the proposed reduction.

It is important, however, to note why McTaggart made his claim about the definitional priority of A-terms over B-terms. He wished to forestall one possible reply to his proof of the unreality of time. After all, he claimed to have proved no more than that the A-series is involved in a paradox. Why, it might be asked, does this force us to renounce the reality of time altogether? Why could we not still retain the B-series? To forestall such an argument, McTaggart claimed to have established the dependence of the B-series on the A-series. Real time consisting of the B-series alone is not possible; as soon as the more fundamental A-series is destroyed, the derivative B-series has no legs to stand on.

(b) From Figure 1, we should be able to see the absurdity of having the A-series alone. It would amount to having a large set of Σ-worlds none of which would contain events or moments. Each Σ-world would then be a completely immutable, lifeless world and hence devoid of events, which are the indispensable elements of time. In fact, none of these worlds could retain its distinct identity. We could no longer distinguish them through the fact that actuality hits each one at a different point along the B-series, since none of them would contain any trace of a B-series. Consequently, it would make no sense to claim that actuality resides in each Σ-world at a different point in the A-series, since all the Σ-worlds being identical, it is impossible to speak of different points in the A-series either. In other words, the A-series on its own can have no existence.

Notes

1. Prior, "Thank Goodness That's Over."
2. Smart, "River of Time"; Paul Needham in reviewing Seddon's book, in *Theoria* 54 (1988); Seddon, *Time;* and Zeilicovici, Essay 23. This last is the only place I know of in which a third theory of time, lying halfway between the dynamic view and the static view, is advanced in rigorous detail. Zeilicovici's bold thesis seems to be immune to a number of objections that have been raised against the dynamic view, as well as to succeed in satisfying some of the demands of those who believe in time's transience.
3. Seddon, *Time,* pp. 9–10.
4. D. J. Farmer, *Being in Time* (Lanham, Md., 1990), p. 114.
5. David J. Buller and Thomas R. Foster, "The New Paradox of Temporal Transience," *Philosophical Quarterly* 42 (July 1992): 357–366.

6. Ibid., p. 361.
7. Seddon, *Time,* p. 16.
8. Smart, "Time and Becoming," p. 6.
9. Here we have considered only two new versions of transientism. In fact, there are indefinitely many more; e.g., the NOW may be taken as the shifting correspondence point, as the changes in position along the B-series are plotted against the changing magnitude of some cosmic variable like the scaling factor or the density of matter in the universe.
10. Cited in P. Yourgrau, *The Disappearance of Time* (Cambridge, 1991), p. 42.

PART III

Time and Experience

Introduction: The Problem of Our Experience of Time

L. NATHAN OAKLANDER

As we have seen in Part II, recent defenders of the new tenseless theory of time have sought to promote acceptance of the theory by defending McTaggart's argument for the unreality of tense. However, regardless of how many arguments Mellor, Smart, I, and other defenders of the tenseless theory offer, we will never persuade tensers to abandon their view unless we can explain those features of our experience of time that seem to require that time be tensed. It is an impression deeply felt by all of us that time, or events in time, flow from the far future to the near future and then, after shining in the spotlight of the NOW or present moment, immediately recede into the more and more distant past. But does this impression reflect a basic truth about time? Do events in time, or moments of time, really have A-properties and change with respect to them; or is the tenseless theory adequate to our experience of the flow of time? This and related questions are debated in Part III.

In addition to our experience of the flow of time, the principal related questions concern two main features of our temporal experience: the presence of experience itself and our different attitudes toward the past and the future. According to the tenseless theory, there are no basic ontological differences between past, present, and future events; all events exist tenselessly in the network of earlier than, later than, and simultaneity temporal relations. According to our experience, however, while we engage in a great number of actions throughout our lives, only a small subset of them are experienced *now*. So how can the tenseless view, which maintains that all events exist tenselessly at their respective dates, explain our knowledge that a certain experience—say, a headache—is occurring *at present*? How can the detenser explain the fact that experiences (or events) can be known to be present? H. Scott Hestevold argues (Essay 32) that our knowing experiences to be present implies that events will occur, are occurring, and have occurred in such a way that "experiences *cannot* be mere tenseless occurrences on the B-series!"

Furthermore, on the tenser's view, our attitudes toward the past and the future are strongly indicative of temporal passage. Concerning our attitudes toward events in the past, we may note that a very painful experience that is known to have happened to us in the past, say a painful operation, will be contemplated, with *relief*—as we sometimes say, "Thank goodness that's over." By contrast, when we know that an equally painful operation will be necessary in the future, our attitude is one of dread and anxiety. Events that are contemplated with relief are events that are not only past but are seen as receding or moving away from us. Conversely, those that are contemplated with dread are events that are not only future but are seen as coming towards us and about to overtake us; events that are about to be in the NOW, the stream of our lived experience. It is attitudes and experiences such as these that, according to the tenser, provide strong evidence against the tenseless theory of time.

The *locus classicus* of these experiential challenges to the detenser is the following passage by Prior:

> One says, e.g., "Thank goodness that's over!," and not only is this, when said, quite clear without any date appended, but it says something which it is impossible that any use of a tenseless copula with a date should convey. It certainly doesn't *mean the same as,* e.g., "Thank goodness the date of the conclusion of that thing is Friday, June 15, 1954," even if it be said then. (Nor, for that matter, does it *mean* "Thank goodness the conclusion of that thing is contemporaneous with this utterance." Why should anyone thank goodness for that?)[1]

Thus, the expression "Thank goodness that's over" depicts an experience that allegedly provides another reason for the idea that time moves, since what we are thanking goodness for is the passage of an unpleasant event from the present to the past.

Mellor, MacBeath, Garrett, Williams, I, and others have found Prior's argument wanting. Admittedly, a tensed sentence does not have the same meaning or the same content as the corresponding tenseless sentence that states its truth conditions; but for the detenser, no metaphysical significance follows from that fact. Thus, even if an utterance of "Thank goodness that's over" conveys something that cannot be conveyed by a tenseless sentence, it does not follow that the two sentences are used to depict different states of affairs, one entailing a tenseless fact, the other a tensed fact.

Furthermore, the experience in question, namely, that an undesirable event is over (or past), has a tenseless explanation. Mellor's tenseless explanation, for example, is that we use the idiom to express the feeling of relief that occurs after the cessation of pain, which relief, we believe, is caused by the ending of the pain. Thus, in order to thank goodness that a pain is over, "pains only need to be causes of later feelings of relief; they do not also need to be in reality at first present and then past" (p. 298).[2] The debates between MacBeath, Mellor, Garrett, and Hestevold center on the adequacy of Mellor's response, as well as the responses of other detensers, to Prior's problem. In my contribution "On the Experience of Tenseless Time" (Essay 33), I reply to Hestevold's critique of Mellor and attempt to show that an adequate analysis of tenseless time is possible.

In Delmas Kiernan-Lewis's essay a different interpretation of Prior's argument is offered. Kiernan-Lewis maintains that the detenser cannot account for the knowledge we have when we are pleased that something (say, a headache) has ceased to exist, for, on the new tenseless theory, nothing *really* does cease to exist. In my reply, I argue that since there is a sense on the tenseless theory in which events do cease to exist, the detenser can account for the knowledge we possess when we are pleased that an unpleasant event is over, and that only a misinterpretation of the tenseless theory would lead Kiernan-Lewis to think otherwise.

The last two essays, by Quentin Smith and Clifford Williams, continue the debate over the correct interpretation of our experience of time. In "The Phenomenology of A-Time" (Essay 34), Smith argues that some basic phenomenological facts about time simply cannot be accounted for by the tenseless theory. He does this by appealing to the commonplace experience of perceiving a cloud passing over the treetops and arguing that whether the detenser tries to analyze the experience solely in terms of an awareness of temporal relations between events and linguistic utterances, events and acts of judging (or thinking), or events and dates, the analysis is incompatible with the experience itself.

In "The Phenomenology of B-Time" (Essay 35), Williams replies to the essays by Smith and Hestevold. He maintains that the question is not whether we have an experience of the passage of time or different attitudes toward events in the future and those in the past, for clearly we do in some sense. It is, rather, whether these experiences support the tensed theory of time. Williams claims that they do not. His strategy for responding to tensers who appeal to the court of experience to undermine the tenseless theory is to draw attention to the analogies between presentness and hereness and argue that just as our experience of space does not support an inference to a theory of space that countenances *hereness* as a property, so our experience of time does not support an inference to a theory of time that countenances *presentness*.

Defenders of the new tenseless theory of time have tried to show that the necessity of tensed discourse is compatible with time being tenseless. To do this successfully, the detenser must deal adequately with the experience of temporal becoming. But can the defender of tenseless time provide an adequate analysis of the presence of experience and the experience of the passage, or flow, of time? At the end of the day, the reader will have to consult his or her own experience to decide whether the tensed or the tenseless theory provides the better fit.

Notes

1. Prior, "Thank Goodness That's Over," p. 17, emphasis added.
2. In addition to the essays included in this volume, other recent discussions of Prior's argument are D. L. Hardin, "Thank Goodness It's Over There," *Philosophy* 59 (1984): 121–125; Delmas Kiernan-Lewis, "Prior's 'Thank Goodness' Argument: A Reply to Hardin," *Philosophy* 61 (1986): 404–407; and Seddon, *Time*.

ESSAY 26

"Thank Goodness That's Over"

D. H. MELLOR

There are two fundamentally opposed views of time. The opposition is over the nature of tense, that is, the distinction between past, present, and future and the seemingly inexorable way everything moves in time from future to past. On one view, this is of the essence of time; on the other, a complete illusion. The difference between the views is best expressed by means of what McTaggart (1908) called the A- and the B-series of temporal positions.[1] The A-series orders events by tense, ranging from the most future events, through the present, to the events of the remotest past. The B-series orders them simply according to which is earlier, that is, by their dates. The difference between the series is that the A-series shows events moving from future to past, whereas the B-series does not. Everything constantly changes its A-series position, whereas B-series positions never change. Eighteenth-century events, for example, were once future and are now past and becoming ever more so. They precede twentieth-century events in the A-series because they were present first, and both are moving all the time from later to earlier A-series positions. In the B-series, on the other hand, they do not move at all: eighteenth-century events are forever just 200 years earlier than twentieth-century ones, and that's that.

The dispute about tense amounts, therefore, to a dispute about the reality of the A-series. I follow McTaggart in thinking that the A-series is a myth, only, unlike him, I deny that this shows that time itself is a myth. I believe that a tenseless view of time, according to which the reality of time consists entirely in the B-series, can be upheld. I cannot argue the whole case for this here; I aim here only to meet one major challenge to the tenseless view. The challenge is to account tenselessly for the seemingly irreducible presence of experience, an aspect that, for many people, provides the strongest proof that tense is real. By the presence of experience I mean the fact that all our experience, our thought, and our action takes place neither in the future nor in the past, but always in the present, the fleeting moment between future and past.

In trying to explain away the presence of experience, I shall rely without argument on another thesis of the tenseless view: namely, that all tensed sentences have tenseless token-reflexive truth conditions. That is, provided that they are alright in nontemporal respects, the truth-values of tokens of tensed sentences are functions of how much earlier or later they are than the events they are about, the functions depending on the tense. Thus, tokens of a present-tense sentence are true only if sufficiently close in time to the event; my saying, for example, "Fred is getting married this week" will be true only if I say it in the same week as Fred's wedding; otherwise it will be false. Similarly, past-tense tokens are true only if they occur appropriately later than the events they are about, and future-tense tokens are true only if they occur earlier. The functions are more complicated for compound tenses like the future perfect, but the upshot is the same: fix the relative dates of token and event, and you fix the token's truth-value. All this is familiar enough and should by now be beyond dispute, since it is easily demonstrable. At any rate, I shall take it all for granted in what follows.

The real question about tense is not whether this token-reflexive thesis is true, but whether it suffices to account for the A-series. In particular, does it enable us to account for the presence of experience? It is by no means obvious that it does. When, for instance, I say that "It is now 1781" is true in 1781 and that "It is now 1981" is true in 1981, I do not seem to have exhausted the relevant facts of tense. Indeed, I seem hardly to have started on them, since I have not said which of these two centuries we are now in, a fact that, it hardly needs saying, makes a great deal of difference to our lives. And what tells us that we are now in the twentieth century, not the eighteenth, is our experience, which is experience of twentieth-century events. A world just like ours except that the present moment lay in the eighteenth century would be perceptibly a very different world from our own. Or so it seems. But if it is, the tenseless view of time I advocate is wrong. So I must try and explain the difference away.

The problem I face is nicely illustrated in a puzzle posed by Lewis Carroll in 1849.[2] He invites us to choose between two clocks, one of which is right twice a day, the other only once a year. Naturally we choose the clock that is more often right—and are disconcerted to get a clock that doesn't go at all, rather than one that merely runs a little slow! Yet we got the clock we asked for; the stopped clock is indeed right twice a day, whereas a slightly slow clock is almost never right. Moreover, as Carroll says, we know *when* the stopped clock is right, namely, at the very time shown on its face, say two o'clock.

Even so, a stopped clock is not quite what we had in mind. Any sane person would prefer one that almost keeps time. But why? Carroll has given tenseless truth conditions for the unchanging token of "It is now two o'clock" which the stopped clock is, in effect, emitting all the time. We can see what the clock says, and we know when it is true. What more could we want to know?

What more we want to know, of course, is whether it is two o'clock *now*. Is two o'clock the date of our present *experience* of looking at the clock in order to see what the time is? We need not look at the clock to see that "It is now two o'clock" is true at two o'clock—*that* is true all day. Those token-reflexive truth conditions never change. In particular, therefore, they do not convey the changing facts of tense that a stopped clock fails to tell us. And once we see that, we can see that they also do not convey what an accurate clock succeeds in telling us. The latter does indeed say "2:15" at 2:15, "3:30" at 3:30, and so on throughout the day. So far, so good; these truth conditions express what makes everything the clock says true. But again, none of these truth conditions ever changes. It is true all day long that the clock says "2:15" at 2:15, "3:30" at 3:30, and so on. So far as their tenseless truth conditions go, there is nothing to choose between any of these tokens at any hour of day or night. Citing them, therefore, never tells us what the clock itself tells us when we look at it, namely, which of all these times is the *present* time. That is what we want a clock to tell us. And a slightly slow clock will nearly always tell us that more accurately than one that has stopped altogether. Assuming—what the absurdity of Carroll's tale needs anyway—that the slow clock is corrected periodically, it will never be more than a few minutes out in its dating of the present moment, whereas the stopped clock will mostly be hours out. So we ought, after all, to prefer the clock that goes.

That is the obvious, tensed solution to Lewis Carroll's puzzle. What is wrong with a stopped clock is that most of the time it is very bad at telling us A-series facts. *My* problem is how to say what is wrong with it without appealing to A-series facts at all. But to do this, I must first tackle another puzzle, set explicitly by Arthur Prior as a problem for the tenseless view of time.[3]

Suppose you have just had a painful experience, for example, a headache. Now that it is over, you say with relief, "Thank goodness that's over." What are you thanking goodness for? On the face of it, the fact that the headache is no longer a present experience, that is, is now past. So what you are thanking goodness for appears to be an essentially tensed fact, that the headache is past. That is presumably why you make your remark after the pain and not during or before it. Can this fact still be explained when tensed facts are traded in for tensed tokens with tenseless truth conditions?

Prior says not. In this case the true or false token is your saying "That's over," referring to the headache, and the tenseless fact which makes it true is that it occurs later than your headache. All this is obvious and not in dispute. The question is whether this is enough to explain your thanking goodness. And the trouble is that this was as much a fact before and during the headache as it is now that the headache is over. It always was and always will be a fact that this particular token of "That's over" occurs later than the headache it refers to. What is more, that fact could have been recognized as such in advance. In particular, you could have decided in advance to say "That's over" after the headache and known about the fact in that way. So if that were the fact you were thanking goodness for, you could just as well have thanked goodness for it before or during the headache as afterwards. Which, of course, is nonsense. So it seems that you must be thanking goodness for some *other* fact, something that was not a fact at all until the headache ceased. The tensed conclusion appears irresistible; the pastness of the headache, for which you are thanking goodness, must be an extra fact over and above the tenseless fact that makes "That's over" true. If your headache had not really had the A-series property of presence and had not now lost it, there would be nothing to thank goodness for at all.

Yet again, as with Lewis Carroll's clocks, our tenseless token-reflexive truth conditions seem to miss the tensed character of experience. Nor is this a feature only of these somewhat contrived examples. Temporal presence seems to be an essential aspect of all experience. By "essential," I mean essential to its being experience. If I gave only the dates of my experiences, without saying which was happening to me *now,* I should on the face of it leave out precisely what makes them experiences. The headache which has just stopped, for example, is really no longer a headache at all, because it is no longer painful. Something can only be a headache, or an experience of any other kind, when it is present. The past event is only a headache in the dispositional sense in which an object in a dark room, though invisible, can be yellow. If the object *were* lit, it would be yellow; if the event *were* present, it would be a pain in the head. But as far as actual pain goes, the event is merely a retired, or emeritus, headache, not something still in business as the genuine experiential article. And that is why I thank goodness for its pastness; by ceasing to be present, it has ceased to be the unpleasant experience it was. Having a headache, in short, inevitably includes knowing—if one thinks about it—that it is present; and similarly for all other experiences.

On the other hand, experiences are also events in tenseless time. They have dates, as do other events. It is true that they are mental rather than merely physical events, but that does not prevent them from having dates. For one thing, they occur at the same time as physical events and acquire dates in that way. My headache, for example, may have started just as the clock struck six, and that fixes the first B-series moment of its date.

So some events with dates, namely, our experiences, we know to be present events and hence located firmly in both the A- and the B-series. But once some events are located in both series, all events are. The tenses of all other things and events follow from how much earlier or later they are than these present events, and thereby arise all the tensed facts that distinguish worlds differing in the date of their present moment.

Our knowledge of tenses, moreover, comes entirely from the presence of experience. Experiences tell us directly of their presence, and the rest of the A-series we fill in from there. We know, for example, how long light takes to reach us from a celestial event we are now seeing, and that tells us how past it is, namely, as far past as it is earlier than our experience of seeing it. Ultimately, therefore, as I remarked earlier, it is the directly perceived presence of experience which tells us what the tensed facts of our world are, that is, that it really is the twentieth century we are living in and not the eighteenth.

The presence of experience is the crux of the matter. Without a tenseless account of it, tenseless truth conditions on their own will never dispose of tensed facts. That account I will now set out to supply. First, let us look again at Prior's puzzle, this time put slightly differently. Before, I acquiesced in the idiom of thanking goodness for facts, but in this case that idiom is tendentious. What a token of "Thank goodness" really does is express a feeling of relief (not necessarily relief from or about anything, just relief). And the real question is when it is natural to have a feeling of relief in relation to a painful experience. The tensed answer to that question is, of course, when the experience is past, rather than present or future. The tenseless answer can only be that it is natural to feel relief *after* a painful experience, that is, at a later date, rather than during or before it. Now this may well seem a rather weak response. Why, after all, should relief be peculiarly natural after pain, if not because the pain is now past and so, as we have seen, no longer pain? To this further question I confess I see no answer. But I also see no answer to the question, Why feel relief only when pain has the A-series position *past,* as opposed to being present or future? The answer is not, as one might suppose, that relief *cannot* be felt while pain is present and is thus still pain. It is not an a priori truth that relief is never felt, in relation to a pain, while the pain is present. Indeed, I believe it not to be a truth at all. For one thing, masochists presumably feel relief when a future pain, for which they have been longing, at last becomes present; "Thank goodness it's started" is what they would naturally say, not "Thank goodness it's over!" And masochism, however deplorable, is certainly possible. No magic in temporal presence prevents relief being felt while pain is present. And short of a priori prevention, which is not to be had, saying that relief normally occurs only when pain is past is no more explanatory than saying that it normally occurs only after the pain.

I conclude that the tenseless description of the phenomenon of relief, as usually following pain rather than preceding or accompanying it, is alright on its own. The tensed description makes the phenomenon no less mysterious, and we have no good reason to insist on it. Nothing about the relation between pain and relief requires us to credit pains with tenses as well as with being earlier or later than other events.

And from the tenseless description, of relief usually coming after pain, I can forge a tenseless solution to Prior's puzzle. We need not, after all, claim to be thanking goodness for the fact that "That's over" is true only after the pain. There is a much more credible tenseless story than that, which goes as follows. Basically, the remark "Thank goodness that's over" is not a single statement but a conjunction, of "That's over" and "Thank goodness." This can be seen in the fact that the conjuncts are just as naturally joined the other way round: "That's over; thank goodness." Now the first conjunct, "That's over," has obvious and undisputed tenseless and token-reflexive truth conditions, and I have just stated the tenseless conditions under which the relief that the second expresses is normally felt. The reason why the two things are usually said together is partly that these two tenseless conditions usually coincide. The relief which "Thank goodness" expresses is usually felt only when "That's over," said of a pain, is true, namely, just after the pain has stopped. However, there is a little more to it than that. The coincidence of these tenseless conditions is not merely that. The ending of the pain is also, we believe, the cause of our relief; and our saying "Thank goodness" in conjunction with "That's over" expresses, among other things, our recognition of this further tenseless fact.

These are the tenseless facts of the matter, and they explain perfectly well why most of us, wishing to tell the truth and not being masochists, will say "Thank goodness that's over" only when our pain has stopped. Here, I believe, is an entirely adequate tenseless account of Prior's case. It does not, after all, compel us to admit tensed facts as well as tenseless ones. Pains need only be causes of later feelings of relief; they need not also be in reality at first present and then past.

I have drawn out the tenseless treatment of Prior's case at some length in order to extract from it the ingredient needed to dispose in general of the presence of experience. That ingredient is a kind of self-awareness. The salient feature of Prior's case is that we not only have painful experiences, we also subsequently remember having had them. However relieved I feel, I shall not thank goodness for the removal of a pain I have forgotten about. The crux of the matter is the recollection of pain, rather than the pain itself. Now this recollection, which is what "That's over" expresses, is in part a token of a past-tense judgment, the judgment that I was in pain in the recent past. But it is also in part a present-tense judgment, the judgment that I am not in pain—or not in as much pain—now. I shall now say "That's over," let alone "Thank goodness," while I still feel the same pain. The present-tense awareness of being relatively free of pain is an essential ingredient in Prior's case; and this is the ingredient I need.

An awareness of being free of pain is, I contend, a token of a present-tense type of judgment about my own experience, namely, that the experience I am now having is painless. This token judgment is itself an experience, an event occurring in my conscious mind, but an event quite distinct from the rest of the experience it is about. I emphasize this distinction, because there is a temptation to identify our experiences with present-tense judgments about them, a temptation that it is essential to my argument to resist. The source of the temptation is that we distinguish experiences from other events, virtually by definition, as those events we are directly conscious of. We may easily seem bound, therefore, both to be aware of our experiences and to be right in our conscious present-tense judgments about what they are. While I might, for instance, overlook or mistake the color of my pen, I can hardly miss or mistake the actual experience of (say) seeing it to be red. My judgment about the experience itself is so closely tied to it that there is a serious risk of confounding the one with the other. Nonetheless, the risk must be avoided, not just for the sake of my argument, but for a number of familiar and independent philosophical reasons which I need not digress here to rehearse. But one, at least, is apparent enough in this example, namely, that I need not be making judgments all the time about every aspect of my experience. In particular, although I can hardly be *in* pain without noticing it, I can quite easily be free of pain without noticing it. Being free of pain does not force me to make the conscious judgment "I am free of pain," even if I am—perhaps—bound to be right if I do so. So if I do make the judgment, that is an extra fact about me, over and above my lack of pain.

In short, to be aware that my present experiences are painless is to have a further experience, namely, that of judging them to be painless. Since this judgment is about the experiences I am having now, it will have the token-reflexive truth conditions characteristic of the present tense. That is, the judgment will be true provided I am having only painless experiences at the very B-series time at which I make it. And as for judgments of painlessness, so for judgments about all aspects of experience. If I judge myself to be seeing a red pen, for example, my judgment about it will likewise be true only in the case that I am actually seeing a red pen at the time I make the judgment.

Grant all this. Now suppose I start making judgments not about my present freedom from pain or about colors I am not seeing, but about temporal aspects of my experience. Specifically, suppose I judge that the experiences I am now having possess the A-series property of being present. Notice that this restriction in the subject matter of my judgment to the experiences I am *now* having does not make the supposition a tautology, at least not in terms of tense. Tenses, after all, must always be ascribed to events at a particular time, because the tense of events is always changing; and the events which happen now to be our experiences are no exception to the rule. Our question, therefore, has to be: What tense do these events *now* have? And it is by no means tautological that they will now all have the same tense as each other, let alone that they will all be in the present. On the face of it, we could now have as experiences events anywhere in the A-series, past, present, or future. Far from being a tautology, it seems in tensed terms to be a striking and impressive fact that events can only be experiences while they are present. It is indeed, as we have seen, the basis for all our knowledge of other tenses. It is what lets us infer from an event's now being an experience that it is now a present event, a conclusion that becomes, in turn, the premise from which the tenses of all other events and things are indirectly inferred.

However, no one actually infers the presence of experience. Rather, presence is itself an aspect of experience, that is, something we are directly conscious of. (How else, after all, would we know that all experience *is* present?) So my judging my experience to be present is much like my judging it to be painless. On the one hand, the judgment is not one I have to make: I can perfectly well have experience without being conscious of its temporal aspects. On the other hand, if I do make it, I am bound to be right, just as when I judge my experience to be painless. The presence of experience, like some, at least, of its other attributes, is something with regard to which one's awareness is infallible.

The real, relevant—and suspicious—difference between judgments of presence and those of painlessness is that whereas only some experience is painless, all of it is present. No matter who I am or when I judge my experience to be present, that judgment will be true. This is the inescapable, experientially given presence of experience which I now have to explain away. And once experience has been distinguished from the tensed judgments we make about it, that is not hard to do.

We are concerned with token judgments to the effect that experiences we are now having possess the A-series property of being present. Now any token which says that an event is present will be true if and only if the event occurs at the same B-series time as the token does. These are the undisputed token-reflexive truth conditions of all such judgments. But in this case the events to which presence is attributed are themselves picked out by the use of the present tense. Not all our experiences, past, present, and to come, are alleged to have this A-series property, only the experiences we are having now. But these, by the same token-reflexive definition of the present tense, are among the events which *do* have the property now ascribed to them: that is, they are events occurring just when the judgment itself is made. So of course these judgments are always true. Their token-reflexive truth conditions are such that they cannot be anything else. In tenseless terms, they are tautologies after all.

That is the tenseless explanation of the presence of experience. And for once it is not merely an alternative to a tensed explanation of the same thing. There *is* no tensed explanation of this phenomenon. If events can in reality have a range of tenses, I see no good reason for experience to be confined as it is to present events. In tensed terms, that is just an unexplained brute fact about experience. The nearest thing to a tensed explanation of the fact is given by the extreme view of St Augustine and Prior that in reality only what is present exists at all.[4] And of course, if only present events exist, then, in particular, real experiences will have to be present. To that extent the phenomenon is explained by this tensed view, albeit in an implausibly Procrustean way. What it does not explain, however, is how experiences differ in this respect from other events. Other events and things at least appear to be spread out throughout the whole of A-series time: the events we see (especially celestial events) all over the past, the events we predict or plan to prevent or to bring about all over the future. Only our experiences, including our judgments (that is, our thoughts) and our intentions, decisions, and actions appear to be restricted to the present. Of that contrast, the token-reflexive account I have just given alone provides a serious explanation.

The tenseless fact is that experiences themselves, like all other events and things, are neither past, present nor future. But we can make past-, present-, and future-tense judgments about them, just as we can about other matters. Indeed, we have compelling reason to do so. In particular, we have compelling reason to make present-tense judgments about our thoughts, actions, and experiences as they occur. Without making such judgments, we should be unable to communicate with each other[5]—and there is nothing tautological about our ability to do that. Nor are most of these judgments tautologies. There is no tautology in my being aware of having a headache. It may be a necessary truth of some kind that I have a headache when I think I have one; but even that is not a trivial truth. The only trivial truth is that the experiences I am having now possess the A-series property of being present. That is not, after all, a profound experiential restriction on our temporal awareness of the realm of tense. It is nothing more than the fact that experiences which occur when we judge them to be occurring now are bound, by the token-reflexive definition of the present tense, to make that judgment true.

What, then, of Lewis Carroll's clocks? Consider first the clock that goes dead right. It is true that nothing tenseless about the clock itself picks out the present position of the hands; but something tenseless does, namely, the time the clock is being looked at, say 2:15. If the clock is right, it will then be emitting what is in effect a token of the sentence "It is now 2:15." Assuming that I believe the clock, that token will generate in me another, mental token of the same tensed type. Ignoring the time this message takes to get through to my brain, this means that the clock will make me think "It is now 2:15" *at* 2:15, so my thought will be true. That, in token-reflexive terms, is the virtue of an accurate clock: it generates in those who look at and believe it true tensed judgments about what time it is—which is, after all, what clocks are for.

A slightly slow clock generates in those who believe it tensed judgments that are not far out. That is, although they are actually false, most of their tensed consequences will be true. If I never need to know the time to more than a minute, a clock which is ten seconds slow will never deceive me in anything that matters. But a stopped clock can deceive people in matters of great moment, for mostly it is hours out. At most times of day, someone who looks at and believes it will make wildly inaccurate judgments about the time, judgments whose inaccuracy could cause him to be hours late for most important occasions. That is really what is wrong with a stopped clock. So even in token-reflexive terms, a slightly slow clock is much to be preferred. Lewis Carroll's puzzle does, after all, have a tenseless solution.

Finally, what of the difference between our twentieth-century world and one with its present moment shifted back 200 years? Actually, this is just the clock writ large, for we might as well ask how a good clock at 2:15 differs from the same clock an hour later. In tenseless terms, the answer is that the clock itself doesn't differ. Similarly, there is no tenseless difference between the two worlds. Indeed, there are not two worlds, any more than there are two clocks. There is only one world, with things and events scattered throughout B-series time as they are throughout space, including both the eighteenth and the twentieth centuries.

But among these things and events are token judgments that people make from time to time, token sentences thought, spoken, and written, including tokens of tensed sentence-types. And since, as a matter of tenseless fact, we are located within the twentieth century, so are all the token sentences we produce. Their tenseless truth conditions therefore differ by two centuries from eighteenth-century tokens of the same types, and many of them will therefore differ also in truth-value. Many eighteenth-century tokens of "The present King of France is Louis XV" are true, therefore, because they occurred during the reign of that French monarch; whereas, as is well known, all twentieth-century tokens of that particular type are false. There is the real objective difference between the eighteenth and twentieth centuries: not a difference of tensed fact, but a difference in truth-value of tensed tokens of the same type located in the two centuries.

Some, I fear, will not be satisfied by this token-reflexive account. Even if I gave tenseless truth conditions for every token sentence and judgment in the history of the world, they would still ask: but which of all these token judgments is being made *now*? To them I can only say that their question is itself a token, with a date that determines of what type an answer must be in order to be true. The judgments that are being made on the date of the question are those that the true answer must give. So that answer too is made true by purely tenseless facts.

Of course, the question can be asked again of any token answer: Is it being given now? An endless regress of such questions and their answers is possible. But the regress is not actual; nor is it vicious. Every question in it has an answer made true by tenseless facts, because every question has a date. Those of us who eschew tensed facts are sometimes accused of trying to take an impossible eternal view of the world, neglecting our own immersion in the stream of time. But the accusation might more justly go the other way. It is those who cling to tense who fail to take seriously that all things are in time—as are all our judgments about them. Things, events, and judgments alike all have dates, dates that suffice to settle, without tensed fact, the truth or falsity of every tensed judgment there ever was or ever will be.[6]

Notes

1. McTaggart, "Unreality of Time."
2. J. Fisher, ed., *The Magic of Lewis Carroll* (London, 1973), p. 25.
3. Prior, "Thank Goodness That's Over."
4. See J. J. C. Smart, ed., *Problems of Space and Time* (London, 1964), p. 58; A. N. Prior, "The Notion of the Present," *Studium Generale* 23 (1970): 245–248.
5. D. H. Mellor, "Consciousness and Degrees of Belief," in Mellor, ed., *Prospects for Pragmatism* (Cambridge, 1980), p. 148.
6. I am indebted to the editor of *Ratio* for suggesting improvements to a first draft of this essay and for supplying the reference to Lewis Carroll. I owe thanks also to the Radcliffe Trust for a Radcliffe Fellowship, during my tenure of which this work was done.

ESSAY 27

Mellor's Emeritus Headache

MURRAY MACBEATH

Hugh Mellor has had a headache all morning. He has been reading Arthur Prior's paper "Thank Goodness That's Over" and trying to see how to deal with the challenge that it poses to a tenseless view of time such as he himself holds. As the clock strikes twelve, the solution strikes him, the headache vanishes, and pausing only to say "Thank goodness that's over," Mellor starts to write an article called "Thank goodness that's over."

Prior's challenge is as follows:

> One says, e.g., "Thank goodness that's over!," and not only is this, when said, quite clear without any date appended, but it says something which it is impossible that any use of a tenseless copula with a date should convey. It certainly doesn't mean the same as, e.g., "Thank goodness the date of the conclusion of that thing is Friday, June 15, 1954," even if it be said then. (Nor, for that matter, does it mean "Thank goodness the conclusion of that thing is contemporaneous with this utterance." Why should anyone thank goodness for that?)[1]

Mellor does not claim that the meaning of tensed sentences can be captured in tenseless ones, so he does not have to provide the kind of translation that Prior claims is impossible. But he does deny that there are any irreducibly tensed facts, and Prior's claims are a challenge to him because they suggest that the fact captured by the words "That's over" is indeed irreducibly tensed, or that the response which consists in thanking goodness for that fact makes sense only on the supposition that the fact is irreducibly tensed.

Mellor connects Prior's argument with the problem of the apparently essential tensedness of experience: "The headache which has just stopped . . . is really no longer a headache at all, because it is no longer painful. . . . so far as actual pain goes, the event is merely a retired, or emeritus, headache, not something still in business as the genuine experiential article. And that is why I thank goodness for its pastness; by ceasing to be present, it has ceased to be the unpleasant experience it was" (p. 296).

I shall not discuss Mellor's treatment of the wider problem of the tensedness of experience, but only his attempt to solve the immediate problem suggested by Prior's argument, namely, that the fact for which one thanks goodness after a headache seems to be an irreducibly tensed fact.

Mellor's solution is this. The remark "Thank goodness that's over" is a conjunction of two phrases—their order, indeed, can be reversed—the second of which "has obvious and undisputed tenseless and token-reflexive truth conditions." "That's over" is true if and only if the end of the event in question is earlier than the relevant token of "That's over." Thus, if he is speaking of the headache induced by reading Prior, Mellor can truly say "That's over" after twelve o'clock but not at any time before. As for the first phrase, the idiom of thanking goodness for facts is, says Mellor, tendentious, and the phrase really serves to express a feeling of relief. Now when is it natural to have a feeling of relief in relation to a painful experience like the headache Mellor had all morning? When the experience is past, of course; but this tensed answer can be put in tenseless terms by saying that it is natural to feel relief *after* a painful experience—in this case, after twelve o'clock. Thus, "the relief which 'Thank goodness' expresses is usually felt only when 'That's over,' said of a pain, is true, namely, just after the pain has stopped. However, there is a little more to it than that. The coincidence of these tenseless conditions is not merely that. The ending of the pain is also, we believe, the cause of our relief; and our saying 'Thank goodness' in conjunction with 'That's over' expresses, among other things, our recognition of this further tenseless fact" (p. 298).

In the first sentence of this quotation Mellor uses the word "usually" because of the existence of masochists, whom, he says, "presumably feel relief when a future pain, for which they have been longing, at last becomes present—'Thank goodness it's started' is what they would naturally say, not 'Thank goodness it's over!'." (The masochist of Mellor's article and of this article is the cardboard figure familiar from jokes, who likes pain and dislikes pleasure, rather than the more complex masochist of real life.) Making an exception of masochists in this way is untidy; but Mellor's account can readily be generalized in such a way as to bring masochists within the fold. Relief of the appropriate kind, we should say, is felt after some disvalued experience or, more generally still, after the end of some disvalued state of affairs. In the case of most of us, disvalued states of affairs will include painful experiences such as headaches; though for the masochist the absence of pain will be disvalued. If relief is tied to the ending of a disvalued, rather than a painful, experience or state of affairs, the connection can plausibly be regarded as a necessary one. There are other kinds of relief, of course—there is, for example, the kind that finds expression in the words "Thank goodness he's alive"—but it does seem plausible to say that the kind of relief that finds expression in the words "Thank goodness that's over" can be felt only after some disvalued state of affairs or, rather, when the state of affairs is believed by the person in question to have ended. The belief clause is essential, for I may feel relief that a war has ended when I am wrong to believe that it has ended. (Mellor's example of a headache makes it possible to overlook the importance of the belief clause, since it is, at the very least, hard to envisage circumstances in which one can wrongly believe that one is or is not experiencing the pain of a headache. The example given by Prior in a later article,[2] of final examinations, is preferable in this respect, and I shall use a similar example later.) It is worth pointing out that, when I introduced the belief clause, I slipped back into a tensed way of expressing the conditions in which relief is felt: "after some disvalued state of affairs or, rather, when the state of affairs is believed by the person in question to have ended." The significance of this will emerge later; in the meantime the reader might try to frame in tenseless terms a statement, including the belief clause, of the conditions in which relief is felt.

The more general account that I have given of the conditions for relief not only gets rid of masochists as constituting an exception; it also provides us with a more explanatory account. For the reason why most of us feel relief on the cessation of pain is because we disvalue pain: this becomes clear when we consider the case of the masochist, who, because he disvalues the absence of pain, feels relief at the onset of an awaited headache.

By generalizing Mellor's account, one would expect to make it stronger. But if we replace his by-and-large connection between relief and the end of a painful experience with a necessary connection between relief and the end of a disvalued state of affairs, we threaten another part of his argument. Mellor considers a question that might be raised against him, namely, why relief should be particularly natural *after* a painful experience if not because the pain is then *past*. As far as he can see, he says, there is no answer to this question; but he goes on to argue that there is equally no answer to the question of why relief should be felt only when the painful experience was past, rather than present or future. In particular, he says that the answer to this second question cannot be supplied by an alleged "a priori truth that relief is never felt, in relation to a pain, while the pain is present." Why not? Precisely because of the existence of masochists. So the untidy role played by masochists, as exceptions to the normal rule that relief occurs after pain, is vital to Mellor's argument, for he needs to claim that the rule is not an exceptionless necessary truth. "Masochism, however deplorable, is certainly possible. No magic in temporal presence prevents relief being felt while pain is present. And short of a priori prevention, which is not to be had, saying that relief normally occurs only when pain is past is no more explanatory than saying that normally it occurs only after the pain" (p. 298).

So my generalizing of Mellor's thesis about the relation between relief and pain has rid it of exceptions and increased its explanatory power, but at the cost of depriving Mellor of an argument that he wants to use against Prior. For if relief (of the appropriate kind) is necessarily felt only after a disvalued state of affairs, there is, after all, a priori prevention of relief's being felt while the disvalued state of affairs is present. And it once again looks as if the best explanation of the fact that relief comes after pain might well be that the pain is then past.

An aspect of Mellor's argument that deserves further examination is his claim that the idiom of thanking goodness for facts is tendentious. I have space only to raise a question and make one comment on it. What if someone were to say—and mean it—"Thank God that's over," or, more explicitly, "Thank you, God, that that's over"? Two different kinds of case can be imagined. In the case of a headache, one usually does not know in advance when the experience will end, and one's relief when it does end at midday is partly relief that it did not go on any longer. And it makes good sense, if anything does, to thank God for bringing to an end at midday the headache that one feared might drag on through the afternoon. In the case of an examination, however, one knows in advance that the experience will end at midday, and the relief that one feels at midday seems to be wholly due to the pastness of the experience. But there is something odd about giving thanks to God for the pastness of the experience, since, given that one knew in advance that the examination would end at midday, the only thing left to thank God for seems to be that he brought midday around, and with it the end of the examination. How much sense it makes to thank God for doing this, I do not know; but what gives it such sense as it has is clearly the fact that, during certain experiences, time seems to crawl, and we wonder whether the end, even though we know when it is due to come, ever will come.

Mellor's treatment of the challenge that he derives from Prior centers on the observation that "Thank goodness that's over" expresses relief. But it is not clear how well he could deal with other cases in which there is what appears at first sight to be an emotional response to a tensed fact or to the tensedness of a fact. Consider, for example, a father who, looking at his daughter on 1 June 1982 as she studies for her finals, says, "Thank goodness I'm never going to sit another examination." Maybe the idiom of thanking goodness for facts is indeed tendentious, but what alternative account would Mellor offer of what the remark amounts to, or of what it expresses? Relief, surely, is not being expressed: the time for that was either just after the father's last examination or just after he discovered that he would never have to sit another examination. Gladness looks like the best candidate; but we cannot say of gladness in this case what Mellor says of relief in relation to his headache when he claims that the relief expressed by "Thank goodness that's over" is "not necessarily relief from or about anything, just relief" (p. 297). Even in relation to relief felt after a headache, Mellor's claim may not seem altogether plausible; but to say of the father in my example that his remark expresses gladness, not necessarily gladness at or about anything, just gladness, would be worse than implausible. It seems hard to resist the description of the father as being glad about the fact that he is never going to sit another examination. And even if the truth conditions of the sentence "I'm never going to sit another examination" can be stated tenselessly, it is not the fact that consists in the obtaining of those truth conditions that the father is glad about. He is not feeling glad that he sits no more examinations after 1 June 1982; nor is he feeling glad that he sits no more examinations after the relevant token of "Thank goodness I'm never going to sit another examination." (Why should anyone feel glad about that?) Prior's challenge, as developed by Mellor, from a claim about irreducibly tensed sentences to a claim about irreducibly tensed facts, seem not have been adequately met.

Mellor's attempt to meet the challenge relied on his denial of the intentionality of the feeling expressed by "Thank goodness that's over." I have suggested that this denial will be very much less plausible in other cases, like that of the father in my example, whose gladness is surely about something. Now once we admit that the father is glad about a fact, it seems impossible to find a tenseless fact that captures what he is glad about, and we seem forced to conclude that the fact that he is glad about is an irreducibly tensed fact.

If I appear so far to have been trying hard to bring out of retirement the headache that Mellor pensioned off at twelve o'clock, let me now supply what I believe to be the remedy. Mellor's treatment of relief as a nonintentional state suggests that he sees it as a direct response to the cessation of pain, a response not mediated by any thought. Where thought does enter, on his view, is between the feeling of relief and the utterance of the sentence that expresses that relief. "However relieved I feel," says Mellor, "I shall not thank goodness for the removal of a pain I have forgotten about." (p. 299). In the case of the father in my example, however, it is clear that not just his utterance, but also the gladness which it expresses, is occasioned by a thought. And it is this observation that may help us to see how Prior's challenge can be met. For the fact for which the father is thanking goodness is what we may call an "intentional fact": that is to say, if we describe the father as thanking goodness for the fact that he is never going to sit another examination, we are not implying that he is indeed never going to sit another examination after 1 June 1982. The "intentional fact" may not be a fact; the father's belief may be false; he may find himself sitting another examination in 1995.

But if the father's belief is true, is it not then indeed a fact that he is never going to sit another examination, and is not that the kind of tensed fact that Mellor wants to disallow? No, for if the father's belief is true, it is true in virtue of the purely tenseless fact that he never sits an examination after the time at which he holds the belief, that is, after 1 June 1982. Now this tenseless fact, as Prior rightly says, is not the fact that the father is thanking goodness for; the fact that he is thanking goodness for is an intentional fact, that is to say, it connects not with what is the case but with what is believed (by the father) to be the case. And, as we have seen, what makes the belief in question true, if it is true, is a tenseless fact. So Prior's argument does not force us to admit the existence of irreducibly tensed *facts,* for the only facts in the case are tenseless ones. What Prior's argument does suggest is that there may be irreducibly tensed *beliefs*. But that is a suggestion that squares well with the conclusions of much recent work on indexicals.[3] It squares well, too, with what I take to be the impossibility of framing in purely tenseless terms a statement of the conditions under which relief is felt, while including the necessary reference to the beliefs of the person in question.

Earlier, we touched on the question of whether the best explanation of the fact that relief comes after pain is not that the pain is then past. Mellor would see his negative answer to this question as being dictated by his denial of the reality of tense, for the pastness of x cannot be invoked in an explanation of y if pastness is not a real property of events. But when we notice that, throughout Mellor's discussion of this question, the crucial reference to belief is missing, we see how to solve his problem. It is not the pastness of x that explains y, but A's belief that x is past; and allowing this belief to explain y is not incompatible with the thesis that pastness is not a real property of events or with the thesis that there are no irreducibly tensed events.

There are those who will not like my talk of intentional facts. In particular, they may disagree with my claim that, when we describe someone as thanking goodness for the fact that *P,* we are not implying that *P*. The word "fact," they may say, functions like the word "know": we cannot say that someone knows that *P* without implying that *P*. To those who have such reservations about my solution to the problem that Mellor derives from Prior, I think I need only offer the following rewording. The father thanks goodness for what he believes to be a fact, namely, that he is never going to sit another examination. That is not to say that he is thanking goodness for his belief. There are circumstances in which people may thank goodness for their beliefs: for example, "Thank goodness I believed you were innocent when I took the lie-detector test" or "Hearing of her reaction to her husband's death makes me thank goodness I believe in life after death." But the father in my example is not thanking goodness for his belief that he is never going to sit another examination: he is thanking goodness that he is never going to sit another examination, because that is what he believes to be the case. Again, if his belief is true, it is made true by the tenseless fact that he sits no examinations after 1 June 1982.

What should be our conclusion? If Prior's argument is simply concerned to show that certain tensed sentences cannot be translated into purely tenseless sentences, then Mellor has not offered any counter-arguments. If Prior's argument is interpreted as being concerned to show that there are irreducibly tensed facts, then Mellor's counter-argument fails, but then so does Prior's own argument. What Prior's argument may show, however, is that there are irreducibly tensed beliefs. And with that Mellor need have no quarrel. Indeed, he advances something like that claim himself towards the end of his essay: "We have compelling reason to make present-tense judgments about our thoughts, actions, and experiences as they occur. Without making such judgments, we should be unable to communicate with each other" (p. 302). If some of our judgments are irreducibly tensed, the same is presumably true of the corresponding beliefs; and the irreducible tensedness of some beliefs is all that need be conceded to Prior. For, though the tensedness of judgments carries over to beliefs, it does not carry over from beliefs to facts: true tensed beliefs are made true by tenseless facts.[4]

Notes

1. Prior, "Thank Goodness That's Over," p. 17.
2. A. N. Prior, "The Formalities of Omniscience," *Philosophy* 37 (1962): 114–129; rpt. in Prior, *Papers on Time and Tense,* pp. 26–44.
3. See, e.g., Perry, "Problem of the Essential Indexical."
4. I am grateful to philosophy colleagues at the University of Stirling for their comments on an earlier version of this essay.

ESSAY 28

MacBeath's Soluble Aspirin

D. H. MELLOR

I agree with most of Murray MacBeath's discussion of my previous essay, and acknowledge that his solution to Prior's "Thank goodness that's over" puzzle supersedes my own. The puzzle is to say what, in saying "Thank goodness that's over," one thanks goodness for. No doubt that that—a headache, say—is over, that is, past. But a headache's being past is a tensed fact, and I believe time to be tenseless. Do I not therefore need a tenseless equivalent of this tensed fact, that is, a tenseless translation of "That's over"? If so, I am stumped, since it has no such translation, as I admitted in my book *Real Time,* though not in my essay. And since many still think that time could only be tenseless if tensed truths had tenseless translations, I should have made clear, as MacBeath does, that this is not so.

When I call time tenseless, what I mean, following McTaggart,[1] is that there are no A-series facts. That is, nothing ever has any A-series position: neither temporal presence nor any degree of pastness or futurity. I say "nothing" because McTaggart ascribed A-series positions to events, and many think his famous argument against them ineffective when A-series predicates are construed not as adjectives of events but as sentential operators. But that just turns A-series positions into properties of present-tense facts, so that, for example, the pastness of a death becomes the pastness of the fact that someone is now dying. And as I remark in *Real Time* (p. 169), this dodge has no effect on McTaggart's argument that A-series properties cannot be ascribed without contradiction; that is true whatever they are properties of. Present-tense facts are no better at both being and not being past, present, and future than events are. Whatever its elements, there is no real A-series.

But (*pace* McTaggart) there is still real time, in the form of a tenseless B-series of things and events, ordered by the relations of simultaneity and of the varying degrees of earlier and later than. In *Real Time* I show how, without recourse to the A-series, one can get a real B-series that meets McTaggart's test for temporality by being the dimension of change. So things and events can indeed be simultaneous with each other, or more or less earlier or later than each other, even though none is really past, present, or future.

These tenseless temporal relations can then be used to give token-reflexive truth conditions for tensed sentences. A token of "That's over," for instance, will be true if and only if it is later than whatever the token "That" refers to. Similarly for other tensed sentences. A tenseless sentence, on the other hand, is not temporally token-reflexive: the truth of its tokens does not depend on when they occur. But the truth of tokens of a tensed sentence generally does. Thus, tokens of the sentence "*E* is past" which are later than *E* are true, while earlier ones are false. Replacing them with tokens of any tenseless sentence would therefore mean replacing some true tokens with false ones, or vice versa. But translations must obviously at least preserve the truth or falsity of the translated tokens. So tensed sentences generally have no tenseless translation. In short, tenseless time does *not* imply that tensed sentences have tenseless translations: on the contrary.

But what, then, does someone saying "Thank goodness that's over" after a headache thank goodness for? Not the A-series fact that the headache is past, for there is no such fact, nor any tenseless equivalent of it. And, as Prior remarks, one certainly doesn't thank goodness for the tenseless facts that make the token true. In Essay 26 I claimed that one doesn't literally thank goodness at all. "Thank goodness," I said, just expresses relief and usually accompanies "That's over" because relief is usually caused by the end of painful experiences and thus occurs when "That's over," said of them, is true; and I added that conjoining "That's over" with "Thank goodness" also expresses *inter alia* one's awareness of this causal connection.

To this, MacBeath raises two objections (Essay 27). The first is to my arguing from the existence of masochists that relief can accompany pain and hence that nothing a priori can prevent it from doing so. In particular, the a priori fact that experience is always present, so that past pain is no pain at all, cannot be what makes relief usually follow pain. And a posteriori we can explain the tenseless fact that relief usually follows pain as readily (for example, by evolution) as the tensed fact that relief usually occurs when pain is past. I concluded that we have no reason to prefer the tensed description of this phenomenon, nor, therefore, to suppose that pains not only generally precede relief but also become past.

MacBeath tries to deprive me of this argument by changing the example from pain to "disvalued experience," which he thinks relief cannot accompany (masochists not being counter-examples, merely people who value pain). If this were indeed a necessary truth, there might be an a priori reason for it, for example, that relief cannot be felt until a disvalued experience is past. But that reason appeals to tensed facts (namely, experiences being past), which would thus have to be admitted after all.

But relief surely *can* accompany disvalued experience. Consider a lusty but inexperienced Puritan relieved to be tasting forbidden fruit at last. An awareness of doing wrong—of enjoying a "disvalued experience"—may well accompany his relief, indeed, add zest to it. Only an implausibly restrictive definition of "disvalued" would rule out that possibility. And the restriction would be no more implausible if it were tenseless, that is, if it prevented relief from being simultaneous with disvalued experience rather than preventing it from occurring while the experience is present. So either way, relief's habit of following disvalued experience no more suggests the existence of tensed facts than its habit of following pains does.

MacBeath's first objection can thus be met. But not the second. My solution to Prior's puzzle requires "Thank goodness" to express a feeling, relief, that is not about anything, and so in particular not about the tensed fact that my headache is past. The solution therefore would not work if someone really believed in Goodness and wanted to thank it; that is, as MacBeath says, "if someone were to say—and mean it—'Thank God that's over'." Here, as in MacBeath's other examples, we have "an emotional response to a tensed fact or to the tensedness of a fact," a response whose fitness or otherwise I have not at all explained.

But MacBeath has. Tensed facts figure only in our responses to tensed facts, and what our responses require is not the facts themselves but beliefs about them, that is, tensed beliefs. But both the content and the truth of tensed beliefs can be fixed, as MacBeath says, by purely tenseless facts. Thus, I thank goodness that my headache is over not because it is over but because I believe it to be over; and the content of this belief is fixed by its token-reflexive truth condition (that the belief occur after the headache) and its truth by the tenseless fact that the condition is satisfied. The alleged tensed fact that my headache is over is not needed after all. Having brought back my emeritus headache as a veritable migraine, MacBeath has cured it again, using ingredients all of which are in my book *Real Time,* and I am duly grateful.

In Essay 26 and *Real Time,* I drew on my solution of Prior's puzzle to account tenselessly for the feeling that experience occurs essentially in the present. Though MacBeath doesn't discuss it, I should remark in conclusion that his solution improves my account considerably. I had located the feeling in a belief, accompanying an experience, that the experience is present—a belief that is, of course, a token-reflexive tautology. But since beliefs aim only at truth, and this truth is trivial, so (arguably) is the belief, which the felt presence of experience seems not to be. Beliefs, moreover, are not feelings, as I argue in an earlier article.[2] For both these reasons, my account was not really adequate. It explained the presence of experience being truly believed, but not its being seriously felt.

MacBeath now enables me to do better, since he has shown how to account tenselessly for tensed feelings and emotions. First, I can now allow the presence of experience to be felt, not just believed. Secondly, since feelings and emotions aim at more than truth, the feeling need not be as trivial as the belief it is based on. If my belief that I live in the present makes me glad to do so, my gladness may be serious, even though my belief is only trivially true.[3]

Notes

1. McTaggart, "Unreality of Time"; idem, *Nature of Existence,* vol. 2, chap. 33.
2. D. H. Mellor, "Conscious Belief," *Proceedings of the Aristotelian Society* 78 (1977): 87–101.
3. I am indebted to my colleague Jeremy Butterfield for comments and corrections to an earlier draft of this essay.

ESSAY 29

"Thank Goodness That's Over" Revisited

BRIAN J. GARRETT

There are two fundamentally opposed views of the nature of time, two accounts of the truth conditions of tensed sentences. According to one view, the *tensed* view, the truth of a tensed sentence-token such as "*E* is past" consists in a particular event, *E,* having a particular property, *pastness*. According to the opposing view, the *tenseless* view, this account of the truth conditions of tensed sentences is illusory. There is no property of pastness (or of presentness or futurity). Tensed sentence-tokens possess tenseless token-reflexive truth conditions; on this view, an utterance "*E* is past" is true if and only if *E* is earlier than the utterance "*E* is past." The account is token-reflexive, since the sentence itself appears in the statement of its own truth conditions, and it is tenseless since it is true at all times that a particular event is earlier than some other event. (Of course, defenders of the tensed view of time do not deny that the above biconditional is true; what they deny is that it gives the correct *analysis* of tensed sentences.)

Arthur Prior has presented the following objection to the tenseless view of time:

> One says, e.g., "Thank goodness that's over," and not only is this, when said, quite unclear without any date appended, but it says something which it is impossible that any use of a tenseless copula with a date should convey. It certainly doesn't *mean the same* as, e.g., "Thank goodness the date of the conclusion of that thing is Friday, 15 June 1954," even if it be said then. (Nor, for that matter, does it *mean* "Thank goodness the conclusion of that thing is contemporaneous with this utterance." Why should anyone thank goodness for that?)[1]

Thus, it appears that the tenseless view of time cannot give utterances of "Thank goodness that's over" their intended content; for such utterances can have that content only if the tensed view of time is correct.

As stated, however, Prior's puzzle appears to admit of an easy solution. For it is no essential part of the tenseless view of time to claim that any particular tensed sentence is *synonymous* with (means the same as) the corresponding tenseless sentence which states its truth condition. The tenseless view of time is simply a view about tensed properties and the nature of the truth conditions of tensed sentences. Hence the fact that an utterance of "Thank goodness that's over" conveys something that it is impossible that any use of a tenseless copula with a date could convey is quite consistent with the tenseless view of time.

However, Mellor has pointed out that Prior's puzzle can be restated in order to avoid this reply:

> Suppose you have just had a painful experience, for example, a headache. Now that it is over, you say with relief, "Thank goodness that's over." What are you thanking goodness for? On the face of it, the fact that the headache is no longer a present experience, that is, is now past. . . . That is presumably why you make your remark after the pain and not during or before it. Can this fact still be explained when tensed facts are traded in for tensed tokens with tenseless truth conditions?[2]

That is, in order to bestow upon utterances of "Thank goodness that's over" their intended content, we must presuppose the reality of tensed facts. But to acknowledge the existence of tensed facts is inconsistent with the tenseless view of time. Hence the tenseless view of time cannot give utterances of "Thank goodness that's over" their intended interpretation.

Mellor's own response to this version of Prior's puzzle is to claim that when I say "Thank goodness that's over" after the ending of a painful headache, I am not thanking goodness for any fact; a fortiori, I am not thanking goodness for the fact that my headache is past but merely expressing my *relief* (not necessarily relief from or about anything, just relief).[3] However, this account of the matter is unconvincing. Contrary to one of Mellor's assumptions, it is plausible to suppose that relief is a mental state which always has an intentional object; if I am relieved, it is surely always appropriate to ask what I am relieved *about*. And no other description of this case appears tenable than that I am relieved about the fact that my headache is past. Furthermore, MacBeath has pointed out (Essay 27) that not all utterances whose correct interpretation apparently presupposes the existence of tensed facts can plausibly be regarded as expressions of relief (for example, "a father who, looking at his daughter on 1 June 1982 as she studies for her finals, says, 'Thank goodness I'm never going to sit another examination.'"

MacBeath himself attempts to solve Mellor's version of Prior's puzzle by claiming that only tensed beliefs, not tensed facts, are required in order to bestow upon utterances of "Thank goodness that's over" their intended content, and that both the truth and the content of tensed beliefs can be fixed by purely tenseless facts. Thus, I thank goodness that my headache is past, not because it is past, but because I believe it to be past, and this belief is true in virtue of the tenseless fact that the belief occurs after the headache.

However, I am not convinced by this solution. Certainly, when I thank goodness after the ending of a painful headache, I thank goodness because I believe that the headache is past (if I didn't have this belief, I wouldn't have thanked goodness). But I do not thank goodness for my *belief* that the headache is past, I thank goodness for the *fact* that my headache is past. That I thank goodness *because* I believe my headache to be past does not serve to undermine the thesis that what I thank goodness *for* is the fact that my headache is past; hence the correct account of the content of utterances of "Thank goodness that's over" does require reference to tensed facts. The problem for the tenseless view of time remains—or so it appears.

However, I think it an illusion to suppose that there is any genuine puzzle for the tenseless view of time in the first place. Contrary to the assumption of Prior and Mellor, there is a perfectly good sense in which, on the tenseless view of time, there are tensed facts. Mellor's claim that "tensed facts are a myth" is, I believe, no essential part of the tenseless view of time.[4] Since the tenseless view of time acknowledges the existence of tensed truths (for example, a 1988 utterance of "Hitler's death is past"), there is a harmless sense in which it acknowledges the existence of tensed facts (the fact, expressed by that utterance, that Hitler's death is past). If we allow that an utterance *P* expresses a truth, there ought to be no objection to the locution "It is a fact that *P*." The tenseless view of time denies the existence only of tensed properties, not of tensed truths or facts, and it is quite consistent to hold that it is a fact that Hitler's death is past *and* that this fact does not consist in a particular event having a particular tensed property.

Examples from other areas may help to illustrate this point. It appears quite consistent for someone to believe that there are no negative, disjunctive, or intensional properties and yet to hold that there are negative, disjunctive, or intensional facts. For example, one could hold that it is a fact that Socrates is believed by Jones to have been a famous Roman philosopher *without* holding that the predicate "is believed by Jones to have been a famous Roman philosopher" denotes a genuine property of Socrates. In general, then, it seems that we can acknowledge the fact that *a* is *F* without thereby incurring any commitment to the existence of a property of *F*-ness. (A defender of the tenseless view of time ought, I suggest, to exploit this result.) Consequently, there is no good reason why a defender of the tenseless view of time cannot agree with Prior that when I say "Thank goodness that's over" after the ending of a painful headache, I am thanking goodness for the fact that my headache is past.

At this point it may be objected that if, on the tenseless view of time, tensed sentences are analyzed in terms of tenseless sentences, then when I say "Thank goodness that's over" after the ending of a painful headache, what I am thanking goodness for must, ultimately, be a *tenseless* fact. But this, as Prior and Mellor rightly stress, is absurd. If I was thanking goodness for the tenseless fact that the utterance "My headache is over" is later than my headache, then, since this tenseless fact was a fact before and during my headache, as well as after it, I could just as well have thanked goodness for it before or during the headache—which is absurd.

However, this objection is, I think, fallacious. The operator "Thank goodness . . ." appears to generate a nonextensional context: I can thank goodness for the fact that *P*, where the fact that *P* is identical with, or logically equivalent to, the fact that *Q*, yet not thank goodness for the fact that *Q*. For example, suppose that, after an accident and suffering from temporary amnesia, I thank goodness for the fact that I am still alive. Then, it might be supposed, I do not thank goodness for the fact that Garrett is still alive, even though—on one plausible view—the former fact *just is* the latter fact under a different mode of presentation. Similarly, on the tenseless view of time, I can thank goodness for the fact that my headache is past *without* thanking goodness for the tenseless fact that my headache is earlier than my utterance "My headache is over."

Hence, once we (1) acknowledge the distinction between tensed facts and tensed properties, (2) appreciate that the tenseless view of time is quite consistent with the existence of tensed facts, and (3) recognize the nonextensionality of the operator "Thank goodness for the fact that . . . ," Prior's puzzle for the tenseless view of time disappears. The tenseless view of time can, after all, bestow upon utterances of "Thank goodness that's over" their intended content.[5]

However, even if this is so, it might be objected that defenders of the tenseless view of time must regard utterances of "Thank goodness that's over" (which are expressions of what Parfit has called "the bias towards the future") as symptoms of an irrational preference structure.[6] It seems irrational, on the tenseless view of time, to thank goodness for the fact that a pain is past. Why should the fact that a pain is past justify caring less about it? On the tenseless view, the fact that a particular pain, *E*, is past simply consists in the fact that *E* is earlier than the judgment that *E* is past. But there appears to be no relevant, intrinsic asymmetry between the relations *earlier than* and *later than* which justifies caring less about pains which are earlier than the time of judgment. On the tenseless view of time, it appears that the bias toward the future cannot be justified.

It might be thought that it is justifiable to care more about future experiences, since we can *control* or *bring about* future states of affairs, whereas we cannot control or bring about past states of affairs, and it is perfectly rational to care more about states of affairs that we can now affect. However, as Parfit points out, if the explanation of why we care more about future pains were simply that we care only about those states of affairs which we can now affect, then *inevitable* future pains ought to have the same value for us as past pains.[7] Yet, clearly, our attitude to inevitable future pains (states of affairs which, *ex hypothesi,* we cannot affect) is not the same as our attitude to past pains. We *are* biased toward the future—we care more about future experiences simply because they are future—and this bias would appear to be unjustified if the tenseless view of time is correct.

However, whether this is an objection to the tenseless view of time depends upon whether the tensed view of time can justify the bias toward the future. But, *prima facie,* it is difficult to see how the tensed view of time is any better placed to provide a rationale for this bias. It is difficult to see how acknowledging the existence of tensed properties is supposed thereby to justify it. For the question still remains: what is it about the property of pastness that justifies caring less about past pains? I am skeptical as to whether this question can be answered, and appeal to time's passage and the metaphor of the moving NOW appears to provide little illumination.

Clearly, this is not the place to go into the general question of the rationality of the bias toward the future. My point is simply that if the tenseless view of time is unable to provide a *justification* for the rationality of the attitude expressed in utterances of "Thank goodness that's over," this is not necessarily an objection to that view of time. And, of course, if the correct thing to say about the bias toward the future is that it is an attitude so deep and so fundamental that it is rational even though it does not admit of justification, this is a response that can be made on either view of the nature of time.

Notes

1. Prior, "Thank Goodness That's Over," p. 17; my italics.
2. Mellor, *Real Time,* p. 48.
3. Ibid., p. 50
4. Ibid., p. 34.
5. Note that this solution is not the same as MacBeath's. Despite its involvement

with intensional notions, MacBeath's solution is one which (like Mellor's) attempts to resolve Prior's puzzle for the tenseless view of time *without* invoking tensed facts. See Essay 27.

6. D. Parfit, *Reasons and Persons* (Oxford, 1984), chap. 8.
7. Ibid., p. 168.

ESSAY 30

Not Over Yet: Prior's "Thank Goodness" Argument

DELMAS KIERNAN-LEWIS

In his article "Thank Goodness That's Over," Prior was concerned with the ontology of substances and their properties. N. L. Wilson had suggested that existence is not datable, but is "a simple something or other which Napoleon simply has and Pegasus (for example) lacks." Prior took issue with this suggestion, because it presupposes the adequacy of a tenseless ontology. According to advocates of tenselessness, all temporal items are stretched out in a tenseless array, and all are on an equal footing with respect to existence. Hence, all past and future items exist in the same way that present items exist—tenselessly and changelessly at some time. Prior's argument is intended to refute this account of temporal existence.

In order to understand Prior's argument, it is important to recognize the implications of the tenseless view. If reality is tenseless, then reality cannot *begin to have* or *cease to have* any feature that it does not tenselessly possess. Hence, on this view, the only available sense of "exist," and the only sense needed to provide a complete description of reality, is the *tenseless* sense. It is simply false, according to the tenseless view, that anything, in a strict, non-Pickwickian sense "begins to exist" or "ceases to exist." Something cannot begin to exist or cease to exist *tenselessly;* it either does or does not exist tenselessly, and that is all there is to say about its existence. Of course, purveyors of tenselessness may certainly continue to employ the locutions "begin to exist" and "cease to exist," but they must regard these as mere figures of speech.

What, then, is the fact that Prior considers so decisive against a tenseless reality? Most philosophers of time have taken it to be a fact about language. Some think that the fact is the untranslatability of tensed discourse into tenseless discourse. Others have taken the fact to be the ineliminability of indexicals or demonstratives from our temporal discourse. Yet in Prior's bare-bones restatement of the argument in "The Formalities of Omniscience," demonstratives such as "that" play no key role. And neither version of Prior's argument seems to be directly concerned with the role of temporal indexicals. Others see in Prior's argument an appeal to the truth-variability of tensed sentences.

I do not deny that Prior accepted any or all of these claims about tensed language. I only deny that they are the basis of his "Thank Goodness" argument against cosmic tenselessness.

I would like to suggest a new reading of Prior's argument. The fact that Prior considers so decisive against a tenseless reality is an epistemic fact, an item of knowledge. It is the sort of knowledge we have when we are pleased that something has ceased. "Thank goodness that's over!" means "Thank goodness that has ceased!" The argument implicit in Prior's admittedly terse remarks is, I believe, analogous to a well-known argument against physicalism, defended, for example, by Frank Jackson and Thomas Nagel.[1] The argument against physicalism runs as follows. Physicalists hold that complete physical knowledge is knowledge *simpliciter,* since the actual world is entirely physical—in particular, entirely colorless. However, I know what it is like to see something, say, red. I could not know this if physicalism were true. Hence, physicalism is false.

The argument I am attributing to Prior has a parallel structure. It runs as follows. Advocates of tenselessness hold that the world is entirely tenseless. So complete tenseless knowledge is complete knowledge *simpliciter.* However, I know what it is like to cease to be aware of a headache (or to be aware that a headache of mine has ceased). I could not know this if the tenseless account of reality were true. Hence, the tenseless account is false.

This is a powerful argument, which contemporary advocates of tenselessness have failed to recognize or address. Consider Mellor, whose treatment of Prior's argument in *Real Time* is typical. Mellor accepts McTaggart's assumption that tense is properly treated as involving the applicability to temporal items of a set of predicates "past," "present," and "future" which, along with verbal tenses, express no more than relative temporal locations. Hence, to determine whether any token of a tensed utterance, belief, judgment, and so on is true, we merely need to determine where in the temporal series the tenselessly existing items talked about stand in relation to the time of the tenselessly existing token. Once these tenseless facts have been established, the truth of the token follows. For Mellor and other advocates of tenselessness, the central ontological issue is whether real objective properties of pastness, presentness, and futurity are needed to account for the truth of tokens of tensed utterances and so forth. Mellor simply misses the point of

Prior's argument, because his perspective blinds him to the ontological issue that Prior's argument is intended to resolve—namely, whether it is true at any time that a past item really has ceased to exist (and is not merely located elsewhere in the temporal series).

To answer Prior, it is not enough to talk about tenseless truth conditions that make it true that a thing or event tenselessly exists at times earlier than, and not including, the time after which it has ceased. Advocates of tenselessness must explain how it is that we are systematically and continually deceived by our awareness that something really has ceased to exist. But the tenseless view does not have the ontological wherewithal to construct such an explanation. At least, so it seems to me now, and did, I suggest, to Prior then.

Note

1. See Frank Jackson, "What Mary Didn't Know," *Journal of Philosophy* 88, no. 5 (May 1986): 291–295.

ESSAY 31

Thank Goodness It's Over

L. NATHAN OAKLANDER

In Essay 30 Delmas Kiernan-Lewis offers a new reading of Prior's much discussed argument against the tenseless theory of time, according to which reality (existence) is tenseless. In this brief essay I shall argue that Kiernan-Lewis's interpretation of Prior's argument does not undermine the tenseless view, since it is either unsound or invalid. I will then offer a diagnosis of Kiernan-Lewis's mistakes.

According to the tenseless theory, all events in the temporal series are equally real; there are no fundamental ontological distinctions between past, present, and future events. Kiernan-Lewis maintains that, according to Prior, this view cannot account for the knowledge we have when we are pleased that something has ceased to exist, for on the tenseless theory nothing *really* ceases to exist. Kiernan-Lewis formulates what he takes to be Prior's argument as follows: "Advocates of tenselessness hold that the world is entirely tenseless. So complete tenseless knowledge is complete knowledge *simpliciter.* However, I know what it is like to cease to be aware of a headache (or to be aware that a headache of mine has ceased). I could not know this if the tenseless account of reality were true. Hence, the tenseless account is false" (p. 323).

Perhaps the best way to see where this argument goes wrong is to reformulate the essence of it in the following three steps:

1. I know what it is like for a headache of mine to cease to exist.
2. I could not know this if the tenseless account of reality were true.
3. Hence, the tenseless account is false.

Pre-analytically, premise 1 is obviously true. If I have a headache and then, after taking an aspirin, say, no longer have a headache, I know what it is for a headache to cease to exist. On the other hand, premise 2 is not obviously true. Indeed, it is not true at all. On the tenseless theory, there are the tenseless facts: (a) I am conscious of having a headache at t_1, (b) I am conscious of taking an aspirin (and having a headache) at t_2, and (c) I am conscious of feeling fine (and not having a headache) at t_3. The succession of these different states of consciousness is the ontological ground of knowing that a headache of mine has ceased to exist. Why would Kiernan-Lewis (or Prior) believe that those facts were not sufficient to explain the knowledge in question?

The issue centers around the correct interpretation of "ceases to exist." On the tensed view that Prior espouses, only the present exists; the past and future have no reality whatsoever. Thus, when a headache ceases to exist, it *really* ("in a strict, non-Pickwickian sense") ceases to exist. Of course, *assuming* that interpretation of "ceases to exist," the detenser cannot know that his or her headache has ceased to exist, since nothing ever does *really* cease to exist. But such an interpretation assumes what needs to be proved. In other words, if we understand "cease to exist" as the tenser would have it, premise 2 is true, but then premise 1 is either false or question-begging, since it assumes that the tensed account of existence is true and that the tenseless account of existence is false.

On the other hand, if we assume the tenseless interpretation of "cease to exist," wherein a thing ceases to exist if and only if there is a time after the existence of all its temporal slices (or after its temporal location), then we *can* know that a headache has ceased to exist; that is, premise 2 is false (although premise 1 is true). And finally, if we confuse the tensed and tenseless senses of "cease to exist," so that (1) is true (in the tenseless sense) and (2) is true (in the tensed sense), then the argument commits the fallacy of equivocation and is thereby invalid. Thus, Kiernan-Lewis's reconstruction of Prior's argument against tenseless reality is either unsound or invalid.

There are two confusions that may underlie Kiernan-Lewis's (Prior's?) argument against the detenser. First, there is the confusion of tenseless existence and eternal (sempiternal) existence. On the tenseless theory, all events exist tenselessly at the moment they do, but this does not imply that they are everlasting or exist at every moment in the time series. If one fails to be cognizant of this distinction, then one might erroneously believe that on the tenseless view we cannot know that an event ceases to exist, since all events always (at every time) exist. Second, Kiernan-Lewis may be confusing God's tenseless knowledge of tenseless facts with our tenseless knowledge of them. If God is outside time, looking at all tenseless facts from a distance, as it were, then for him no event would appear to come into existence or cease to exist. But for us, beings who are in time, there is the experience of events coming into being and ceasing to exist, and that experience can be known simply by (tenselessly) having different experiences at different times.

In short, it does not seem to me that I am aware that something *really* ceases to exist (in the tensed sense, whatever exactly that may be); so advocates of the tenseless view need not explain, in Kiernan-Lewis's words, "how it is that we are systematically and continually deceived by our awareness." There is no item of knowledge that the tenseless view cannot explain, and Kiernan-Lewis could only think that there was by falling prey to the confusions I have discussed.

ESSAY 32

Passage and the Presence of Experience

H. SCOTT HESTEVOLD

The doctrine of "transitory time" ("tensed time") is the view that events undergo "temporal becoming," "passage" from the future to the present to the past:

TT Events will occur, events are occurring, and events did occur.[1]

According to TT, the event *Reagan's becoming President* was in 1979 such that it would occur, in January 1981 such that it was occurring, and since then such that it did occur. In McTaggart's familiar terms, TT is the view that events "flow" along the A-series.

Defended by Mellor, among others,[2] the doctrine of "static time" ("tenseless time") is the view that events do *not* undergo "temporal becoming." This doctrine implies that my birth *is* occurring *and* my death *is* occurring in the *only* sense in which it can be said of any event that it is occurring:

ST Events can only occur tenselessly.[3]

Using McTaggart's terminology, ST is the view that although there is no A-series, events occur tenselessly on the B-series such that their "passage" is illusory.

Schuster has argued against ST on the grounds that it cannot adequately explain the illusion of "passage," since ST cannot account for why the illusion is of "forward passage" rather than "reverse passage." In Section 1, I explicate Schuster's argument and conclude that it is not a telling objection to ST. The serious problem for ST is not whether it can explain why we are deluded by "forward" rather than "reverse passage," but whether it is reasonable to believe that "passage" is an illusion at all. More specifically, since talk of the "flow" or "passage" of time can only be metaphorical if ST *is* correct,[4] the defender of ST must explain why it is reasonable to believe that the past, present, and future are illusory, given the seeming presence of experience.

Although TT should not be defended by claiming that *being present* is a phenomenal property of which one can be directly aware, I argue in Section 2 that TT is supported by the doctrine that experiences, necessarily, may be known to be present. In Section 3, I formulate and defend against objections Prior's defense of TT, which involves the appropriateness of certain of our attitudes toward future and past events. Finally, both these arguments for TT are defended against Mellor's tenseless analysis of the presence of experience, which, I argue, is inadequate.

1. Schuster on the Illusion of "Passage"

Referring to defenders of ST as subjectivists (since they hold that "passage" is but a subjective illusion), Schuster offers the standard objection that subjectivists have no plausible tenseless account of "passage." More specifically, he argues that "passage" cannot be explained tenselessly in terms of the view that events (tenselessly) are *first* future, *then* present, *then* past in some tenseless sense of "first future," "then present," and "then past."[5]

Although the subjectivist may lack a tenseless analysis of "passage," Schuster acknowledges that ST could be defended against his objection by "denying completely that the temporal relationship establishing the futurity of an event is actually prior to the relationship establishing its status as past."[6] That is, the subjectivist could deny that there is any tenseless sense in which an event's being "first future" precedes its being "then present" and "then past." Schuster rejects this response, claiming that discounting "passage" as a mere illusion is not itself an adequate reply to the original objection; what the subjectivists owe, he claims, is an explanation of why we are deluded by an appearance of "forward passage" rather than an appearance of "passage" in the other direction: "Why then should we not suffer an illusion the reverse of what we do, wherein t_2's relation to t_3 would be prior to its relation to t_1? Of course the actual sequence is from future to past—that is acknowledged to be a fact—but the point is that the subjectivists' account does not explain it."[7] Schuster demands of the subjectivist an account of why, if there is no "flow" of time at all, it appears to you that you read the first page of this essay prior to the present page instead of it appearing to you that you read the present page prior to the first page.

Adolph Grünbaum and Lynn Rudder Baker have attempted to explain the illusion of "forward passage" by the observation that "passage" appears to follow the direction of increase with respect to both memory and entropy.[8] Their explanation, claims Schuster, is insufficient:

> What Grünbaum and Baker might claim is that . . . entropy and memory increase in the direction of later time, and . . . that the illusion of time

> flow is also toward later time. The conjunction of the two facts is left unexplained. The question we have posed is why time should appear to move from earlier to later, and Grünbaum and Baker . . . merely indicate certain other conditions that are also correlated with earlier and later. . . . Why then should time not appear to flow from more information to less instead of from less to more?[9]

Offering further support for his point, Schuster asks the reader to imagine that time *does* appear to flow toward earlier events and then asks just how subjectivists would explain this illusion of "reversed passage." Schuster writes:

> They would merely point out that time appears to move in the direction of entropy decrease and from more to less stored information, as water flows from a fuller to a less full container. It would seem perfectly natural, and they might even take their observation as an explanation. . . . Their theory would remain intact, exactly as it presently is, with nothing added, nothing subtracted, and nothing altered. . . . A theory which purports to explain some phenomenon but which actually provides an account that applies equally well to the reverse phenomenon . . . must be considered fatally inadequate and therefore philosophically unacceptable.[10]

Schuster's claim that the subjectivist cannot explain why we have the illusion we have is not a telling criticism of ST. First, the subjectivist would resist Schuster's construing ST as the view that every event is (tenselessly) first future, then present, and then past in some tenseless sense of "first future," "then present," and "then past." ST implies that all events *are* tenselessly occurring in the *only* sense in which events can possibly occur. To insist that the subjectivist attribute to each event the properties of being first future, then present, then past is to misrepresent ST.

Second, the subjectivist could take it as a "brute fact" that the tenseless occurrence of events just *does* appear to "flow" toward later events in the very same way that it is a "brute fact" that, say, a fluttering butterfly appears motionless to one whose neurophysiology has been altered in a particular way by mescalin.

Third, the defender of ST could explain why we have the illusion we have by appealing to what subjectivist Mellor says about causation and the direction of time: "To be affected is to be later, to be the cause, earlier. The direction of time is the direction of causation."[11] And again: "So much for our perception of the flow of time, which we see to be nothing more than an accumulation of memories. The fact that a memory is an effect, not a cause, of what is remembered is the real reason the flow of time takes us into the future rather than the past."[12]

If Mellor is right that the direction of static time *is* the direction of causation, then it should not be too surprising that time appears to "flow" in the direction of causation if we experience an illusion of "time flow" at all. Schuster may object that ST remains inadequate, since it cannot explain exactly why the illusion is of "passage" in the direction of causation rather than "passage" in the opposite direction. This challenge is one that the subjectivist cannot meet, but it is a kind of challenge which no theory of time can meet. After all, if time *is* transitory, then why *does* it appear to "flow" forward as it does? The TT theorist can, at this point, only appeal to its being a "brute fact" that the "forward passage" of time just *does* veridically appear to us as "forward flowing" rather than as "backward flowing" or "static."

Thus, the serious problem for ST is *not* its inability to explain why we experience the illusion of "forward passage" rather than "reverse passage." It is whether it is reasonable to believe that the appearance of events becoming present and then past is an illusion at all—whether the ST defender can explain in tenseless terms what we know about the presence of some events and the futurity and pastness of others.

2. Experiences May Be Known to Be Present

A naive defense of TT is that one can be directly aware of an event's having the nonrelational transitory temporal property *being present*. For example, one could claim that you are now directly aware that *your reading Section 2* and *your remembering having read Section 1* have the phenomenal property *being present;* and your direct awareness of the presence of these events justifies your belief that the event *your reading Section 1* has the property *being past*. In short, the view is that we literally see that there are transitory temporal properties.[13]

Mellor rejects this defense of TT:

> Suppose I am looking through a telescope at events far off in outer space. I observe a number of events, and I observe the temporal order in which they occur: which is earlier, which later. I do *not* observe their tense. What I see through the telescope does not tell me how long ago those events occurred. That is a question for whatever theory tells me how far off the events are and how long it takes light to travel that distance. . . . So, depending on our theory, we might place the events we see anywhere in the A-series from a few minutes ago to millions of years ago. Yet they

> would *look* exactly the same. What we see tells us nothing about the A-series positions of these events. It does not even tell us that the events are past rather than future. Someone who claims to see the future in a crystal ball cannot be refuted by pointing to some visible trace of pastness in the image. Our reason for thinking that we cannot observe the future rests on theory, not on observation.[14]

Mellor's point is that since past events appear through the telescope to an observer in the same way that present events appear to the observer, *being present* is not a phenomenal property.[15]

There are two possible responses to Mellor. First, one could continue to hold that *being present* is a phenomenal property, concluding that Mellor's telescope example does not prove otherwise: "Although the telescope example shows that we are *not* directly aware of the transitory temporal properties had by *some* events (namely, events in the distant past), it does not show that I never perceive transitory temporal properties. In fact, I *am* aware of the transitory temporal properties had by more immediate events; for example, I am directly aware that *my experiencing an event* —an event which may itself be past, present, or future—has the property *being present*." Mellor anticipates this response, admitting that he can "only produce the paradoxical reply that, although we observe our experience to be present, it really isn't."[16]

Although Mellor's doctrinaire reply will hardly satisfy his critic, he grants too much to the defender of phenomenal temporal properties when he admits that, at least seemingly, "we observe our experience to be present." First, although I may (at present) see red patches or may (at present) be appeared to redly, I do not see "presentness," nor am I appeared to presently. Hence, I simply lack phenomenological evidence that *being present* is a phenomenal property.

Second, if there are transitory phenomenal temporal properties, then TT is true; but TT does not obviously entail that there are such temporal properties, phenomenal or otherwise. As formulated, TT is simply the view that some events will occur, others are occurring, and still others did occur. Thus, a version of TT which avoids commitment to phenomenal temporal properties may be ontologically simpler than the phenomenal-property version. Third, if one can defend TT without commitment to phenomenal temporal properties, then TT may skirt the serious objections which have been raised to those theories of time that explain temporal passage in terms of transitory temporal properties.[17]

Finally, phenomenal properties serve as a means of distinguishing some experiences from others. For example, *being red* is the property which allows me to distinguish one part of my visual field from, say, another part which is blue. (Alternatively, *being appeared to redly* allows me to discriminate *this* way of being appeared to from *being appeared to bluely.*) *Being present,* however, serves no such discriminatory function. Since all my (present) experiences *are* present, the property *being present* does not allow me to distinguish certain of my present experiences from other present experiences.

Consider the second way in which the TT defender can respond to Mellor's telescope example and his claim that "although we observe our experience to be present, it really isn't." The stronger response to Mellor's telescope example is that, although *being present* is not a phenomenal property, an experience is necessarily such that it can be known to be present:

PE For any experience E and time t, E is necessarily such that if E occurs at t, then there exists someone S who knows at t that E is (presently) occurring if S considers (entertains) at t that E is occurring.

Informally, to *consider* that an event is occurring is to *contemplate* whether it is occurring or to *entertain* the possibility that it is occurring. One can consider an event's occurrence without the event's occurring and without believing that it occurs. By "experience," I mean any "conscious state," any "mental state." Although distinguishing "mental states" from "nonmental states" lies well beyond the scope of this essay, I assume that the distinction can be drawn in terms of incorrigibility.[18] Thus, one's being depressed, wishing for snow, seeing a red, round sense-datum, and being appeared to bluely would all count as experiences. PE implies that if an experience is [presently] occurring, then there is someone who can know that it occurs, even though there may be no phenomenal property *being present*. This is analogous to the fact that if one contemplates a perceived object's being self-identical, then one knows that it is self-identical, even though *being self-identical* is not a phenomenal property.

PE is not a trivial thesis, like the claim that, necessarily, every present event is present, an expression of modality *de dicto*. And it is not on a par with the trivial expression of modality *de re,* that all things are necessarily self-identical. The claim that experiences, essentially, can be known to be present implies that there cannot occur an experience that occurs only tenselessly; experiences *cannot* be mere tenseless occurrences on the B-series!

Two points can be made on behalf of PE. First, one can cite PE's obvious intuitive appeal: experiences just *are* the sorts of events which occur presently and can be known to be so. Second, consider an implication of denying PE. If PE is false, then it is possible for one to (a) have a headache at a time t, (b) contemplate at t whether the headache presently occurs, and (c) believe that the headache is not presently occurring. This is impossible; it would be absurd for one to claim, "I have a headache and contemplate its presently occurring, but I believe that my headache is past." If one has a headache and considers whether this is so, then one *knows* that the headache occurs presently and cannot believe that the pain is past.

Consider this Mellorian reply to the second defense of PE: "PE is false: when one claims to know that one's headache is present, one is merely claiming that one knows (tenselessly) one is having a headache at the same time that one utters (tenselessly) 'I have a headache.' This tenseless explanation captures the view that one can know that one has a headache at the time it is occurring without implying that it *presently* occurs." The problem with this reply is that it fails to distinguish between my knowing I had a headache at the time when, years ago, I uttered "I have a headache" and my knowing I have a headache *now*. It is not simply the case that I know I have a headache on the date that I write this sentence; rather, I know that the headache is not past but *present*—it *does* hurt.

After formulating a second defense of TT, I shall, in Section 4, entertain Mellor's rebuttal that experiences *can* occur tenselessly.

3. Attitudes Toward Future and Past Events

Prior has argued that one mark in favor of TT is that only a theory of time which allows for "passage" can account for the appropriateness of certain of our attitudes toward future and past events.[19] Consider the following version of Prior's argument: "On Monday I dread the painful tooth extraction scheduled for Tuesday, and on Wednesday I am relieved that the extraction is over. Dread on Monday and relief on Wednesday *are* appropriate attitudes to have toward the Tuesday tooth extraction. If ST is correct, however, then dread and relief are never appropriate, since there are *no* future and past events toward which to direct them! That Wednesday *follows* the day of the tooth extraction is a tenseless fact which is true before, during, and after the extraction. Thus, it is as appropriate to feel relief that Wednesday follows Tuesday before or during the extraction as it is to feel such relief on Wednesday. But this is absurd! Apparently, dread on Monday and relief on Wednesday are appropriate attitudes because I believe correctly that the extraction is on Monday such that it will occur and is on Wednesday such that it did occur. Hence, TT must be adopted to make sense of the appropriateness of our attitudes toward the future and the past."[20]

Acknowledging the significance of this argument by admitting that the "presence of experience is the crux of the matter" such that, without "a tenseless account of it, tenseless truth conditions on their own will never dispose of tensed facts," Mellor resists the argument's conclusion, offering a tenseless explanation of the appropriateness of relief: "What a token of 'Thank goodness' really does is express a feeling of relief: not necessarily relief from or about anything, just relief. So when is it natural to have a feeling of relief in relation to a painful experience? The tensed answer is: when the experience is past. The tenseless answer can only be *after* a painful experience, that is, at a later date, rather than during or before it."[21] Mellor's point is that relief on Wednesday is appropriate because it (tenselessly) occurs *after* the tooth extraction.

Mellor's explanation—which even he admits "may well seem weak"—is unacceptable, because he has omitted an account of why on Wednesday relief, not dread, is appropriate. After all, on Wednesday, there is a sense in which the extraction is not over; on Wednesday, the extraction is "eternally" and tenselessly occurring on Tuesday. Thus, Mellor must explain how Wednesday's bearing the temporal relation *follows* to Tuesday renders relief on Wednesday appropriate.

Mellor offers this reply:

> To this further question I confess I see no answer. But I also see no answer to the question: why feel relief only when pain has the A-series position *past*? The answer is not, as one might suppose, that relief *cannot* be felt while pain is present and thus still pain. It is no a priori truth that relief is never felt, in relation to a pain, while the pain is present. Indeed it is no truth at all. Masochists, for a start, presumably feel relief when a future pain they have been longing for at last becomes present. . . . . . There is no a priori magic in temporal presence therefore to prevent relief being felt while pain is present. So saying that relief normally occurs only when pain is past is no more of an explanation than saying that normally it only occurs after the pain.[22]
>
> The relief "Thank goodness" expresses is usually felt only when "That's over," said of a pain, is true, namely just after the pain. However, there is a little more to it than that. The coincidence of these tenseless conditions is not just a coincidence. The ending of the pain is also, we believe, the cause of our relief; and saying "Thank goodness" in conjunction with "That's over" expresses *inter alia* our recognition of this further tenseless fact.[23]

In short, Mellor's solution to the Prior puzzle is that relief on Wednesday is appropriate because (a) the ending of the pain causes the relief, and (b) the relief occurs after the Tuesday extraction. And, Mellor concludes, saying that it is appropriate to feel relief after a painful experience is no more arbitrary than saying, according to the tensed account, that relief is appropriate because the pain is past.

Although the cessation of pain is not the "sole cause" of one's relief,[24] Mellor is correct that, with most people, the cessation of pain is a *partial* cause of relief. But *if* masochists *do* desire that which I call "pain," then it is still not obvious that the masochist's relief can possibly be directed toward present pain. One could argue that when the masochist enjoys relief while simultaneously reveling in his agony, he is relieved not that pain is present but that, say, his sexual frustration is past.[25] Overlooking this problem, Mellor's twofold reply to the Prior puzzle still fails to explain the appropriateness of some dread and relief.

First, since *I* am not a masochist, I *do* dread painful tooth extractions and experience relief when they are over. Thus, although my directing dread toward future pain may be a contingent matter, Mellor must explain why my dread on Monday is appropriate given that I *do* loathe pain. Second, Mellor's reply does not demonstrate that the appropriateness of relief and dread fails to be explained by TT or that it can be explained instead by ST. The Prior puzzle works *not* because there is a necessary connection between future *pain* and dread, but because there is a necessary connection between dread and the futurity of something which I loathe and between relief and the cessation of that which I loathe. Dread is appropriate *because* I knowingly "move closer" to an unpleasant event, and relief is appropriate *because* I knowingly "move away" from an unpleasant event; and such temporal "movement" cannot be captured in terms of ST. The point of the Prior puzzle is that to make sense of why the masochist feels relief toward the presence of pain, one must suppose that he anticipated pain with desire and that the pain is finally occurring. Thus, since Mellor has failed to explain the appropriateness of certain attitudes toward past and future events, and since the tensed account of their appropriateness is *not* arbitrary as he claims it is, the appropriateness of these attitudes counts in favor of TT.

4. Mellor's Tenseless Account of the Presence of Experience

Recall the conclusion from Section 2 that Mellor's rejection of the argument appealing to our direct awareness of phenomenal temporal properties is, by his own admission *not* entirely satisfactory, since he omits a tenseless account of the "inescapable, experientially given presence of experience." Mellor believes that his own discussion of the Prior puzzle affords him the machinery to offer such a tenseless account. To appreciate his effort, assume that his response to the Prior puzzle does not suffer from the problems raised in the previous section. He claims that what is needed to dispose of the presence of experience is "a kind of self-awareness. The salient feature of Prior's case is that we not only have painful experiences, we also remember them. . . . Now this recollection, which is what 'That's over' expresses, is in part a token of a past-tense judgment, the judgment that I was in pain in the recent past. But it is also in part a present-tense judgment, the judgment that I am not in pain—or not in so much pain—now. . . . A present-tense awareness of being relatively free of pain is an essential ingredient in Prior's case; and this is the ingredient I need."[26]

By revealing more about present-tense self-awareness and the judgments one makes about it, Mellor's tenseless analysis of the presence of experience unfolds:

> We are concerned with token judgments to the effect that experiences we are now having possess the . . . property of being present. Now any token which says that an event is present will be true if and only if the event occurs at the same B-series time as the token does. Those are the undisputed token-reflexive truth conditions of all such judgments. But in this case the events to which presence is attributed are themselves picked out by the use of the present tense. Not all our experiences, past, present, and to come, are alleged to have this A-series property, only the experiences we are having now. But these, by the same token-reflexive definition of the present tense, are among the events which *do* have the property now ascribed to them: that is, events occurring just when the judgment itself is made. So of course these judgments are always true. Their token-reflexive truth conditions are such that they cannot be anything else. In tenseless terms they are tautologies after all.[27]

There are two possible interpretations of Mellor's tenseless analysis of the presence of experience.

On the first interpretation, my saying, for example, "My experience of pain is present," is to say that that judgment occurs contemporaneously with the experience of pain. Generally, the analysis would go like this:

(A) My experience E is present = Df The judgment "My experience E is present" is occurring at the same time (B-series position) as E.

Interpretation A faces two problems. First, to elicit sympathy, I could utter on Wednesday, "My experience of being in pain is present," when in fact my pain ended on Tuesday. Thus, the judgment that an experience is occurring can occur at a time other than that at which the said experience occurs. Hence, the *definiens* of the analysis is not a necessary truth, and this implies that the *definiendum* cannot be the tautology that Mellor claims it must be.

Second, interpretation A blatantly fails to capture the presence of experience. Desperately telling the dentist drilling your tooth that pain is *present* involves something more than telling him that an utterance about an experience is occurring tenselessly at the same time as the experience. After all, he may not be moved to administer more anesthetic if he is told merely that *your reporting pain* and *your being in pain* occur (tenselessly) at the same time; unless he is told that that time is *now,* you may be denied relief.

The second interpretation of Mellor's tenseless account of the presence of experience is more complex. Presumably, when I correctly report on Tuesday that I am in pain, I do not mean simply that *my being in pain* is tenselessly occurring. After all, *my being in pain* may be tenselessly occurring on a B-series date six months after today's date. Thus, when I report on Tuesday that my pain is present, I have in mind some particular pain. Which pain? The pain which is occurring while I report it. And I do not mean to imply that *that* pain is occurring on just any date at all; I mean that it is occurring on the date that I am reporting its occurrence. Thus, to say that my pain is present is to say that the pain that is occurring while I report its occurrence is occurring while I report its occurrence; and this *is* tautologous, as Mellor says it must be. The general form of this second interpretation of Mellor's analysis is as follows:

(B) My experience *E* is present = Df The experience *E* which is (tenselessly) occurring while I report that *E* is (tenselessly) occurring is (tenselessly) occurring while I report that *E* is (tenselessly) occurring.

In short, to say of an experience that it is present is to say that an experience occurring when reported is occurring when reported.

Mellor offers a defense of (B), claiming that his tenseless explanation of the presence of experience

> is not merely an alternative to a tensed explanation of the same thing. There is no tensed explanation of this phenomenon. If events can in reality have a range of tenses, I see no good reason for experience to be confined as it is to present events. In tensed terms, that is just an unexplained brute fact about experience. . . . What [the tensed account of the presence of experience] does not explain, however, is how experience differs in this respect from other events. Other events and things at least seem to be spread out throughout the whole of A-series time. . . . Only our experiences, including our judgments (that is, our thoughts), and our intentions, decisions, and actions appear to be restricted to the present. Of that contrast, the token-reflexive account I have just given alone provides a serious explanation.[28]

Explicitly, Mellor's defense of (B) is the following:

(C) (1) Experiences are events that are present.

(2) If experiences are events that are present, then their presence is explained either by TT or by tenseless analysis (B).

(3) TT does not explain the presence of experience (since it does not explain why experiences are the only sorts of events which can have only the temporal property of being present).

(4) Therefore, the presence of experience is explained by the tenseless analysis (B).

There are several problems with (C), beginning with Mellor's defense of (3).

Mellor argues that TT does not provide an acceptable explanation of the presence of experience, by claiming (a) that if TT is correct, then we must accept as an "unexplained brute fact" that experience is confined to the present, and (b) that this "brute fact" does not explain why only experiences and no other events are such that they can only occur presently. First, that we must take as an "unexplained brute fact" the thesis that experiences occur presently is hardly a reason for rejecting the obvious, when no tenseless alternative has been offered. Moreover, what could it possibly be like to have an experience that does *not* occur presently? (*If* the precognitive psychic can glimpse future events, his glimpsing does not itself lie in the future; rather, he glimpses presently the future.)

Mellor's second point, that TT offers no explanation of why only experiences can only occur presently, is misguided; TT does not entail that only experiences can only occur presently. After all, *every* event that occurs presently will be such that it *occurred,* and experiences are simply a kind of event. Thus, experiences *are* "spread out throughout the whole of A-series time" insofar as any event is. The defender of TT, then, is not committed to the thesis for which Mellor claims TT offers no explanation. By appealing to PE, however, the TT theorist *can* explain the special relation that obtains between experiences and the present that does not obtain between other events and the present: the former are necessarily such that they can be known by someone to be presently occurring, while the latter are not.

The major objection to (C) is that (B) is an unacceptable account of the presence of experience and, thus, either (C2) is false, or the presence of experience *can* be explained by the TT theorist after all. My first objection to (B) is that it is counterintuitive to suppose that the proper analysis of, say, "Excruciating pain is present" is a tautology. If it *is* a tautology, then one can only wonder why a dentist promptly administers more anesthesia when told by his patient that excruciating pain is present. Presumably the dentist would *not* move as swiftly were he told "All pain is pain" or "If pain is present, then pain is present."

Second, along similar lines, suppose that I am not in pain and report this by saying "Pain is not present." Since the analysis of what is expressed by "Pain is present" is a tautology, then the analysis of its denial is a necessary falsehood. Thus, when, in fact, no pain is present, it is necessarily false that pain is present. But it is *not* necessarily false that pain is present when no pain is present, since I could have then been in pain.

To understand (B)'s third shortcoming, assume that, not being a masochist, I *am* distraught that the excruciating pain caused by the dentist's drill is present. According to (B), I am distraught that the pain that is occurring while I judge that it is occurring is occurring while I judge that it is occurring. Clearly, however, I am *not* distraught about this mere triviality. I am distraught that the pain is occurring *now,* and this essential element is omitted from Mellor's tenseless analysis of the presence of experience.

The Mellorian reply to these objections is that although there are no tensed facts, there is tensed belief and that "[a]ction is what really makes tensed belief indispensable";[29] tensed beliefs are "the psychological reality behind the myth of tense, the myth of the flow of time."[30] Thus, Mellor could admit that, although the patient's tensed belief about his pain is indispensable to communicating his distress to his dentist, his having the tensed belief in no way implies that there are tensed facts, since the truth conditions of the tensed belief are themselves tenseless.

Even if Mellor is right that tensed beliefs do not entail that there are tensed facts, he cannot explain how the dentist's hearing his patient's affirmation of a tautologous tenseless fact can lead the dentist to acquire a belief that this patient is in pain at the time that the dentist hears the patient's utterance. Perhaps the patient who reports an occurring pain is, on Mellor's view, to be understood to be uttering the following:

(D) I believe-now that my excruciating pain is (tenselessly) occurring.

(D), however, is not strong enough to convey the message that the pain is present, since (D) could be used to express the tensed belief that pain is tenselessly occurring at some time or the other, and this tensed belief can be held when one is not then in pain. Perhaps (D) could be replaced by the following more complex statement:

(E) I believe-now that my excruciating pain which is (tenselessly) occurring while I report its (tenseless) occurrence is (tenselessly) occurring while I report it.

(E) merely implies that its utterer has a present-tense belief that a certain tautology is true and does not thereby imply that its utterer is then in pain. After all, I could believe-now on Friday that my pain that is tenselessly occurring on the previous Monday is then simultaneously occurring with my tenselessly reporting its occurrence.

What is missing in both (D) and (E) is the message that the pain in question is occurring at the very same time as the utterance; the following may capture this:

(F) I believe-now that time t_1 is (tenselessly) occurring and that my excruciating pain is (tenselessly) occurring at t_1.

Again, (F) fails to convey to the dentist the requisite information, since I could, at time t_3, have the tensed belief that t_1 is tenselessly occurring and that a pain is tenselessly occurring at that time.

Apparently, the only way to convey to the dentist that anesthesia is needed *now* is to utter something like the following:

(G) I believe-now that time t_1 is (tenselessly) occurring and that my pain is now occurring at t_1.

If, however, as you read this sentence, (G) is uttered by someone to report a *true* tensed belief, then what makes that tensed belief a true belief is the tensed fact that the person's pain is *now* occurring. Thus, Mellor is right that tensed beliefs are indispensable with respect to coordinating our actions and communicating with others, but tensed facts are also indispensable.

For two reasons, ST should not be rejected on the grounds that it cannot explain why we suffer the delusion of "forward passage" rather than "reverse passage," if the subjectivists are correct that "passage" is illusory. First, Mellor's appeal to causation may provide the subjectivist with the needed explanation. Second, the demand for an explanation of why temporally located events appear to us as they do may be a demand that no theory of time can meet.

Although TT should not be defended by claiming that *being present* is a phenomenal property of which one may be directly aware, TT is rendered plausible by PE, the doctrine that experiences are necessarily such that they can be known to be presently occurring when they occur. A second mark in favor of TT is that the appropriateness of certain of our attitudes toward past and future events can be explained only by a theory of time that allows for "temporal becoming."

Finally, Mellor's tenseless analysis of the presence of experience offers us no reason to abandon either defense of TT.[31]

Notes

1. There are theories of time other than TT which allow for "temporal becoming," and what is said on behalf of TT in this essay can also be said on behalf of these other transitory time theories. See C. D. Broad, "Ostensible Temporality," in Gale, ed., *Philosophy of Time,* pp. 117–142; M. Capek, "The Inclusion of Becoming in the Physical World," in Capek, ed., *The Concepts of Space and Time* (Dordrecht, 1976), pp. 501–524; Chisholm, *First Person,* pp. 126–128; M. M. Schuster, "Is the Flow of Time Subjective?" *Review of Metaphysics* 39 (1986): 695–714; Sosa, "Status of Becoming"; and Taylor, *Metaphysics,* pp. 71–79.
2. Mellor, *Real Time.* See also Grünbaum, "Status of Temporal Becoming"; Russell, "On the Experience of Time"; and Smart, *Philosophy and Scientific Realism.*
3. ST does *not* imply that my birth occurs simultaneously with my death; many other events have temporal locations between those two events. Time, according to ST, can be likened to a yardstick: the numeral "6" lies to the right of the "2" and to the left of the "9," but it makes no sense to say that the "2" rests in the past or that the "9" looms in the future. Similarly, although today's events *follow* yesterday's and *precede* tomorrow's, ST implies that it makes no sense to say that today's events will be past tomorrow, having been present today and future yesterday.
4. Cf. Smart, "Time and Becoming," p. 4; P. Gassendi, "Reality of Absolute Time," in Capek, ed., *Concepts of Space and Time,* p. 195.
5. Schuster, "Is the Flow of Time Subjective?" pp. 700–708.
6. Ibid., p. 710.
7. Ibid., p. 711.
8. Adolf Grünbaum, "The Meaning of Time," in Eugene Freeman and Wilfred Sellars, eds., *Basic Issues in the Philosophy of Time* (La Salle, Ill., 1971), pp. 195–228; L. R. Baker, "On the Mind-Dependence of Temporal Becoming," *Philosophy and Phenomenological Research* 39 (1979): 341–357.
9. Schuster, "Is the Flow of Time Subjective?" p. 712.
10. Ibid., p. 713.
11. Mellor, *Real Time,* p. 150.
12. Ibid., p. 171.
13. Cf. Henri Bergson, *Time and Free Will: An Essay on the Immediate Data of Consciousness,* trans. F. L. Pogson (New York, 1910), p. 130: "[if] to-day's impression were absolutely identical with that of yesterday, what difference would there be between perceiving and recognizing, between learning and remembering?"
14. Mellor, *Real Time,* p. 26.
15. Cf. Smart, "Time and Becoming," p. 6.
16. Mellor, *Real Time,* p. 26.
17. See, e.g., Sosa's criticism of "OPD" in "Status of Becoming," pp. 26–28.
18. See William Alston, "Varieties of Privileged Access," *American Philosophical Quarterly* 8 (1971): 223–241, and Roderick M. Chisholm, "Self-Profile," in

Radu J. Bogdan, ed., *Roderick M. Chisholm* (Dordrecht, 1986), pp. 36–37. On p. 37, Chisholm writes that one could "characterize the *mental* in terms of the purely psychological," which "are those properties to which we have privileged access."

19. Prior, "Thank Goodness That's Over." For Mellor's discussion of Prior's puzzle, see *Real Time,* pp. 48–55. Cf. Schlesinger, *Aspects of Time,* pp. 34–38; Taylor, *Metaphysics,* p. 73.
20. I use "dread" in a way such that a dreaded event may or may not be an event which will in fact occur. Thus, nuclear war may be dreaded even though, say, there will never be a nuclear war. What make the dread appropriate are the logical possibility that *there being a nuclear war* has the transitory temporal property *being future* and one's (perhaps false) belief that it does have that property. Similar remarks can be made about my use of "relief" and those events believed to be past.
21. Mellor, *Real Time,* p. 50.
22. Ibid.
23. Ibid., p. 20.
24. Cf. Joel Feinberg, *Doing and Deserving* (Princeton, N.J., 1970), p. 146.
25. Norvin Richards brought this point to my attention.
26. Mellor, *Real Time,* pp. 51–52. Mellor offers a different account of relief in Essay 28.
27. Ibid., p. 54.
28. Ibid.
29. Ibid., p. 82.
30. Ibid., p. 116.
31. For helpful comments, I thank Robert Coburn, Nita Nestevold, Philip Quinn, Norvin Richards, Ernest Sosa, and the referees for *Philosophy and Phenomenological Research,* in which this essay was originally published. I also thank the University of Alabama Research Grants Committee for its support of this work.

ESSAY 33

On the Experience of Tenseless Time

L. NATHAN OAKLANDER

The status of temporal becoming, temporal passage, or the transitory aspect of time is a paradigmatic metaphysical problem. It involves a *prima facie* conflict between reason and experience. The experience in question involves the passage of time: the "perception" of events flowing from the future into the present and from the present into the past. This experience is reflected in such statements as "I can't wait until the basketball season comes around again," "Hurray, I am finally graduating," and "Thank goodness the exam is over!" When we rationally reflect upon these statements and wonder what reality must be like in order for them to be true, we find that logical difficulties, such as McTaggart's paradox, emerge. Faced with this conflict, the goal of the metaphysician is to provide an ontology of time that fits the experience in question and is logically consistent. Broadly speaking, two theories of time have taken up the challenge to realize that goal: the tensed and the tenseless theories. According to the tenseless view, the logical problems surrounding temporal becoming are real and can only be avoided by recognizing, in Donald C. Williams's words, "the myth of passage."[1] According to the tensed view, the experience of passage and the presence of experience are real and can be accounted for only by accepting the tenses as reflecting basic ontological distinctions.[2]

The debate between proponents of the two camps has been fought on several fronts. Until the early 1980s the question of translatability was of central importance. If tensed discourse could be translated without loss of meaning into tenseless discourse, then, it was argued, tenses lacked ontological significance. More recently, defenders of the so-called new tenseless theory of time have sought to demonstrate that the necessity of tensed discourse is compatible with time being tenseless.[3] To do this successfully, it is necessary for the detenser to deal adequately with the experience of temporal becoming. The issue centers on whether the defender of tenseless time can provide an adequate analysis of the presence of experience and the appropriateness of certain of our attitudes toward future and past events. In Essay 32, Hestevold argues that the tenseless theory of time cannot account for our experience of time. In what follows, I shall attempt to show that his objections to the tenseless theory can be overcome and that an adequate analysis of tenseless time is possible.

The Presence of Experience

According to the tenseless theory of time, there are no basic ontological differences between past, present, and future events. All events exist tenselessly in the network of earlier than, later than, and simultaneity, or temporal relations. If, however, all events exist tenselessly, then how can the detenser explain our knowledge that a certain experience, say a headache, is (now) occurring? How can the detenser explain the fact that experiences can be known to be present? According to Hestevold, no explanation is possible, because "the claim that experiences, essentially, can be known to be present implies that there cannot occur an experience that occurs only tenselessly; experiences *cannot* be mere tenseless occurrences on the B-series!" The reasoning underlying the implication in this passage is open to two interpretations. First, since none of the terms of the B-series (the series of events generated by the earlier–later relation) is intrinsically present, no experienced events on the B-series can be known to be present. Second, if the detenser defines the presence of an experience in terms of its occurring at a certain date or its being simultaneous with some temporal item, then it follows that all experiences are (tenselessly) present. In that case, however, the detenser cannot explain the knowledge we possess of which experiences are happening *now*. For if all our experiences exist tenselessly at the moment they do, what is the explanation for the phenomenological fact that certain of those experiences are known to be occurring *now,* while others are not known to be occurring *now* or are even known *not* to be occurring now?

I think the detenser has a reasonably good response to this question. It begins with the truism that whenever we are aware of an object (or have an experience), we are conscious of being aware of that object (or of having that experience). Thus, one aspect of our knowledge of the present is grounded in the consciousness of our experiences at the time they are occurring. If we combine this thesis with the claim by Thomas Reid that "consciousness . . . is an immediate knowledge of the present,"[4] we arrive at the required result: that for an individual, every experience he or she is conscious of is one known to be present. Of course, Hestevold may ask,

"Doesn't this claim from Reid posit something that *really is* present? If you need that claim to finish off your argument, haven't I made my point?" Not necessarily, because Reid's reference to "the present" can be understood to designate the cross-section of experiences that are simultaneous with one's consciousness of them. There is nothing more, ontologically speaking, to the presence of experience than our being conscious of our experiences when they are happening.

To this explanation of the presence of experience it may be objected that merely being (tenselessly) conscious of an experience when it is (tenselessly) occurring does not give knowledge of which experiences are (presently) occurring. But I do not think that this objection can be sustained, for the knowledge we seek *can* be explained tenselessly, and the argument to the contrary is a non sequitur.

According to a detenser, if I am conscious at t_1 of an experience that occurs tenselessly at t_1, and if, as a matter of tenseless fact, it is t_1, then I know the experience is present. Of course, tensers use the same antecedent to infer that detensers cannot know which experiences are present, but such an inference is based on a misinterpretation of the tenseless view. The tenseless view gives rise to several different images. One is that of experiences in the B-series existing "eternally" or totally outside time. Another is of experiences as existing sempiternally, or at every time; and still another is of experiences seen from a point of view outside time, looking down and seeing them as parts of a never changing present. Each of these images falsifies the detenser's view in a fundamental way. On the tenseless theory, experiences and events are not eternal or sempiternal, and they do not all exist at once, *totum simul*. Rather, experiences, like our consciousness of them, exist in time, in succession, one after another. We are in time and therefore conscious of our experiences from a temporal point of view. The significance of this last point can be clarified by means of a spatial analogy. We are in space and so experience things from a spatial point of view. I am *here*, hence distant from some places and near to others. Accordingly, the answer to the question "Which things are existing here?" depends on the place at which the question is asked. Similarly, the answer to the question "Which events are existing now?" depends on the time at which the question is asked. Right now, as I look at the clock on my desk, it is 10:00 A.M., 11 September 1993, and so the experience of my looking at the clock (of which I am conscious) is known by me to be present. There is no need to suppose that there is some special property of events that are present, or objects that are here, that enables us to know which events are present or which objects are here. Admittedly, if we were somehow outside time and so nontemporally conscious of all our experiences (as God might be of the history of the world), then no experience could be known to be present to the exclusion of others. But our consciousness of experiences, like the experiences themselves, is in time, and at any given time we can know what experiences are present simply by being conscious of them, as opposed to remembering or anticipating them.

To all this Hestevold makes the following reply. If, at t_1, I record the presence of my experience of, say, an excruciating toothache, by telling the dentist "I am now in pain," then, on the tenseless theory, that *means* that "t_1 is tenselessly occurring, and my excruciating pain is (tenselessly) occurring at t_1." However, that judgment is true *at any time* and so would not be sufficient to convey to the dentist the requisite information, namely, that I am in pain *now*. In order to convey that information, the *tensed fact* that my pain is occurring *now* is indispensable.

Once again, Hestevold's argument is a non sequitur. It proves that the tenseless sentence ("t_1 is tenselessly occurring, and my excruciating pain is (tenselessly) occurring at t_1") does not have the same *meaning* as the tensed sentence ("I am now in pain"); but it does not prove that the two sentences are used to describe different *states of affairs,* one involving a tenseless fact, the other a tensed fact (see Essay 9). Admittedly, the tensed sentence conveys more information than the tenseless one, but it does not follow that it does so because of the reality of tense. The dentist who hears the tensed sentence-token "I am now in pain" knows that I am using that sentence to describe a state of affairs existing simultaneously with my utterance, and so he or she administers an anesthetic; whereas if I uttered the tenseless sentence, the dentist would not know that I needed relief now unless he also knew what time it was. Thus, the two sentences do not convey the same information and so do not have the same meaning. Nevertheless, it does not follow that they do not describe the same state of affairs, and, more generally, the indispensability of tensed discourse does not imply the indispensability of tensed facts.

Before leaving the topic of the presence of experience, I want to consider another phenomenological datum that allegedly supports the tensed theory. Schlesinger has defended the tensed theory by appealing to the experience of the NOW as "the point in time at which any individual who is temporally extended is alive, real, or Exists with a capital *E*" (p. 214). More recently he claimed that "our attitude toward the present may be described as regarding it as distinct from every other temporal position, for while the future is yet to be born and the past is rapidly fading, the present is palpably real."[5] I suggest that we can make sense of Schlesinger's phenomenological claims without countenancing transitory temporal properties. Again, a spatial analogy may help. I am here and so experience space differently from the way I would if I were outside space. I can know what goes on in distant places, but, given causal laws, I can affect what goes on elsewhere less surely and can reasonably regard what takes place there as less important, because it affects my life much less. Similarly, I am now (at this time), so those events that are at temporally distant times are less affected by me and have less effect upon me than those in the present. Thus, I may reasonably regard what is happening now as being more important or more real, and that is the only (harmless) sense in which the present is "palpably real" or Exists with a capital *E;* the reality of tense has nothing to do with it.

Our Attitudes Toward the Future and the Past

Another argument intended to demonstrate that the tenseless theory of time cannot be squared with our experience is based on the claim that dread and relief are inexplicable attitudes on the tenseless theory. Hestevold's point in relation to his tooth extraction example (p. 334) is that since the fact *Wednesday is later than Tuesday* is a fact that exists before Wednesday, then if *that* fact is what explains relief, it is just as sensible to feel relief on Monday or Tuesday on account of a painful experience that is taking place on Tuesday as it is to feel relief on Wednesday as regards the same painful experience.[6]

The mistake in this argument is the assumption that the tenseless fact that renders relief appropriate exists before, during, and after the extraction. On the tenseless view, the fact in question does not exist before, after, or during the extraction. The pain exists before the relief, and the experience of the relief exists after the cessation of the pain, but the fact that *the pain occurs before Wednesday* (or that *the relief occurs after the pain*) does not exist in time at all. Thus, while it is "always" true to assert that "Wednesday follows Tuesday," it does not follow that *Wednesday's following Tuesday* always exists; so Hestevold ought not to conclude that relief is justified before the pain or during it. To think that it *is* justified is to confuse tenseless facts with sempiternal things. Further evidence that Hestevold succumbs to this confusion is his statement: "After all, on Wednesday, there is a sense in which the extraction is not over; on Wednesday, the extraction is 'eternally' and tenselessly occurring on Tuesday" (p. 335).

Again, this way of viewing the matter is fraught with difficulties. To say that an extraction is tenselessly occurring on Tuesday (t_2) is to say, assuming that time is relational, that the extraction is simultaneous with each member of the set of simultaneous events that constitutes t_2. This fact, however, does not exist on Tuesday or on Wednesday or on any other day; it is eternal. But to say that an event's occurring at a certain time is an eternal fact does not imply that the event in some sense is always occurring, although looked at from an external God-like perspective, it may appear as if this is so. But from the inside, and in reality, my painful experiences are (hopefully) short-lived, and as they are succeeded by more pleasant experiences, my awareness of the painful ones become a mere memory. Indeed, it is just this succession of different psychological attitudes toward the same event (first anticipation, then consciousness, then memory) that gives rise to the impression of time's flow, and it is this impression that provides the basis for our different attitudes toward the future and the past.

Accepting all this, a critic may wonder why treating facts as outside time helps to resolve Hestevold's problem. If it is always true that the dread occurs before the painful experience and if the fact of the dread occurring before the pain never changes, why should I be happy *now* that the toothache is over? Of course, if the painful experience will occur, but is not yet occurring—that is, if it will move from the future to the present with the passage of time—then, so the critic alleges, we can easily understand an attitude or feeling of dread. We cannot understand that attitude on the tenseless theory, where all events exist and nothing really moves through time at all. In short, the tenseless theory never explains why dread is "appropriate" *before* a bad event rather than after or during it.

One way of responding to this objection is to question the premise upon which it is based, namely, the assumption that the feeling of dread *is* appropriate when the dreaded event is in the future. Perhaps we should say that dread is an appropriate attitude to have before an unwelcome event in that it is a *rational* attitude to take. However, it might plausibly be argued that dread is often not rational, in that it does not make us more efficient in meeting the problems we face. Dread of a visit to the dentist does nothing but make one's life miserable beforehand, and it may even stop us from keeping our appointment. It serves in no way to direct one's actions when the visit is necessary for good health. So, although it is *natural* enough, perhaps, at least for a certain sort of personality, to dread certain sorts of events, it is not clear that it is appropriate in the sense of being rational.

Maybe it is *in general* useful to dread bad events, because dread in general motivates us to prepare for or avoid such events in ways that we would not employ if we did not experience dread. So it is easy to see how dread might evolve biologically. But, like many biologically evolved defense systems, this one often does harm, preventing us from acting efficiently, and so must often be controlled or suppressed in the interests of behaving more reasonably. Dread before an event is functional (when dread is functional at all) because one can still do something about it. The same feeling after the event is never functional, so never appropriate. Thus, our attitude toward dread is like the attitude we have toward preparing, say, for an exam; it makes sense *before* but not *after* the exam, because preparation affects the outcome. Similarly, if dread spurs us to preparation, it may have survival value and thus be appropriate before an event, but not after. In other words, the *causal* efficacy of dread and the direction of causality in time are what explain its appropriateness before, but not after, an event.[7] Since this explanation works perfectly well for a detenser, I conclude that neither the presence of experience nor our attitudes toward the past, present, or future pose insurmountable difficulties for an adequate analysis of tenseless time.[8]

Notes

1. See Williams, "Myth of Passage."
2. The most elaborate defense of the tensed theory of time is found in Smith, *Language and Time*.
3. Proponents of the new tenseless theory of time include Beer (see Essay 7), Murray MacBeath (see Essay 27), Mellor (see his *Real Time*), and Oaklander (see his *Temporal Relations* and Essay 3).
4. Thomas Reid, *Essays on the Intellectual Powers of Man* (1785; rpt. Cambridge, Mass., 1969), p. 359.
5. George Schlesinger, "E Pur Si Muove," *Philosophical Quarterly* 41 (1991): 427.
6. Recent discussions of this type of argument are found in Essays 27, 29, and 31 and in Kiernan-Lewis, "Prior's 'Thank Goodness' Argument."
7. Also, of course, the word "dread" contains "before the event" in its meaning or usage; after the event, one may regret or rue it or look back on it with horror, but one cannot dread it. But looking back on it with horror is close enough to the feeling of dread that we can get away from mere grammar here and ask why such a feeling of horror is not appropriate after it is over. If one holds that dread is appropriate even when it is dysfunctional, merely because the event really is awful, then looking back on it with horror would also be appropriate.
8. I wish to thank the anonymous referee for the *Journal of Philosophical Research*, where this was first published, and Robert Audi for helpful comments and the University of Michigan–Flint for an award that partially funded the research.

ESSAY 34

The Phenomenology of A-Time

QUENTIN SMITH

B-Relations and A-Properties

One of the central debates in current analytic philosophy of time is whether time consists only of relations of simultaneity, earlier than, and later than (B-relations), or whether it also consists of properties of futurity, presentness, and pastness (A-properties). If time consists only of B-relations, then all temporal determinations are permanent; if at any one time it is the case that Dante's birth is *later than* Homer's birth, then it is ever after the case that Dante's birth is later than Homer's. The temporal position of Dante's birth vis-à-vis Homer's is permanently fixed. Moreover, if B-relations are the only temporal determinations possessed by events, then each event, regardless of when it occurs, is just as real as each other event. Each event sustains B-relations to other events and thus, in respect of its temporal determinations, is ontologically undistinguished from each other event. Why should Dante's birth be "more real" than Homer's just because it is later than it? But if time also consists of A-properties, then some events are ontologically distinguished by virtue of their temporal determinations; the events that *are* or *exist* in the tensed sense, the events that possess the A-property of presentness, have a reality not possessed by other events. All other events are no longer (are past) or are not yet (are future) and in this respect are deprived of the being, the presentness, possessed by events that *are*. The A-properties possessed by events are impermanent temporal determinations; if an event possesses a certain A-property at one time, then there is another time at which the event will not have that A-property but some other A-property instead. First an event is future, then it is present, and finally it is past.

This shows that the issue between the defenders of the B-theory and the defenders of the A-theory is of fundamental ontological importance. But analytic philosophers discuss this issue almost exclusively in terms of the *language* we use to describe the temporal determinations of events. They engage in what Quine calls "semantic ascent," that is, they redirect their concern from the "things themselves" to the words we use to describe things. Defenders of the A-theory argue that tensed sentences and their tokens, sentences containing tensed copulas like "is," "was," and "will be" and adverbs like "now" and "at present," are untranslatable or unanalyzable into tenseless sentences about B-related events and therefore that tensed copulas and adverbs refer to A-properties of events. Defenders of the B-theory argue that tensed sentences or their tokens are translatable or analyzable into tenseless sentences about B-related events and hence that tensed sentences or tokens refer only to B-related events. While this semantic ascent is not without its advantages, it seems to me that additional light can be thrown on this subject if it is approached from a nonlinguistic *phenomenological* perspective. This approach is all the more needed since this particular issue in the philosophy of time has not been explicitly addressed by any of the practitioners of "phenomenology" (in the wide sense), such as Husserl, Scheler, Heidegger, Sartre, and Merleau-Ponty. Indeed, phenomenologists have generally seemed to be unaware of the debate between A-theorists and B-theorists. This issue has been a concern exclusively of analytic philosophers (with the exception perhaps of the early twentieth-century British Idealist John McTaggart[1]).

In what follows I will point to a number of phenomenological facts that are pertinent to the debate. These facts all favor the A-theory. I shall make the case that the basic phenomenological truths about time simply cannot be accounted for by the B-theory.

A-Properties and B-Relations to Linguistic Utterances

There are three versions of the B-theory, one of which I shall criticize in this section, the version which states that A-properties are reducible to *B-relations between events and linguistic utterances*. This version of the B-theory has been defended by Hans Reichenbach, J. J. C. Smart, Bernard Mayo, Milton Fisk, Paul Fitzgerald, and others.[2] According to this doctrine, the presentness of an event is not an irreducible A-property that it possesses in addition to its B-relations; the presentness of an event is just its relation to a linguistic utterance of being simultaneous with it. The pastness of an event is just an event's relation to an utterance of being earlier than it, and an event's futurity is its relation to an utterance of being later than it. A-properties, accordingly, are not aspects of time distinguishable from B-relations; they are a certain class of B-relations, the class of B-relations that obtain between events and linguistic utterances.

If this theory were consistent with the phenomenological facts about time, it would be the case that for an event to appear to be present is for it to appear to be simultaneous with a linguistic utterance. But this is not what appears. *Contra* the linguistic thesis, there are many instances in which events appear to be present without there appearing to be any utterances simultaneous with the events in question. I perceive a cloud presently passing over the treetops, and during this perceptual experience no speech act is apprehended: I utter a sentence neither out loud nor to myself and do not hear anybody else uttering a sentence. I am beholding the present event in a wordless silence. Moreover, I am neither reading nor writing any sentence.

It might be objected that these phenomenological facts show not that the linguistically reductive theory is wrong but merely that it is incomplete, that it accounts for the temporal awareness involved in our linguistic experience but not for the awareness involved in our nonlinguistic experience.

Two responses can be made to this. First, the proponents of this B-theory do not regard the linguistic thesis as a partially adequate reductive account of our seeming awareness of A-properties but as sufficient to establish this reduction. For example, Smart first makes some linguistic observations like "When we say that the boat '*was* upstream, *is* level, *will be* downstream,' we are saying that occasions on which the boat is upstream are *earlier than* this utterance, that the occasion on which it is level is *simultaneous with* this utterance, and that occasions on which it is downstream are later than this utterance."[3] He then concludes: "This shows how misleading it is to think of the pastness, presentness, and futurity of events as properties, even as relational properties."[4]

Not only do our nonlinguistic apprehensions of A-determinations show that such an account is insufficient to establish the requisite reduction; they also reveal that this account is wholly mistaken. For if in our nonlinguistic experiences we apprehend A-determinations and the apprehended A-determinations are not B-relations of events to linguistic utterances, then it follows quite simply that A-determinations *are not* these B-relations and cannot plausibly be said to be taken or rather mistaken to be such in our linguistic experiences. The phenomenological facts bear this out: if after wordlessly perceiving for a few seconds the cloud presently passing over the treetops, I say (while the cloud continues to pass over the treetops) "The cloud is passing over the treetops," I do not cease to apprehend the presentness as a property of this event and begin to apprehend it as a B-relation to the linguistic utterance; rather, I continue to apprehend the presentness of the event while *also* discerning the said B-relation.

A-Properties and B-Relations to Psychological Events

A second version of the B-theory reduces A-properties to B-relations of events to psychological occurrences like sensations or acts of awareness. Proponents of this version include, among others, Russell, Grünbaum, and to some extent Mellor, who reduces A-properties to B-relations of events to utterances or acts of judging or thinking.[5] This version of the B-theory seems on the face of it to be more consistent with our experience of time than does the linguistic version, for it coheres with the fact that we are always aware of A-properties; while we are not always apprehending utterances of sentences, we are always undergoing some psychological experience. ("Always" should be taken to mean here "at every moment of our waking life.") However, this version runs into difficulties once the distinction between introspective, or reflexive, states of mind and extrospective, or unreflexive, states of mind is made. For proponents of this thesis, apprehending an A-determination requires a reflexive act of consciousness in which I turn my attention back onto myself and discern that my psychological experiences stand in some B-relation to some other event(s). Grünbaum writes, for example, that to be aware of an event as present, one must be "conceptually aware of the following complex fact: that his having the experience of the event *coincides temporally* with an awareness of the fact that he has it at all."[6]

Definitions such as these do not square with our many *unreflexive* awarenesses of events as present, past, or future; I perceive the cloud to be passing over the treetops at present without at the same time reflexively grasping my own perceptual experiencing of the event. I am not attending to my perceiving but to what I am perceiving, the cloud passing over the treetops. If somebody asks me, "What are you experiencing right now?," I may then reflect upon my perceptual state, but until then my attention is directed elsewhere.

It is possible to modify the psychological thesis to say that I need not be *attentionally* aware of some B-relation between my experience and some other event but need only be implicitly or marginally aware of it. And surely even in the unreflexive experience under discussion I am implicitly aware that my perceiving is simultaneous with the cloud passing over the treetops.

This may be granted, but it is still unsatisfactory. For I am explicitly and unreflexively aware that the cloud is presently passing over the treetops, and if I am only implicitly and reflexively aware of a B-relation between this event and my perceiving, this B-relation is *ipso facto* different from presentness. That which currently has a relational property of being an object of my explicit and unreflexive awareness is nonidentical with that which currently has a relational property of being an object merely of my implicit and reflexive awareness.

A-Properties and Dates

The third version of the B-theory reduces A-properties to dates, that is, to B-relations of events to some historical event (such as Christ's birth) that serves as the origin or zero-point of a calendar system. This version is adopted by Russell in his early writings on time and by Nelson Goodman, W. V. O. Quine, Clifford E. Williams, and others.[7] According to this view, if at 2:00 P.M. E.S.T. on 26 August 1985 I apprehend an event to be present, I am apprehending the event to occur at 2:00 P.M. E.S.T. on 26 August 1985.

This version of the B-theory overcomes both the problems with the linguistic version and the problems with the psychological version, for it reduces awareness of A-properties neither to awareness of linguistic utterances nor to reflexive awarenesses of psychical events. Clearly I can be unreflexively and nonlinguistically aware that the cloud passes over the treetops at 2:00 P.M. E.S.T. on 26 August 1985.

But is this what our awareness of A-properties amounts to? I think not, for I can very well be aware that the cloud is presently passing over the treetops, without knowing what o'clock it is or whether it is 25 or 26 August. Indeed, in cases of severe amnesia a person can be aware that an event is presently occurring without even knowing the century in which the event is located. In fact, in most cases of awareness of the A-properties of events, we are not aware of the dates of the events, either because we don't know the dates or because a knowledge of the dates is irrelevant to our involvement with the events. Moreover, we can have correct beliefs about the A-properties of events at times when we have incorrect beliefs about their dates; I can at 2 P.M. E.S.T. correctly judge that the cloud is passing over the treetops *at present* while erroneously believing that it is passing over the treetops at 3 P.M. E.S.T.

A-Properties, B-Relations, and the Experienced Causes of Emotions

I shall now present some phenomenological facts that are inconsistent with all three versions of the B-theory, facts concerning the experienced causes of emotions.[8] At least some of our emotional reactions are elicited or caused by what we believe to be the case, and the kinds of belief that elicit these emotions differ from one kind of emotion to the next. The B-theory of time is inconsistent with these facts, for it is a consequence of this theory that different kinds of emotion are caused by the same kinds of belief.

Nostalgia and eagerness are different emotions and are caused by different beliefs. According to the A-theory of time, nostalgia is caused by the belief that some joyous or happy occasion *is past* (that is, possesses an irreducible A-property of pastness), and eagerness by the belief that such an occasion *is future* (that is, possesses an irreducible A-property of futurity). A defender of the B-theory might try to explain this by saying that nostalgia is elicited by the belief that the happy occasion is *earlier than* the experience of the nostalgia and that eagerness is elicited by the belief that the happy occasion is *later than* the experience of the eagerness. But this attempted explanation is unsuccessful, for there is no difficulty in supposing that *both* these beliefs are held at both these times. At the time I experience nostalgia, I not only believe the nostalgia to be later than the happy event but also believe the eagerness to be earlier than the event, and at the time I experience the eagerness, I not only believe the happy event to be later than the eagerness but also (supposing I reasonably expect myself to be subsequently nostalgic) that the happy event is earlier than the nostalgia.

Referring to dates will not help the defender of the B-theory explain the difference in the causes of these emotions, for the belief on 3 September 1984, when eagerness is being felt, that *the happy event occurs on 4 September 1984* is also held on 5 September 1984, when nostalgia is being felt.

Nor will reference to linguistic utterances solve the problem; for if before the event I utter "The happy event is imminent" and believe that *the happy event is later than my utterance of "The happy event is imminent,"* I will not be believing anything different from what I can reasonably be said to believe *after* the event.

The problem is that any facts about the B-relations of the event that I could believe before the event I could just as well believe afterwards, and vice versa, so if B-facts are the only kind, it is impossible to explain the difference in belief that causes the different emotions of eagerness and nostalgia. This difference is explicable only if it is assumed that before the event I believe that the event *is future* and after the event I believe that it *is past.*

The Viability of the Phenomenological Account of Time

It is conceivable that the defender of the B-theory may concede my point that his theory is inconsistent with the phenomenological facts but claim that this is not detrimental to his theory, since the phenomenological facts concern *what merely appears to be the case* and his theory is about time *as it really is*. The proponent of the B-theory may draw an analogy between A-properties and color-properties: just as physical bodies appear to us to be colored but really are not, so time appears to us to be A-dimensional but really is not. Bodies as they really are possess only primary qualities, and time as it really is, is composed only of B-relations. Colors and the distinction among future, present, and past belong to the "manifest image of man-in-the-world" but not the "scientific image of man-in-the world."[9] Time *as it really is* is not known phenomenologically but only by science or a scientifically based philosophy.

A phenomenologist may respond to this objection in at least two ways. I shall call the first response that of the "phenomenological idealist," whom I define as follows. The phenomenological idealist believes that the scientific realm no less than the manifest realm (the "life-world") is projected by, or essentially dependent upon, man. The realm of scientific entities is nothing real in itself but a mere "theoretical construct" fashioned from materials in the manifest realm, such that the latter realm is the foundation of the former. Accordingly, scientific time is not only as human-dependent and human-relative as manifest time but is an abstract construct fashioned from the latter. Given this, the phrase "time as it really is" is more suitably applied to the original phenomenologically manifest time than to the derivative time of the sciences. It follows that time *as it really is* is A-dimensional, inasmuch as manifest time is A-dimensional, and that the B-theory accordingly is false, or at least restricted to an abstract scientific image of real time.

The position underlying a response of this sort is approximately consistent with that of phenomenologists like Husserl, Heidegger, Sartre, and Merleau-Ponty, although of course they would not have expressed themselves in precisely this way, and they did not use the phrase "phenomenological idealism" to characterize their position. The essential idea underlying this response is that (in Heidegger's way of putting it) "time produces itself only insofar as man is" and that scientific time is "produced" derivatively from the time of prescientific being-in-the-world.[10]

A second, considerably different response is also open to a phenomenologist. This second response is that of a "phenomenological realist," whom I characterize in this way. The phenomenological realist holds that time and physical bodies exist nondependently upon and without being constituted, projected, or produced by man. This time is studied by the sciences and is also manifest to us in prescientific experience. But there is no dimensional dissimilarity between time as scientifically known and time as phenomenologically manifest, or between time as it really is and time as it phenomenologically appears. Time *really is* A- and B-dimensional, and it is represented as such in the sciences and manifest as such in prescientific experience. The task of the phenomenologist is to elucidate the A- and B-dimensionality of time as it intuitively appears, and the task of the scientist is to formulate and test hypotheses about this dimensionality in respect of its nonintuitively manifest features. Both approaches support the same conclusion, that the B-theory of time is inadequate to the (phenomenological or scientific) data.

The phenomenological realist position underlying this second response has not (to my knowledge) been defended by any mainstream phenomenologist, but I have defended it at length in my book *The Felt Meanings of the World.*[11] The specific claim that the sciences represent time as both A- and B-dimensional I have defended in my article "The Mind-Independence of Temporal Becoming."

But it is not necessary to adopt the realist position I defend in order to be convinced by the phenomenological studies of time presented in Sections 2–5. It follows from both the phenomenological idealist and realist positions that the phenomenological elucidation of the A-dimensionality of time reflects time *as it really is* and hence that the A-theory of time is to be preferred.[12]

Notes

1. Cf. McTaggart, *Nature of Existence,* vol. 2.
2. Reichenbach, *Elements of Symbolic Logic;* Smart, "River of Time"; Bernard Mayo, "Events and Language," in M. Macdonald, ed., *Philosophy and Analysis* (Oxford, 1954); Fisk, "Pragmatic Account of Tenses"; Fitzgerald, "Nowness and the Understanding of Time."
3. Smart. "River of Time," p. 492.
4. Ibid., p. 493.
5. Russell, "On the Experience of Time"; Grünbaum, *Modern Science and Zeno's Paradoxes;* Mellor, *Real Time.*
6. Grünbaum, *Modern Science and Zeno's Paradoxes,* p. 17.
7. Russell, "Review of MacColl's *Symbolic Logic*"; Goodman, *Structure of Appearance;* Quine, *Elementary Logic;* Clifford E. Williams, "'Now', Extensional Interchangeability and the Passage of Time," *Philosophical Forum* 5 (1974): 405–423.
8. The subsequent discussion is a development of some suggestions made by Prior in "Thank Goodness That's Over" and Schlesinger in *Aspects of Time* (see esp. pp. 34–36).
9. W. Sellars, "Philosophy and the Scientific Image of Man," in Robert G. Colony, ed., *Frontiers of Science and Philosophy* (Pittsburgh, 1962).
10. Martin Heidegger, *An Introduction to Metaphysics,* trans. Ralph Manheim (New York, 1961), p. 71.
11. By "phenomenologist" I mean anybody who practices the method of intuition

(not necessarily Husserl's method of intuition) or who is usually associated with the so-called phenomenological movement.

12. For some discussions of the A-theory/B-theory controversy that employ the analytic method of "semantic ascent," see Smith, "Sentences about Time"; idem, "Impossibility of Token-Reflexive Analyses"; and Essays 2 and 12.

ESSAY 35

The Phenomenology of B-Time

CLIFFORD WILLIAMS

Advocates of the A-theory of time have sometimes argued that the A-theory is true because it conforms to our experience of time, whereas the B-theory does not. A fatal defect with the B-theory, they say, is that it cannot account for the fact that our experience of time includes something more than the experience of the B-relations of earlier than, simultaneous with, and later than. This something more is the experience of the mind-independent A-properties: pastness, presentness, and futurity.

I shall argue, contrary to A-theorists, that a correct account of our experience of time confirms the B-theory and not the A-theory. We do not experience the mind-independent A-properties that the A-theory says that events possess.

My strategy will be first to clear away some confusions about the debate between the A- and the B-theories and consequently some confusions about the experiential component of the debate. Then I shall argue that our experience of presentness is like our experience of hereness; in neither case are we aware of a mind-independent property over and above the events or objects to which we ascribe presentness or hereness. After doing this, I shall reply to two objections that two recent A-theorists who discuss the phenomenology of time, Quentin Smith and H. Scott Hestevold, have raised (see Essays 34 and 32). My conclusion will be that, insofar as experience is a valid appeal, it supports the B-theory, not the A-theory.

Some Confusions

The difference between the way in which reality is characterized in the two theories has sometimes been described as the difference between a moving, vibrant universe on the one hand and a block universe on the other. Time in the first universe flows out of the past and into the future, whereas time in the second universe is static. In the A-theorist's universe, the only events that exist are present ones; past events are gone, and future ones are yet to come. In the B-theorist's universe, all events exist; past, present, and future events have an equal status.

If these descriptions of the two theories are correct, one would expect a definite experiential difference between them. If the A-theory is true, one would expect to experience flow of time—the goneness of the past and the not-yetness of the future. If the B-theory is true, one would expect to experience time as not having this kind of flow, as not having the division into past, present, and future, and as somehow being static. And if this experiential difference between the two theories is correct, it is evident which theory is right—the A-theory. Our experience of time does indeed involve the experience of some kind of flow and of some kind of distinction between past, present, and future.

However, these ways of characterizing the difference between the A- and the B-theories are misleading. The trouble with them is that they describe the B-theory in inconsistent ways—ways that no B-theorist would accept. B-theorists do not say that we experience time as static, since the very idea of time being static is self-contradictory. Nor do B-theorists say that we experience the universe—past, present, and future—as a big block, the parts of which coexist equally. This concept too—the concept of temporally separated parts being coexistent—is incoherent.

Several recent A-theorists have characterized the B-theory in these anomalous ways. William Lane Craig writes: "according to the B-theorist, temporal becoming is mind-dependent and purely subjective. Time neither flows nor do things come to be except in the sense that we at one moment are conscious of them after not having been conscious of them at an earlier moment."[1] This description distorts the B-theory of time, because there are senses in which, according to it, temporal becoming is objective, time flows, and things come to be. B-theorists do not assert that things do not come to be unless we are conscious of them. On the contrary, they assert that things do come to be even though we are not conscious of them. This coming to be is the coming to be at a B-time, which differs from the coming to be that A-theorists ascribe to events. Craig's confusion here becomes explicit in his statement that the B-theory of time cannot account for the Christian doctrine of creation *ex nihilo,* because in the B-theory there "is no state of affairs in the actual world which consists of God existing alone without creation." The reason for this is that on the B-theory, "the creation as a whole is co-eternal with God in the sense that it exists as tenselessly as He."[2] This assertion confuses tenseless occurrence with permanence. For B-theorists the occurrence of an event at a particular time is neither static nor permanent, and this is a conceptual truth about time, whether it be B-time or A-time. A tenseless truth about an event may be permanently true, but that does not make the event itself permanent, and this, too, is true whether the B-theory or the A-theory is true.

Quentin Smith also identifies tenselessness with permanence. "If time consists only of B-relations," he says, "then all temporal determinations are permanent" (Essay 34). Hestevold makes similar remarks. He labels the B-theory (or what he calls the tenseless view) the static view of time, according to which "talk of the 'flow' or 'passage' of time can only be metaphorical" (essay 32). Since we think of time, by definition, as not being static or permanent, we will have trouble on Smith's and Hestevold's characterizations figuring out why anyone would be tempted to adopt the B-theory.

These ways of describing the B-theory are not new. C. D. Broad writes that "the theory seems to presuppose that all events, past, present, and future, in some sense 'coexist,' and stand to each other timelessly or sempiternally in determinate relations of temporal precedence."[3] G. J. Whitrow mirrors this description: "the theory of 'the block universe' . . . implies that past (and future) events coexist with those that are present."[4] And M. Capek refers to "the preposterous view . . . that . . . time is merely a huge and chronic hallucination of the human mind."[5]

The general principle that should be invoked here is that whenever there is an obvious tilt against a theory, we should reexamine our understanding of it, because the dispute between two opposing philosophical theories is hardly ever so easily decidable. Invoking this principle in the present case means that we must take the obvious unintelligibility of the coexistence of past, present, and future events (which Broad notes) and the preposterousness of time's illusoriness not as reasons to reject the B-theory but as motivation to reexamine our understanding of what the B-theory in fact is. In particular, the A-theorist must not set up a bogus distinction that makes one wonder how anyone could possibly be a B-theorist. But, equally, the B-theorist must describe how our experience of time's flow differs from the A-theorist's conception of that flow.

The model that I shall use to distinguish the two theories and their corresponding experiential claims is the analogue between presentness and hereness. I shall use this analogue both to define the difference between the A- and the B-theories and to illuminate my claim that our experience of time accords with the B-theory, not the A-theory.

The difference between the two theories is that in the A-theory presentness is a mind-independent property of events in addition to the times at which events occur and in addition to the events being simultaneous with our experience of them. In the B-theory presentness is not such an extra property.

It is either the simultaneity of the events with our experiences of them or the times themselves at which events occur—times that are simultaneous with our experiences of events. The B-theory's conception of presentness is analogous to our common conception of hereness, which we think of as objects being in the proximity of the places we occupy or as the locations themselves of the objects—the locations that are in the proximity of the locations we occupy. We do not think of hereness as a mind-independent property of objects in addition to their being in the proximity of the places we occupy or in addition to their locations. The A-theorist claims that presentness differs from the common conception of hereness, because presentness is an extra, mind-independent property, whereas the B-theorist claims that presentness does not differ from this conception of hereness because neither presentness nor hereness is an extra, mind-independent property.

What exactly is this extra, mind-independent property that A-theorists ascribe to events? When we search what A-theorists have written for an answer to this question, we find them generally silent. They tell us what presentness is not: it is not simply the occurrence of an event; it is not the simultaneity of an event with the utterance or thought that the event occurs; it is not the time at which the event occurs; and it is not the simultaneity of an event with an experience of the event's occurrence. But they do not tell us what presentness actually is. This silence suggests that presentness, like yellow, is a simple, indefinable property that we can know only by direct experience. Though we can say what it is not, we cannot say what it is. We simply know it when we encounter it. If this is correct, and it is reasonable for A-theorists to say that it is, then presentness is the something more we experience when we experience an event's occurring, something above and beyond the event's occurring simultaneously with our experience of the event. A-theorists assert, accordingly, that we have such experiences, and B-theorists deny that we do. What we experience, B-theorists say, is one of the things that A-theorists say presentness is not.

To show that B-theorists are right, I shall compare the A- and the B-theories to what I shall call the A- and B-theories of space. My aim in making this comparison is to make it evident to introspection that we do not experience the something more that A-theorists say we experience.

Our Experience of Here and Now

The A-theory of space says that hereness and thereness are mind-independent properties of objects over and above the places at which the objects are located and over and above the fact that the objects are in the proximity of the places we occupy. The B-theory of space denies that hereness and thereness are extra, mind-independent properties. All that exist are the objects, their spatial locations, and their spatial relations to other objects. Space does not consist of anything other than locations and relations. (It is another matter whether these are absolute or relative.) And our experience of space consists of nothing more than experiences of these locations and relations. When we experience an object being here, says the B-theorist of space, we experience it being in the proximity of the place we occupy. The A-theorist of space disagrees: we experience something more than simply the object being in the proximity of the place we occupy. We also experience the object's hereness.

This analogue to the A- and B-theories of time helps us to see that we do not experience an extra, mind-independent property of presentness. I say "helps us to see," for I am using the analogy only as an aid to illumine our experience of presentness, not as the premise of an argument. I shall consider four features of our experience of time that A-theorists of time have appealed to in order to show that our experience of presentness is of an extra, mind-independent property: the inexorability involved in our experience of presentness, the privilegedness we sense about presentness, the movement of presentness, and the extra-ness of which we are aware. Each of these has a spatial analogue that throws light on its temporal counterpart.

Inexorability

One feature of our experience of time that A-theorists of time appeal to is the inexorability of the movement of time. We have the sense of being swept along against our wills. We do not say about times, as we do about places, "I think I'll stay here for awhile" or "Let's go over there." We feel ourselves being taken to later times, and when we get to them, we feel ourselves taken to still later ones. We cannot choose which times we will be at, as we can with places.

Richard Gale once used this sense of inexorability to support the A-theory. He wrote: "This difference between here and present or now is due to the fact that there is no spatial analogue to temporal becoming: the present (now), unlike here, shifts inexorably, independently of what we do."[6]

The question here is not whether we experience inexorability, for the answer is that we surely do. The question is whether the experience of inexorability supports the A-theorist's contention that we also experience an extra, mind-independent property of presentness. The answer to this question is, I believe, that the experience does not support the contention. The B-theorist can readily admit that the sense of inexorability is involved in our experience of B-time, because the very nature of time—the B-theory's time—involves an inexorability that is lacking in space.

To see this point, consider the A- and the B-theories of space, but with one emendation: imagine that we are carried from place to place without a choice of where we are taken. We would, in this condition, experience inexorability—we could not choose what places we were at; the hereness of objects would exist independently of our voluntary actions. But we would not, because of this condition, think of hereness as an objective property of objects over and above their being in the proximity of the places we occupy or over and above the locations they occupy. Nor, except for the inexorability, would we experience anything more than the objects being in the proximity of the places we occupy. We would not, in short, adopt an A-theory of space because of our experience of spatial inexorability. Nor should we adopt an A-theory of time because of our experience of temporal inexorability. That we experience temporal inexorability does not mean that we also experience presentness as a mind-independent property over and above our awareness of the simultaneity of an event with our experience of that event. All that we experience is this simultaneity plus the inexorability involved in the movement of B-time.

Privilegedness

Another feature of our experience of time that the A-theorist appeals to is the privilegedness we sense when we experience events as present. This sense manifests itself in the feeling that we occupy the central place in the temporal spectrum. Past and future events, we feel, do not occupy this central place; they recede from it as they become more and more past and approach it as they become less and less future.

The question here again is not whether we possess an experience of privilegedness, for we clearly do in some sense. The question is, rather, whether the existence of the experience supports the A-theory of time. That it does not can be seen by looking at the A- and the B-theories of space. We also possess a sense of privilegedness about space, the sense that we occupy the central place in the spatial spectrum. Objects not in the proximity of the place we occupy do not reside in this central place; they recede from it as we move away from them and approach it as we approach them. Surely, though, we do not think that this sense of privilege shows that we experience objects as having an extra, mind-independent property of hereness. We still accept the B-theory of space, even though we regard certain objects as privileged. Similarly, we should not adopt the A-theory of time just because we regard certain events as privileged. Our experience of privilegedness is not the experience of an extra, mind-independent property of presentness; nor does having the former entail having the latter.

The A-theorist of time may object that there is a difference between our experience of temporal privilege and our experience of spatial privilege. Though we regard objects in the proximity of the place we occupy as being central in the spatial spectrum, we nevertheless view objects elsewhere as equally real. But we do not view events occurring in the past or future as equally real to present ones. Present events possess a certain kind of privilege that objects that are here do not possess.

The response to this objection is to point out that "equally real" is one of those phrases, like the ones I mentioned above, that invite confusion. What the A-theorist of time seems to mean is that we experience all objects, both those here and those there, as existing at the same time, whereas we do not experience all events, past, present, and future, as occurring at the same time. This asymmetry, the A-theorist argues, gives us an experience of the centrality of present events that is lacking in our experiences of objects that are here. Time possesses an inegalitarian quality that space does not possess, a quality that is prominent in our experience of time but absent from our experience of space.

If the A-theorist of time means these things, then the B-theorist of time will reply that the asymmetrical experiences do not support the A-theory of time. For they are derived from the difference between space and B-time; spatially separated objects can exist at the same time, but temporally separated events cannot occur at the same time. If this difference between space and time means that we experience a privilegedness with time that we do not experience with space, then so be it. But the privilegedness does not entail the existence of presentness as an extra, mind-independent property. All that it entails is the asymmetry of time and space, which B-theorists of time are happy to admit—indeed, must admit if they are to maintain that there is a difference between time and space at all, which clearly they do maintain.

Movement of the Present

Another asymmetry that the A-theorist of time appeals to is our experience of time as a movement. We have a vivid sense of the present first characterizing one set of events and then characterizing another. But we do not similarly sense *here* as a movement. What counts as being here is often stationary, and when it does move, it is really our movement that we sense, not the movement of space. Our experience of the movement of the present, though, is an experience of time itself moving.

It is not the inexorability of the present's movement that the A-theorist appeals to here, or else the point would be a duplicate of the inexorability argument. Nor must we take the A-theorist's reference to motion literally, or we will fall into the conceptual absurdities that have been pointed out in the literature on the subject.[7] The A-theorist's claim is that we possess an awareness of time's movement that is not matched by any analogous sense of space's movement.

However, this claim too can be accounted for by the B-theorist. If anything is true of time and space, it is that they are intrinsically different. So one would expect an asymmetry in our experience of them. But this asymmetry does not justify the postulation of presentness as an extra feature of time. To see this, think again of the A- and B-theories of space. Imagine that we are moving from place to place voluntarily (so that the idea of inexorability does not get entangled with the idea of movement). We would then experience the here moving. But we would not infer that because we do, we should adopt the A-theory of space. The reason why we would not make this inference is that we are already convinced that the B-theory of space is true. Our not making this inference shows that we distinguish between our experience of the moving here and the experience of here being an extra, mind-independent property of objects. And this fact in turn suggests that we should make a similar distinction with regard to time. Our experience of the moving present is different from the experience of the present being an extra, mind-independent property of events. The latter experience would support the A-theory of time if the experience existed; but the experience does not exist. The former experience exists, but its existence does not support the A-theory of time; for it is compatible with the B-theory of time, in the same way that our experience of the moving here is compatible with the B-theory of space.

Extra-ness of the Present

The fourth feature of our experience of time that A-theorists of time appeal to is our direct experience of presentness as a characteristic that events possess in addition to their occurrences at certain times and in addition to their simultaneity with our experiences of them. If all that we experience is this occurrence or this simultaneity, A-theorists assert, then we do not know when the events occur, namely, now. We might be experiencing their past or future occurrences. To know that the events occur now, we must either experience their presentness directly or infer their presentness from our direct awareness of the presentness of our *experiences* of the events (which are themselves simultaneous with the events). Since we do in fact know that events occur now, our experience of their presentness is something more than the experience of the events occurring and also something more than the experience of the events occurring simultaneously with our experiences of the events.

A linguistic analogue to show this argument has commonly been used by

A-theorists to show that tensed sentences are not equivalent to tenseless date-sentences. The tensed sentence

(1) Laura is outside now

is not equivalent to

(2) Laura is (tenseless) outside at 11:45 A.M. on 9 July 1991,

the A-theorist claims, because we can know that (2) is true without knowing that (1) is true—we might know that (2) is true when the event it reports is past. So if we know only that (2) is true, we still do not know when Laura is outside. This means that there is an added element in (1), namely, the ascription of the A-property of nowness to the event reported. The A-theorist may be making only a linguistic claim in this argument, but she may mean to be making an experiential claim as well. In any case, my response to the experiential claim has an equally valid linguistic analogue (see Essay 9).

Consider the following argument that an A-theorist of space might use in response to the claim made by the B-theorist of space that, when we experience an object as here, we are experiencing merely the object's being in the proximity of the place we occupy and not an extra, mind-independent property of hereness. If this is all we experience, the A-theorist of space might respond, then we do not know that the object is here; for having this experience is compatible with our not being here, as would be the case if we clairvoyantly observed both ourselves and the object being somewhere else at a future time. To know that an object is here, we must either experience its hereness directly or be aware of the hereness of our experience of the object. Since we do know that objects are here, our experience of their hereness must be something more than simply the experience of their being in the proximity of the places we occupy.

We would reject this argument, because we are already convinced that its conclusion is mistaken; we do not experience hereness as an extra, mind-independent property of objects. All that we experience is the object being near our spatial location. This experience is what tells us that the object is here; no additional experience of hereness is needed for us to know this.

This fact about our experience of space suggests that no additional experience beyond the experience of simultaneity is needed for us to experience presentness. If this is so, our experience of an event's presentness would be the awareness of the simultaneity of the event with our experience of the event. And the something more we are aware of when we step back and experience the experience itself as being present is the following complex experience: the awareness of our awareness of our experience of an event being simultaneous with the experience of the event.

To clarify, when we experience an event as being present, four items are involved: the event, the experience of the event, the simultaneity of the two, and our awareness of this simultaneity, thus:

Awareness of
experience of event simultaneous with the event

The A-theorist of time objects that this is not enough to give us an experience of presentness, for we do not know in this schema either when the event occurs or when the experience of the event occurs. They could occur at some past time. But since we do in fact know when the event occurs and when our experience of the event occurs, the A-theorist continues, namely, in the present, we must, in addition to these elements, possess a direct experience of presentness.

My B-theorist response is that there may, indeed, be something more than what I have just described, but it, too, is limited to our awareness of further simultaneity, thus:

Awareness of
awareness of experience of event
simultaneous with
experience of event

This awareness is what we possess when we experience the presentness of our experience of an event (and not just the presentness of the event itself). We are, therefore, not obliged to introduce presentness as an "objective" property of mental events in order to account for our experience of the presentness of external objects.[8] And in general we are not obliged to introduce an experience of presentness as an extra property, since our experience of simultaneity takes care of all that we know, in the same way that our experience of objects being in the proximity of our locations takes care of our knowledge of hereness.

The A-theorist of time may object that my account of our experience of an event's presentness and of our experience of the experience's presentness introduces an infinite regress. My answer is that if there is a regress with our experience of presentness, then there is a regress with our experience of hereness. But since there is no regress with the latter, there is none with the former. We have an experience of an object being here (which consists of our awareness of the object being in the proximity of our location), and we may or may not also have an experience of the experience's being here. The existence of the former does not entail the existence of the latter. Similarly, that we experience an event's presentness does not entail that we experience the experience itself as present, though, of course, we may and often, do.

Objections

I shall now consider two objections to my claim that we do not experience the A-theory's presentness.

First, Quentin Smith asserts that we sometimes have an unreflexive awareness of events being present. By this he means that we are aware of events being present without also being aware of our experiences of the events. He writes:

> I perceive the cloud to be passing over the treetops at present without at the same time reflexively grasping my own perceptual experiencing of the event. I am not attending to my perceiving but to what I am perceiving, the cloud passing over the treetops. If somebody asks me, "What are you experiencing right now?," I may then reflect upon my perceptual state, but until then my attention is directed elsewhere. (p. 354)

If Smith is right, then B-theorists of time cannot claim that our experience of presentness is the awareness of the simultaneity of an event with our experience of the event, for this awareness is reflexive in Smith's sense.

The question here is not whether we have unreflexive temporal experiences, for Smith is surely right in pointing out that we do. The question is rather whether these unreflexive experiences are of the presentness of events or simply of the occurrence of events. B-theorists of time assert that when we experience events as occurring, we do not also unreflexively experience them as present. And their court of appeal is introspection. They fail to find the extra, unreflexive experience of presentness. They do, of course, find numerous occasions when we unreflexively experience events occurring, but this fact is compatible with their claim that the events occur in B-time.

The analogy to our experience of *here* helps us to see these points. There are occasions, perhaps quite numerous, when we unreflexively experience, simply as existing, objects that happen to be in the proximity of our locations, without also experiencing them as being here. To experience them as being here would require reflexively experiencing them as existing in the proximity of our location. With space, then, we distinguish between our unreflexive experience of an object existing and our reflexive experience of an object existing here. Introspection reveals a similar distinction in our temporal experiences: we sometimes unreflexively experience present events occurring, without experiencing their presentness, and at other times we experience the presentness of present events. The former experiences do not require A-time, and the latter are reflexive experiences compatible with B-time.

Second, both Smith and Hestevold (Essays 34 and 32) claim that the different attitudes we have toward the past and the future are incompatible with the B-theory of time.[9] We dread a future tooth extraction and experience relief regarding a past one. "Dread is appropriate," Hestevold writes (Essay 32) "*because* I knowingly 'move closer' to an unpleasant event, and relief is appropriate *because* I knowingly 'move away' from an unpleasant event, and such temporal 'movement' cannot be captured in terms of ST [the static view of time.]" Smith and Hestevold reject Mellor's B-theory response that the time relations of before and after are sufficient to explain the appropriateness of dread and relief.[10] If Mellor's response were true, Smith asserts, the two emotions would be caused by the same fact that they are caused by different beliefs; when we experience dread, we would believe (if the B-theory's explanation were correct) that the tooth extraction is later than the dread and before the relief, and at the time we experience the relief, we would believe exactly the same thing (Essay 34).

Mellor's B-theory response can be reinforced by noting once more the analogy between our experience of presentness and our experience of hereness. It is appropriate to dread being under a tree during a thunderstorm, and it is also appropriate to experience relief after having moved from beneath the tree if moments later it is hit by lightning. Dread in this case is connected with the experience of the tree's being here, and relief is connected with the experience of the tree's being there. No one is tempted, though, to say that hereness and thereness are extra, mind-independent properties of the tree because of these facts. The existence of space relations is sufficient to explain the dread and the relief. Similarly, we do not need the extra, mind-independent characteristics of past and future to account for these emotions. Thus, to Smith's assertion that the B-theory of time makes dread and relief derive from the same beliefs, we may respond that the B-theory of space does so as well, contrary to the fact that the two emotions come from different beliefs; when we experience dread, we believe, according to the B-theory of space, that the lightning strike occurs at the same location as the dread and at a different location from the relief, and when we experience the relief, we believe exactly the same thing. But since we all accept the B-theory of space, we are not persuaded of its falsity by this reasoning and, consequently, should not be persuaded of the falsity of the B-theory of time by Smith's analogous reasoning.

I conclude that we do not experience the presentness that the A-theorist of time claims characterizes events. The basis for this conclusion is inspection of our inner states. I have used our experience of hereness not to prove this, but to aid in our inspection of temporal experiences. What the A-theorist of time must do to counter this conclusion is show that our experience of presentness differs from our experience of hereness in such a way as to require the postulation of an extra, mind-independent property for the former experience but not for the latter one. There are, of course, differences between the two experiences, but the question is whether they make a difference to the A- and the B-theories of time.

The inference from my conclusion is that the B-theory of time is more likely to be true than the A-theory of time. It is possible, of course, for the A-theory's presentness to exist even though we do not experience it, and thus it is possible that the A-theory might be true in spite of all I have said. However, if the A-theory were true, one would naturally (though not logically) expect that we would experience the phenomenon it asserts to exist. The precritical and unanalyzed experience of presentness is, perhaps, the source of most A-theorists' conviction that the A-theory is true. If what I have said is correct, this precritical experience turns out to be compatible with the B-theorist's conception of time. I infer that, insofar as experience as a source of support for any theory, it supports the B-theory and not the A-theory of time.

Notes

1. Craig, "God and Real Time," p. 336.
2. Ibid., pp. 337–338.
3. Broad, *Examination of McTaggart's Philosophy,* vol. 2, p. 137.
4. Whitrow, *Natural Philosophy of Time,* p. 228.
5. M. Capek, *The Philosophical Impact of Contemporary Physics* (Princeton, N.J., 1961), p. 337.
6. Gale, *Language of Time,* p. 214.
7. See, e.g., Williams, "Myth of Passage."
8. I mean here to be addressing Craig's charge that "no B-theorist . . . has successfully answered, in my estimation, the charge that his theory is incoherent because the mind-dependence of physical becoming requires a real becoming in the subjective contents of consciousness" ("God and Real Time," p. 337). If Craig's charge is unanswerable, then so also is an analogous charge against the B-theorist of space: the mind-dependence of hereness requires a real hereness in the subjective contents of consciousness.
9. Their argument derives from Prior, "Thank Goodness That's Over."
10. Mellor, *Real Time,* p. 50.

Acknowledgments

D. H. Mellor, "The Need for Tense," from *Real Time* © 1981 by Cambridge University Press. Reprinted by permission of Cambridge University Press and the author.

Quentin Smith, "Problems with the New Tenseless Theory of Time," *Philosophical Studies* 52 (1987): 371–392. © 1987 Kluwer Academic Publishers. Printed in the Netherlands. Reprinted by permission of Kluwer Academic Publishers.

L. Nathan Oaklander, "A Defense of the New Tenseless Theory of Time," *Philosophical Quarterly* 41 (1991): 26–38. Reprinted by permission of Blackwell Publishers.

L. Nathan Oaklander, "The New Tenseless Theory of Time: A Reply to Smith," *Philosophical Studies* 58 (1990): 287–293. © 1990 Kluwer Academic Publishers. Printed in the Netherlands. Reprinted by permission of Kluwer Academic Publishers.

Michelle Beer, "Temporal Indexicals and the Passage of Time," *Philosophical Quarterly* 38 (1988): 158–164. Reprinted by permission of Blackwell Publishers and the author.

Quentin Smith, "The Co-reporting Theory of Tensed and Tenseless Sentences," *Philosophical Quarterly* 40 (1990): 213–222. Reprinted by permission of Blackwell Publishers.

Clifford Williams, "The Date Analysis of Tensed Sentences," *Australasian Journal of Philosophy* 70 (1992): 198–203. Reprinted by permission of the editor and the author.

David Kaplan, "Demonstratives," from *Themes from Kaplan,* edited by Joseph Almog, John Perry, and Howard Wettstein (Oxford University Press, 1989), pp. 489–507, 520–521, 523–524 © by David Kaplan. Reprinted by permission of the author.

Quentin Smith, "Temporal Indexicals," *Erkenntnis* 32 (1990): 5–25. © 1990

Kluwer Academic Publishers. Printed in the Netherlands. Reprinted by permission of Kluwer Academic Publishers.

D. H. Mellor, "The Unreality of Tense," from *Real Time* © 1981 by Cambridge University Press. Reprinted by permission of Cambridge University Press.

Quentin Smith, "The Infinite Regress of Temporal Attributions," *Southern Journal of Philosophy* 24 (1986): 383–396. Reprinted by permission of the editor.

L. Nathan Oaklander, "McTaggart's Paradox and the Infinite Regress of Temporal Attributions: A Reply to Smith," *Southern Journal of Philosophy* 25 (1987): 425–431. Reprinted by permission of the editor.

Quentin Smith, "The Logical Structure of the Debate about McTaggart's Paradox," *Philosophy Research Archives* 24 (1988–89): 371–379. Reprinted by permission of the editor of the *Journal of Philosophical Research*.

George Schlesinger, "Temporal Becoming," from *Aspects of Time,* pp. 23–26, 30–33, 140–141 © 1980 by Hackett Publishing Company, Inc. Reprinted by permission of Hackett Publishing Company and the author.

L. Nathan Oaklander, "McTaggart, Schlesinger, and the Two-Dimensional Time Hypothesis," *Philosophical Quarterly* 33 (1984): 391–397. Reprinted by permission of Blackwell Publishers.

George Schlesinger, "How to Navigate the River of Time," *Philosophical Quarterly* 35 (1985): 91–92. Reprinted by permission of Blackwell Publishers and the author.

L. Nathan Oaklander, "A Reply to Schlesinger," *Philosophical Quarterly* 35 (1985): 93–94. Reprinted by permission of Blackwell Publishers.

David Zeilicovici, "Temporal Becoming Minus the Moving Now," *Noûs* 23 (1989): 505–524. Reprinted by permission of Blackwell Publishers and the author.

L. Nathan Oaklander, "Zeilicovici on Temporal Becoming," *Philosophia* 21, nos. 3–4 (1992): 329–334. Reprinted by permission of the editor of *Philosophia,* Asa Kasher.

George Schlesinger, "The Stream of Time," from *Timely Topics* © The Macmillan Press Ltd. (forthcoming). Reprinted by permission of the publisher and the author.

D. H. Mellor, "Thank Goodness That's Over," *Ratio* 23 (1981): 20–30. Reprinted by permission of Blackwell Publishers and the author.

Murray MacBeath, "Mellor's Emeritus Headache," *Ratio* 25 (1983): 81–88. Reprinted by permission of Blackwell Publishers and the author.

D. H. Mellor, "MacBeath's Soluble Aspirin," *Ratio* 25 (1983): 89–92. Reprinted by permission of Blackwell Publishers and the author.

Brian J. Garrett, "'Thank Goodness That's Over' Revisited," *Philosophical Quarterly* 39 (1988): 201–205. Reprinted by permission of Blackwell Publishers and the author.

J. D. Kiernan-Lewis, "Not Over Yet: Prior's 'Thank Goodness' Argument," *Philosophy* 66 (1991): 242–243. Reprinted by permission of Cambridge University Press and the author.

L. Nathan Oaklander, "Thank Goodness It's Over," *Philosophy* 67 (1992): 256–258. Reprinted by permission of Cambridge University Press.

H. Scott Hestevold, "Passage and the Presence of Experience," *Philosophy and Phenomenological Research* 50 (1990): 537–552. Reprinted by permission of the publisher and the author.

L. Nathan Oaklander, "On the Experience of Tenseless Time," *Journal of Philosophical Research* 18 (1993): 159–166. Reprinted by permission of the editor.

Quentin Smith, "The Phenomenology of A-Time," *Diálogos* 52 (1988): 143–153. Reprinted by permission of the editor.

Clifford Williams, "The Phenomenology of B-Time," *Southern Journal of Philosophy* 30 (1992): 123–137. Reprinted by permission of the editor and the author.

Name Index